Canadian

Professional Engineering and Geoscience

Practice and Ethics

Canadian

Professional Engineering and Geoscience

Practice and Ethics

FOURTH EDITION

Gordon C. Andrews
University of Waterloo

NELSON EDUCATION

NELSON / EDUCATION

Canadian Professional Engineering and Geoscience: Practice and Ethics, Fourth Edition
by Gordon C. Andrews

Associate Vice President, Editorial Director:
Evelyn Veitch

Editor-in-Chief, Higher Education:
Anne Williams

Executive Editor:
Paul Fam

Senior Marketing Manager:
Sean Chamberland

Senior Developmental Editor:
Elke Price

Photo Researcher:
Beth Yarzab

Permissions Coordinator:
Beth Yarzab

Production Service:
Macmillan Publishing Solutions

Copy Editor:
Elizabeth Phinney

Proofreader:
Barbara Storey

Indexer:
Maura Brown

Manufacturing Coordinator:
Ferial Suleman

Design Director:
Ken Phipps

Managing Designer:
Katherine Strain

Interior Design:
Katherine Strain

Cover Design:
Jennifer Leung

Cover Image:
© All Canada Photos/Alamy

Compositor:
Macmillan Publishing Solutions

Printer:
RR Donnelley

Library and Archives Canada
Cataloguing in Publication

Andrews, G. C. (Gordon Clifford),
1937-
    Canadian professional
engineering and geoscience :
practice and ethics / Gordon C.
Andrews.—4th ed.

First published under title:
Canadian professional
engineering.

Includes bibliographical references
and index.
ISBN 978-0-17-644134-0

    1. Engineering ethics—
Textbooks. 2. Earth scientists—
Professional ethics—Textbooks.
3. Engineering—Canada—
Textbooks. I. Title.

TA157.A68 2008   174'.962
C2008-906777-0

ISBN 13: 978-0-17-644134-0
ISBN 10: 0-17-644134-4

ABOUT THE COVER AND FRONTISPIECE

The cover and frontispiece illustrate recent Canadian engineering and geo-science achievements, which range from the ocean floor to outer space.

COVER—THE CONFEDERATION BRIDGE

The cover photograph shows the Confederation Bridge, which opened to traffic in May 1997 and links New Brunswick and Prince Edward Island across the Northumberland Strait. The bridge is 12.9 km (8 miles) long, which makes it the world's longest saltwater bridge subject to ice hazards.

The Confederation Bridge was designed by Strait Crossing Inc. of Calgary, Alberta, and will survive the harsh climate of the Northumberland Strait for the next century. It has two highway traffic lanes (plus emergency shoulders). The distinctive arches provide a clearance of 60 m (197 ft.) for seagoing vessels. The bridge approach on Prince Edward Island has 7 spans (580 m; 1,903 ft.); the central bridge portion has 44 spans (11 km; 7 miles); the bridge approach in New Brunswick has 14 spans (1300 m; 4,265 ft.). The water is as deep as 35 m (115 ft.) at the support piers, which are protected by conical ice shields to resist the severe scouring of floating ice.

The bridge's piers, spans, and other components were prefabricated on land, moved to a jetty by a special slider system, and positioned on the site by a huge floating crane. The bridge is equipped with electronic devices that monitor stress, strain, and motion, and transmit this data to universities and research facilities. In this way, the bridge's responses to ice, traffic, and earth movements can be constantly observed.

Further information about the bridge's operation is available from the Confederation Bridge website at <www.confederationbridge.com>, including rates, current weather and wind speed conditions, and a live Web camera.

FRONTISPIECE—CANADARM2

The frontispiece is a photograph of the Canadian robot arm on the International Space Station (ISS), which is in orbit about 400 km (250 miles) above the earth. The technical name for the robot arm is the Space Station Remote Manipulator System (SSRMS); its working name is Canadarm2, and it is a larger version of the Canadarm—the robot arm installed on the American space shuttles. The Canadarm2 is part of Canada's contribution to the International Space Station and was designed for the Canadian Space Agency (CSA) by MD Robotics of Brampton, Ontario.

The robot arm is essential for assembling the space station and identifies Canada as a key partner in the project. Canada contributed the Mobile Servicing System (MSS), which is made up of three parts: a movable base, called the Mobile Base System (MBS); the Canadarm2; and the Special Purpose Dextrous Manipulator (SPDM), known as Dextre. Canada also contributed the Canadian Space Vision System (CSVS), which permits objects outside the space station to be located accurately by the Canadarm2 operator. Canadian Space Agency astronaut Chris Hadfield installed the Canadarm2 on the space station in April 2001.

The photograph shows Canadian astronaut and mission specialist Dave Williams, representing the Canadian Space Agency, anchored on the foot restraint of the Canadarm2 as he participated in the second session of the Extra Vehicular Activity (EVA) for mission STS-118 on August 13, 2007. Assisting Williams was Rick Mastracchio (who is out of the frame). During the 6-hour, 28-minute space walk, the two removed a faulty control moment gyroscope (CMG-3) with a mass of 600 kg (1,320 lbs.), and installed a new CMG. The new gyroscope is one of four CMGs that are used to control the orbital attitude of the station. Williams took part in three of the four space-walks, the highest number of spacewalks performed so far in a single mission, and spent a total of 17 hours and 47 minutes in EVA activity, a Canadian record.

The space station is a truly international project, constructed by the United States, Russia, 11 European countries, Japan, Brazil, and Canada. In return for Canada's contribution, Canadian researchers will have proportional use of the station. Once all of the components and solar panels have been installed, the space station will have a mass of 450 tonnes and will cover an area equal to a Canadian football field.

The International Space Station has been occupied by astronauts since November 2000 and orbits the earth about every 92 minutes. It is visible with the naked eye as a moving star in the night sky. The CSA website <www.space.gc.ca> provides links to orbit data and sighting times (under the International Space Station heading), and much more information about Canada's satellites and space program.

PREFACE TO THE FOURTH EDITION

This textbook introduces engineers and geoscientists to the structure, practice, and ethics of their professions and encourages them to apply ethical concepts in their professional lives. It is a comprehensive reference for engineers and geoscientists in any branch of these professions, in any province or territory of Canada. The book is intended for practising professionals, recent graduates, senior undergraduates, and immigrants who wish to practise engineering or geoscience in Canada. The book is an excellent study guide for the practice and ethics part of the Professional Practice Examination (PPE) required for licensing in every province and territory.

ORGANIZATION AND OUTLINE

This text is organized into five parts, covering four key aspects of practice and ethics, and a final part describing the professional practice exam.

Part One—Professional Licensing and Regulation

The first five chapters describe the history, structure, and administration of engineering and geoscience in Canada. A licence is required to practise these professions, and this part describes the licensing Acts (or laws), the Associations established to enforce those Acts; the licensing process, and the academic and experience requirements for admission. The Acts require licensed professionals to maintain their competence, and every Association has a (mandatory or voluntary) competence assurance program. The Associations also have the authority to discipline unethical, negligent, or incompetent practitioners. Associations are not technical societies, but societies are important, as they assist professionals by stimulating research and publishing new theories, techniques, and standards. In addition, these chapters include the following:

- **Chapters 1 and 2 (Introduction and Regulation)** describe the inspiring achievements of both professions in Canada, marred only by the tragic case histories that created public demand for regulation—the Quebec Bridge collapse and the Bre-X fraud.

- **Chapter 3 (Continuing Professional Development)** explains the importance of professional competence, and how to achieve it through continuing professional development (CPD).
- **Chapter 4 (Disciplinary Powers and Procedures)** defines professional misconduct and reviews the disciplinary process. The chapter concludes with the case history of the Burnaby supermarket roof collapse, which led to changes in the licensing laws.
- **Chapter 5 (Technical Societies)** discusses the key role of technical societies, lists many societies by discipline, including charitable, honorary, and student societies, and describes the Iron Ring and Earth Sciences Ring rituals to welcome new members into our professions.

Part Two—Professional Practice

These five chapters give essential, basic, practical knowledge needed by professionals. A wide variety of relevant topics are grouped under appropriate headings.

- **Chapter 6 (Basic Concepts of Professional Practice)** describes professional working conditions, salary expectations, responsibility levels, promotion options, the significance of the professional seal, and how and when to use the seal.
- **Chapter 7 (Private Practice, Consulting, and Business)** describes the benefits of private practice; encourages the professional to consider private practice as a career; and explains the basic steps in doing so, including licensing, business formats, assistance available, and the Quality-Based Selection (QBS) process for consultants.
- **Chapter 8 (Hazards, Liability, Standards, and Safety)** gives important advice for avoiding hazards and liability by using standards to ensure safety. Safety is a critical matter for professionals, so this chapter includes a comprehensive section on Occupational Health and Safety (OHS) legislation, and two important case histories—the Rivtow Marine and the Westray Mine—in which unsafe practices led to financial loss and death.
- **Chapter 9 (Computers, Software, and Intellectual Property)** explains liability and ethical problems related to computers, and gives advice for using commercial software and avoiding software piracy. The chapter summarizes intellectual property laws, including copyright, patents, industrial designs, integrated circuits, and trademarks. Case histories on the Hartford Arena collapse and patent infringement illustrate the concepts.
- **Chapter 10 (Fairness and Equity in the Professional Workplace)** explains that harassment and discrimination are not acceptable in our profession. Such behaviour is illegal under the Canadian Criminal Code and human rights legislation. The chapter includes three case studies concerning discrimination and harassment.

Part Three—Professional Ethics

These four chapters explain the basic principles of ethics and justice, apply them to common employment, management, and consulting situations, and illustrate them with 16 case studies and three case histories.

- **Chapter 11 (Principles of Ethics and Justice)** explains four well-established ethical theories and the basic principles of justice. The chapter discusses and compares the Codes of Ethics mandated by the licensing Associations, and proposes a six-step strategy for solving ethical problems. Most readers find this chapter enlightening, as it links theory to practice.
- **Chapter 12 (Ethics in Professional Employment)** examines ethical issues in professional employment, such as unethical managers, labour activities, and conflicts of interest. The chapter includes five case studies, and closes with the Challenger space shuttle case history—a disaster caused when an engineer's advice was over-ruled by a manager.
- **Chapter 13 (Ethics in Management)** examines typical issues in management, such as adhering to the licensing Act, hiring, dismissal, reviewing performance, and conflict of interest. The chapter includes six case studies, as well as the case history of the Vancouver Second Narrows Bridge collapse, illustrating the importance of checking plans and calculations thoroughly.
- **Chapter 14 (Ethics in Private Practice and Consulting)** examines ethical issues in consulting, such as client–consultant relationships, advertising, competitive bidding, confidentiality, conflict of interest, and more. The chapter includes five case studies, as well as the inspiring case history of consultant William LeMessurier and the Citicorp Tower.

Part Four—Environmental Practice and Ethics

These three chapters discuss the professional's duty to protect the environment, and illustrate the importance of ethical decisions by describing present and potential environmental threats, and examining several well-known environmental disasters. Sustainable thinking is the key to ensuring the future quality of our environment:

- **Chapter 15 (Environmental Ethics)** reviews the professional's duty to protect the environment, the various laws and guidelines that apply, and the duty to report unethical behaviour, often called "whistle-blowing."
- **Chapter 16 (Environmental Threats and Disasters)** describes many threats to Canada's environment. The chapter closes with two very thorough case histories on toxic pollution and nuclear safety, showing how unsafe practices can easily lead to tragedy.
- **Chapter 17 (Environmental Sustainability)** defines sustainability and gives a brief history of sustainable thinking; reviews climate change,

the depletion of fossil fuels, and how to make our lifestyle sustainable; and closes with the Ladyfern Natural Gas Field case history.

Part Five—Exam Preparation

- **Chapter 18 (Writing the Professional Practice Exam)** describes the Professional Practice Examination (PPE) syllabus and format, but should be of value to anyone preparing to write any ethics exam. The chapter suggests a general technique—the EGAD! strategy—for writing essay-type ethics exams, and includes about 30 practice questions, many with answers.

FEATURES

This comprehensive textbook is a reference for practising engineers and geo-scientists, and is suitable for individual study or classroom use. The Fourth Edition has the following features:

- The writing style is logical and readable.
- The coverage includes every province and territory in Canada.
- The topics have been completely reorganized to group them logically.
- All chapters have been revised, updated, and thoroughly rewritten.
- A new chapter on the important topic of environmental sustainability has been added.
- The chapter on ethical theories has been expanded to include principles of justice.
- Thirteen case histories of actual events illustrate that unethical practices can lead to personal tragedy or disaster.
- Over 20 case studies pose realistic ethical problems, ask readers to suggest the appropriate course of action, and then recommend a solution.
- About 30 typical examination questions, from several provinces, assist readers who are preparing for the Professional Practice Examination.
- Personal advice is included to guide young professionals in planning their careers.
- Professional practice is illustrated from several perspectives, with the engineer or geoscientist as employee, as manager, or in private practice.
- Topics for further study and discussion are at the end of each chapter, and many more are available on the textbook's website.

The website accompanying the textbook (www.andrews4e.nelson.com) contains over 300 additional pages of material, including:

- A—Links to all provincial and territorial licensing Associations;
- B—Important excerpts from the Acts that regulate engineering and geo-science, including Codes of Ethics, admission criteria, definitions of professional misconduct, and disciplinary powers for all provinces and territories;

- C—Codes of Ethics from many technical societies;
- D—NSPE guidelines on working conditions for professional employees;
- E—Over one hundred more assignments and discussion topics, organized by chapter;
- F—Twenty-five detailed case studies, with the author's recommended solutions; and
- G—Two articles: "Reducing Hazards in Design" and "Getting Started in Consulting."

ACKNOWLEDGMENTS

I would like to thank the many creative people who provided assistance or advice in writing this textbook, as well as everyone who gave permission to publish copyrighted material.

I would also like to acknowledge many others who assisted me in writing the three earlier editions of this textbook. I am indebted to the following for past assistance: Scott Duncan, Sarah Duncan, Semareh Al-Hillal, Su Mei Ku, John D. Kemper, Ken Nauss, Grant Boundy, Stephen Jack, Georges Lozano, Wendy Ryan-Bacon, G.A. Bernard, E.R. Corneil, Dennis Brooks, John Gartner, C. Peter Jones, Harold Macklin, Gordon Slemon, Al Schuld, Kent Fletcher, Elvis Rioux, Richard Furst, Anita Direnfeld, Larry Gill, Laurie Macdonald, Chris Lyon, Jerry M. Whiting, Norman Ball, Dick van Heeswijk, Judith Dimitriu, David Frost, Gilles Y. Delisle, Richard Thibault, Andrew Latus, Deborah Wolfe, Karen Martinson, Marie Carter, Marc Bourgeois, David Burns, the late Alan Hale, Beth Weckman, Herb Ratz, Roydon Fraser, Dwight Aplevich, the late Kathy Roenspiess, Evelyn Veitch, Anthony Rezek, Joanne Sutherland, Terri Rothman, Susan Calvert, and Matthew Kudelka.

I am indebted to many reviewers who read the Third Edition and contributed advice, opinions, and evaluations that were critically important in shaping the chapter organization and content of this Fourth Edition. The reviewers were Judith Dimitriu of Ryerson University, David Frost of McGill University, Gilles Y. Delisle of the University of Ottawa, Brian Orend of the University of Waterloo, A.O. Abd El Halim of Carleton University, and N.S.W. (Norm) Williams. I also appreciated the extensive input from Milt Petruk, Manager, Examinations, APEGGA, and the members of the Professional Practice Exam Committee, who reviewed the draft manuscript.

Sincere thanks are directed to the licensing Associations and *Ordres* for engineering and geoscience in every province and territory for their kind assistance, and to Engineers Canada, which provided useful statistics. In particular, I would like to thank Deborah Wolfe, Director, Education, Outreach and Research; Samantha Colasante, Manager, Research; and Marc Bourgeois, Manager, Communications. Much-appreciated information on the history of the Canadian Council of Professional Geoscientists was received from Dr. Gordon D. Williams, former president of CCPG. Wanda Howe of CCOHS and Stephanie Gray of ACEC gladly responded to requests for information.

Richard Furst of PEO and Harry McBride, formerly of APEGBC, provided generous advice for past editions, included again in this edition.

Chapter 10 was contributed by Dr. Monique Frize, P.Eng. O.C., Professor in the Faculty of Engineering, University of Ottawa and Carleton University, and formerly the NSERC/Nortel Chair for Women in Science and Engineering (Ontario). Dr. Roydon Fraser of the University of Waterloo collaborated in writing Chapter 17, and Dennis Burningham, a petroleum engineer in Britain, contributed many useful insights for that chapter. I am very grateful for their help. I received advice and encouragement from Waterloo colleagues Dwight Aplevich and Carolyn Macgregor, and from colleagues Dr. Harold Davis, P.Eng. (UBC), Steven Brown, P.Geo. (Calgary), Toivo Roht (Manotick), and Jon Legg (Ottawa).

The contribution of Nelson Education Ltd. staff members was critically important and much appreciated. Although I have not personally met all of the people involved in acquisition, development, photographs, permissions, copy editing, cover design, production, and marketing, I thank them for their creative and professional work. I send special thanks to Elke Price, Senior Developmental Editor, who provided excellent communication and support during the unexpectedly lengthy rewriting, and to Paul Fam, Executive Editor, Higher Education, who monitored and resolved several key problems during the gestation of this edition.

Finally, I would like to express my thanks and appreciation to my wife, Isobelle, for her companionship and unwavering support during the writing and revision of the manuscript.

Gordon C. Andrews
July 1, 2008

CONTENTS

PART FOUR—ENVIRONMENTAL PRACTICE AND ETHICS

Chapter 15   Environmental Ethics

Web Appendixes
The website accompanying this textbook (www.andrews4e.nelson.com) consists of over 300 pages of additional information in the following appendixes:

**APPENDIX A—PROVINCIAL AND TERRITORIAL
         ENGINEERING/GEOSCIENCE ASSOCIATIONS**
**APPENDIX B—EXCERPTS FROM THE ACTS, REGULATIONS, AND
         CODES OF ETHICS**
**APPENDIX C—CODES OF ETHICS FOR VARIOUS TECHNICAL
         SOCIETIES**
**APPENDIX D—NSPE GUIDELINES TO EMPLOYMENT FOR
         PROFESSIONAL ENGINEERS**
**APPENDIX E—ADDITIONAL ASSIGNMENTS AND DISCUSSION
         TOPICS**
APPENDIX F—ADDITIONAL CASE STUDIES
APPENDIX G—ADDITIONAL ARTICLES OF INTEREST

Chapter 1
Introduction to the Professions

Whether you are entering engineering or geoscience, welcome to a challenging, creative, and rewarding career! Engineering and geoscience are highly respected professions that guard our health and safety, improve the quality of our daily lives, and generate great wealth. Chapter 1 begins with a review of many engineering and geoscience achievements, and gives a brief overview of the professions in Canada. The chapter concludes with a description of the tragic 1907 collapse of the Quebec Bridge, a key event in the history of the professions.

Today, good engineering and geoscience design are found everywhere, from the sleek lines of a new automobile, to the digital accuracy of electronic equipment, to the graceful structure of a bridge and in the wealth that flows from our mines, oil, natural gas, and other resource developments. The importance of engineering and geoscience is especially evident in the infrastructure of the civilized society that surrounds us, making Canada such a great place to live. Most Canadians enjoy reliable electricity; a secure supply of natural gas; safe vehicles and aircraft; and a dependable supply of pure and abundant tap water. These devices, structures, and systems were designed by engineers and/or geoscientists, and illustrate the ingenuity, competence, professional attitude, and diverse interests of the people who enter these professions.

AN INSPIRING LEGACY

Our history is an exciting chronicle of great achievement in a harsh climate. Canada is a huge country, and early settlers were faced with dense forests, rough and rocky terrain, and vast distances that obstructed travel, trade, settlement, and agriculture. Engineers and geoscientists responded willingly to the challenge. For example, the Rideau Canal, built in 1832 to connect Ottawa and Kingston, was Canada's first megaproject. The canal is an ingenious linkage of dams, locks, waterways, rivers and lakes, extending for 202 km (126 miles). The Canal was built by manual labour in only five years—a remarkable achievement, considering the primitive tools of the time.

In 2007, the United Nations Educational, Scientific and Cultural Organization (UNESCO) designated the Rideau Canal and the fortifications in Kingston (Fort Henry and the Martello towers) as a World Heritage Site on the 175th anniversary of their completion. The Rideau Canal joins 800 other sites on the world heritage list, such as the pyramids of Egypt and the Great Wall of China. The Rideau Canal is the only North American canal system of its era that remains in use today, with almost all of its original structures intact.[1]

The "Top Ten" Canadian Engineering and Geoscience Achievements

The Rideau Canal is only part of a great legacy that still inspires us today. A jury of prominent Canadians identified the "top ten" engineering and geoscience achievements in 1987, the centennial year of the Engineering Institute of Canada (EIC). The ranking was based on the significance of the achievement, the contribution to Canadian well-being, international recognition, management required, and, of course, originality, ingenuity, and creativity or uniqueness.[2] Over 110 projects were nominated, and each project selected has a social or political significance that goes far beyond mere technical success. Using any measure of impact, cost, or size, these projects are critically important, are uniquely Canadian, and most are simply massive!

Photographs of a few of them are in this text, and many more are in the impressive pictorial history book by Norman Ball.[3] Also, more information is readily obtained from *The Canadian Encyclopedia*.[4] These successes are only a sample of thousands of projects in which we may all take pride.

THE TRANSCONTINENTAL RAILWAY NETWORK (CPR)

The importance of (and the huge investment in) the transcontinental railway cannot be overemphasized. The railway connected the country from Atlantic to Pacific, making Canada a viable social, economic, and political entity. The federal government signed a contract with the Canadian Pacific Railway (CPR) in 1880, and by 1885, the "last spike" had been hammered in. The first train from Montreal arrived in Port Moody, British Columbia, in July 1886.

THE ST. LAWRENCE SEAWAY

The St. Lawrence Seaway (also known as the Great Lakes Waterway) permits ocean-going ships to travel up the St. Lawrence River and through the Great Lakes, an almost unbelievable distance of almost 3,800 km (2,400 miles). Construction began in August 1954, and the Seaway was open for commercial ships in April 1959. The Seaway made the Great Lakes accessible to industry and trade, and is a major route for shipping bulk products, such as grain, iron ore, coal, petroleum, cement, and rolled iron and steel.

THE POLYMER/POLYSAR SYNTHETIC RUBBER PLANT IN SARNIA

During the Second World War (1939–1945) the shortage of natural rubber led to the rapid construction of the Polymer synthetic rubber factory in Sarnia, Ontario. The plant was incredibly successful and efficient, and helped the Allied forces win the war. It became a Crown corporation after the war ended, was renamed Polysar in 1976, and was eventually sold. Bayer AG of Germany now owns it.

THE ATHABASCA COMMERCIAL OIL SANDS DEVELOPMENT

The Athabasca oil sands in Alberta contain bitumen, the heaviest, thickest form of petroleum. In fact, Alberta has more petroleum than the entire proven oil reserves of the Persian Gulf. However, in its natural state, bitumen is so dense and viscous that it is suitable only for paving roads. Creating oil from bitumen is expensive and complicated. Compared to conventional crude oil, bitumen's carbon-to-hydrogen ratio is too high, so special refining processes had to be invented to remove the mineral content and to adjust the carbon–hydrogen ratio. The successful research in separation methods in the 1950s, and the first successful large-scale commercial plant at Fort McMurray in the 1960s, proved that oil extraction was viable. These methods opened the deposits to satisfy oil demands that will become even more intense in the 21st century.

THE HYDRO-QUÉBEC VERY-HIGH-VOLTAGE TRANSMISSION SYSTEM

Electric power is critically important to our standard of living, but transmitting electricity over long distances from remote hydroelectric generators can be costly and wasteful. Energy loss is proportional to the square of the current, so voltage must be high and current low. When the 735 kV Manicouagan transmission line was electrified in 1965, Hydro-Québec became the first electrical power producer to transmit electricity at voltages over 500 kV (AC). Proving that power lines are safe and stable at such high voltages is very promising for future electrical distribution.

THE CANDU NUCLEAR POWER SYSTEM

The CANDU nuclear power system produces electricity using natural uranium, moderated by heavy water. The uranium does not need enrichment, thus making the system safer in many respects than other nuclear generators. The first CANDU Nuclear Power Demonstration (NPD) reactor was built in 1962; a 200 MW Douglas Point prototype was built in 1966, which delivered electricity to the grid; and four commercial 500 MW Pickering-A units came into service from 1971 to 1973. Although nuclear power is controversial, as discussed later in this text, it has many advantages over coal-fired generators.

THE DE HAVILLAND BEAVER DHC-2 AIRCRAFT

The Canadian-designed and built De Havilland Beaver DHC-2 first flew in 1947. It could carry six passengers and cargo, in addition to the pilot, and could take off and land in very short distances. When equipped with floats or skis, it could fly into remote locations. The Beaver was a crucial aid to developing northern Canada.

THE ALOUETTE I ORBITING RESEARCH SATELLITE

In 1962, Canada became the third nation to have a satellite in orbit (after the Soviet Union and the U.S.A.). Alouette I was designed and built in Canada, and was launched on a U.S. rocket. It completed its mission successfully, investigating the ionosphere as part of ISIS (International Satellites for Ionospheric Studies) from 1963 to 1969.

THE BOMBARDIER SNOWMOBILE

Joseph-Armand Bombardier started experimenting with snow machines in 1922, and in 1937, he invented the endless-track vehicle that we know as the snowmobile. Its speed over snow-covered ground makes it essential in remote parts of Canada, and it has changed the hunting methods of the Inuit. The snowmobile changed winter life in most of Canada, although concerns about noise, environmental damage, and safety led to laws restricting snowmobile use.

THE TRANS-CANADA TELEPHONE NETWORK

The Trans-Canada Telephone System (now known as Stentor) is an association of telephone companies that was formed in 1931 to integrate national telephone service. Previously, Canada relied on U.S. transmission facilities for most cross-Canada communication. Stentor was expanded in 1969 to include Telesat Canada (Canada's sole domestic satellite carrier). In 1972, Anik A-1 was launched, followed by Anik A-2 in 1973, and Canada became the first country in the world to use satellites for domestic communications.

Significant Achievements of the 20th Century

The five most significant Canadian engineering achievements of the 20th century, chosen using objective criteria by the 1999 National Engineering Week committee, are listed below.[5]

THE CPR ROGERS PASS PROJECT (1989)

The CPR transcontinental railway, built in 1885, crosses the Rocky Mountains via the Rogers Pass, but avalanches (since snow reaches 15 m [50 ft.] per year) and steep grades were constant problems. In 1989, the CPR completed the 34-km (21-mile) Rogers Pass project, including six bridges and two tunnels. The Mount MacDonald Tunnel under the Rogers Pass is 14.7 km (9 miles) long, making it North America's longest railway tunnel.

THE CONFEDERATION BRIDGE (NORTHUMBERLAND STRAIT)

The Confederation Bridge links New Brunswick and Prince Edward Island across the Northumberland Strait. The bridge opened to traffic in May 1997 and is 12.9 km (8 miles) long, which makes it the world's longest saltwater bridge subject to ice hazards. The distinctive arches provide a clearance of 60 m (197 ft.) for seagoing vessels.

THE CANADARM REMOTE MANIPULATOR SYSTEM

Canadian engineers, from many disciplines, designed and developed the Canadarm for the U.S. space shuttle. The Canadarm was used to deploy and retrieve satellites from the shuttle cargo bay. The arm was installed and flew on the second shuttle flight in 1981. A larger version, the Canadarm2, is part of Canada's contribution to the International Space Station and was installed on the space station by Canadian Space Agency astronaut Chris Hadfield in April 2001.

Photo 1.1 — The Mountain Creek Bridge, BC. *The Mountain Creek Bridge on the Canadian Pacific Railway line in the Selkirk mountain range in British Columbia was erected in 1884, and is a tribute to the ingenuity, skill, and determination of the builders of the transcontinental railway. The bridge was 50 m (164 ft.) high and 331 m (1,086 ft.) long, and contained an immense amount of locally cut timber. It was the longest bridge on the CPR transcontinental railway, but within two decades, the wood had deteriorated, and was replaced by steel.*

Source: photos.com/JUPITERIMAGES.

THE IMAX SYSTEM (MOTION PICTURE PHOTOGRAPHY AND PROJECTION)

Canadian mechanical engineer William Shaw perfected the "rolling loop" film transport mechanism in the IMAX projector, which photographs and projects images that are 10 times the size of conventional motion-picture images onto huge screens, without vibration or streaking.

THE HOPPS PACEMAKER

Canadian electrical engineer Dr. John A. Hopps, working with medical colleagues at the Banting and Best Institute in Toronto in 1949, discovered that a gentle electrical stimulus would restart a heart that had stopped beating. In 1950, Hopps developed the first heart pacemaker at the National Research Council. The pacemaker helps millions of people to lead normal, healthy lives. Dr. Hopps was in the vanguard of biomedical engineering in Canada.

The Challenge of the 21st Century

Today's engineers and geoscientists have knowledge and assistance that previous generations could barely imagine. Every design office has computer hardware and software for analysis, design, and visualization that are unmatched in history. The tools to carry out the designs are also more powerful and versatile than ever before, from giant excavators and cranes, to numerically controlled machine tools, to satellite communication. However, your generation will face one of the greatest challenges ever: countering the effect of greenhouse gas emissions on global warming. This challenge is discussed in Part 4 of this text, but a brief overview is appropriate here.

In 2007, the International Panel on Climate Change (IPCC) published their fourth report.[6] The report concludes that global warming is unmistakable and inevitable; moreover, most of the increase in global temperature since the mid-20th century is very likely due to human-caused greenhouse gas emissions. The effects of global warming occur slowly, but will alter life as we know it.

Unless changes are made in our greenhouse gas emissions, the gradual rise in surface air temperatures will increase severe weather events, such as heavy rainfalls, heat waves, droughts, hurricanes, and so forth, and will eventually melt Arctic and Antarctic ice, and raise sea levels. Insect-borne diseases will be able to move further north, and many animal species will face extinction. Fish stocks may be seriously depleted, creating food shortages in some nations. To avoid the most serious effects, the report indicates that the average global temperature rise must be kept below 2°C. The challenge will be to see if we can reverse the trend without experiencing a devastating impact on our standard of living.

Every nation in the world is affected, so political, economic, and social changes will also be necessary if populations are to be convinced to reduce their consumption. However, a huge task will likely fall on engineers and geoscientists—key professionals who know how to reduce emissions by

Photo 1.2 — The Blackberry® Handheld. *The Blackberry® is a well-known Canadian communications invention that converges the features of a wireless tele-phone, computer, and geographical position sensor (GPS) into a single, hand-held personal digital assistant (with calendar, to-do list, and telephone book). The Blackberry sends and receives secure e-mail immediately, permitting users to remain instantly available, wherever they may be. The handy device proliferates in all levels of industry and government. Research in Motion (RIM) the Waterloo, Ontario inventor and manufacturer, reports that over 21 million Blackberry owners subscribe to the RIM service (as of February 2008).*

Source: MAGNUM/Peter Marlow.

increasing efficiency of existing processes, machinery, buildings, and infra-structure. We must reduce energy waste, especially energy wasted in trans-portation, in heating buildings, and in electrical production and delivery, and we must increase efficiency by re-using materials and recycling waste into useful energy or products.

The IPCC report gives an extensive list of areas where engineers and geoscientists should change technology or where new technology should be developed. Part 4 of this textbook discusses these topics in more detail. A challenging task awaits the engineers and geoscientists of the 21st century, but with the knowledge and tools at your disposal, you are well equipped to meet that challenge!

ENGINEERING AND GEOSCIENCE IN CANADA

Engineering and geoscience are recognized as professions in Canada. Engineering was first regulated as a profession during the 1920s, and geoscience (specifically, geology and geophysics) was first regulated in Alberta in 1955. All provinces and territories now recognize, license, and regulate both professions, with a few minor exceptions. The two professions are closely related, so most provinces and territories regulate them under the same laws (as explained in detail in Chapter 2).

Provincial and Territorial Licensing Laws

Each province and territory of Canada has passed a law or "Act" that establishes engineering and/or geoscience as a profession. Each Act, in turn, creates an Association of Professional Engineers and/or Geoscientists (in Quebec, the *Ordre des ingénieurs du Québec* or the *Ordre des géologues du Québec*). These Associations are the licensing bodies that are responsible for setting and enforcing high standards of practice in engineering and geoscience. That is, the Associations enforce the qualifications for admission into the profession, set standards of professional practice, and discipline members who fail to meet these standards. They also prevent the misuse of titles and/or the illegal practice of the profession by unqualified individuals.

Entering the Professions

Academic and experience requirements are very high to enter either engineering or geoscience. Obtaining a licence in Canada typically requires both a four-year university degree and an internship—usually four years of acceptable experience. (See Chapter 2 for a detailed explanation of admission procedures.)

Distribution of Engineers and Geoscientists in Canada

In 2006, about 160,000 practising professional engineers were licensed in Canada, according to *Engineers Canada*.[7] The number of professional geoscientists increased sharply in recent years, from about 6,500 in 2001 to approximately 7,700 in 2007, and is expected to reach 10,000 in a few years, according to the Canadian Council of Professional Geoscientists (CCPG).[8]

The numbers of both engineers and geoscientists are both approximate: Many individuals (about 12 percent) are licensed in more than one province or territory, or are licensed as both engineers and geoscientists, and professionals born in the 1945 "baby boom" are now rapidly retiring. The distribution of professionals is also changing, as resource developments in Alberta, British Columbia, and Saskatchewan attract them.

The distribution of professional engineers across Canada, shown in Table 1.1, is not uniform: Most are clustered in the industrially developed regions of Ontario (39 percent) and Quebec (23 percent), with the next-largest number in the resource-rich province of Alberta (17.6 percent). When we compare the coasts, we find that British Columbia (with 9.09 percent) has many more engineers than the four Atlantic Provinces combined (5.9 percent).

TABLE 1.1 — Licensed Professional Engineers and Geoscientists in Canada

Professional Engineers	Engineers Licensed in 2006	Percentage of Total Engineers
Ontario	60,987	39.22%
Quebec	35,368	22.75%
Alberta	27,422	17.64%
British Columbia	14,137	9.09%
Saskatchewan	3,559	2.29%
Manitoba	3,556	2.29%
Nova Scotia	3,548	2.28%
New Brunswick	3,391	2.18%
Newfoundland & Labrador	1,814	1.17%
Northwest Territories	894	0.58%
Prince Edward Island	413	0.27%
Yukon	396	0.25%
Total—Professional Engineers	**155,485**	
(See note 1)	**100.0%**	
Total—Professional Geoscientists	**7,700 (Estimate for 2007)**	
(See note 2)		

Note 1: Professional Engineers: The number of professional engineers does not include non-practising engineers, engineers-in-training, or engineering students, but includes some duplicate members, since about 12 percent of professional engineers are licensed to practise in more than one jurisdiction.

Note 2: Professional Geoscientists: The number of professional geoscientists is approximate because three jurisdictions (Ontario, Quebec, and Nova Scotia) recently passed legislation regulating geoscientists, licensing has increased sharply in recent years, and the distribution of geoscientists is changing.

Source: Engineers Canada, *2006 Membership Survey*, June 2007. Data reproduced with permission of Engineers Canada.

Saskatchewan and Manitoba together (4.6 percent) have about a quarter of those in Alberta, the adjacent province. The Yukon and Northwest Territories have very few (about 0.8 percent), even though Nunavut is included in the Northwest Territories' total.

Branches of Engineering and Geoscience

Universities offer degree programs in recognized disciplines (or branches) of engineering and geoscience. These programs have well-identified course requirements, but more importantly, the programs are accredited (for most disciplines, in Canada), so they are guaranteed to be acceptable for licensing. However, licensed engineers and geoscientists may practise in any discipline for which they can justify their competence, regardless of which discipline they studied in university. (The responsibility for competence is explained later in this text—for example, see "Responsibility for Continuing Competence" in Chapter 3 and "Ensuring Competence" in Chapter 14.) Many branches or disciplines are accredited for education and licensing:

- **Engineering.** The most general branches are civil, electrical, mechanical, industrial, and chemical engineering, but many more exist, from Aerospace Engineering (at Carleton and Ryerson) to Water Resource Engineering (at Guelph). In fact, the 2007 Canadian Engineering Accreditation Board (CEAB) report lists about 80 accredited engineering programs that are currently offered at Canadian universities (including about 24 French engineering programs offered in Quebec). Some newer disciplines, such as Mechatronics and Nanotechnology Engineering, are awaiting CEAB accreditation, and are not yet listed.[9]
- **Geoscience.** The Canadian Council of Professional Geoscientists (CCPG) recognizes that acceptable geoscience education and experience is usually obtained in three main areas and recognizes professional practice in Geology, Environmental Geoscience, and Geophysics.[10]

Employment Prospects

In spite of recent economic turmoil (in 2008–2009), two factors show employment prospects for graduate engineers and geoscientists are still very positive. First, the retirement of the "baby boom" generation (born after the end of the Second World War in 1945) will begin to peak in 2010. Many senior engineers and geoscientists will be retiring, creating opportunities for promotion. Second, a drive for increased efficiency, new energy sources, and alternative (less-polluting) energy sources will stimulate research, investment opportunities, and jobs. Engineers and geoscientists are essential if we are to sustain and improve our standard of living while competing with other countries and adapting to the evolving threats of climate change and peak oil. The employment history of the past few decades is also reassuring, although history is not necessarily a predictor of future trends.

- **Engineering.** The documented employment history for engineers is fairly stable. The unemployment rate for engineers was typically around 1 percent in the decade prior to 1982, indicating very secure employment. During the recession of 1982, unemployment reached a peak of 7,000 engineers, or about 6 percent of registered professional engineers.[11] It declined gradually until it was typically below 2 percent by 1997 (when the national average Canadian unemployment rate was 4.8 percent).[12] The collapse of the "dot.com bubble" in 2000, the terrorist attacks of 2001, and the recessionary effects of scandalous bankruptcies in several giant American companies, including Enron and WorldCom, had an impact on the entire global economy. As a result, the unemployment rate for engineers rose to about 3 percent in 2002 (when the national average was 7.5 percent).[13] The outlook for engineering in 2009, as estimated by *Job Futures*, the Government of Canada's National Career and Education Planning Tool, shows that most engineering disciplines have unemployment rates in the range of 1 to 4 percent, as compared to the national average of 7 percent.[14]

- **Geoscience.** Geoscience has fewer practitioners than engineering, so historical employment data are not easily found. The *Job Futures* statistics categorizes geoscientists, with several unrelated professions, under the heading of Physical Science Professionals. The 2007 unemployment rate is only 2 percent for this category, as compared to the national average of 7 percent.[15] In view of the rising demand for fossil fuels, the demand for geoscientists should continue to grow for the foreseeable future. Professional engineers and geoscientists therefore have good employment prospects. However, if economic turmoil claims your job, remember that the technical societies (discussed in Chapter 5) provide good contacts. Alternately, consider upgrading your qualifications (as discussed in Chapter 4) to prepare for a better job.

A Brief Discussion of Professional Status

The general public holds the engineering and geoscience professions in high regard. In Canada and the United States, opinion surveys consistently show engineers near the top for honesty and integrity (and this ranking would include geoscientists, although they are rarely identified as a separate group). But are engineering and geoscience really professions?

To answer this question, we need to define the term "profession" more precisely. What is a profession? How does it differ from a job? The following dictionary definition of a profession helps to answer these questions:

> **Profession:** A calling requiring specialized knowledge and often long and intensive preparation including instruction in skills and methods as well as in the scientific, historical, or scholarly principles underlying such skills and methods, maintaining by force of organization or concerned opinion high standards of achievement and conduct, and committing its members to continued study and to a kind of work which has for its prime purpose the rendering of a public service. (Copyright © Webster's Third New International Dictionary)[16]

Engineering and geoscience certainly require "specialized knowledge," "intensive preparation," and "instruction in skills and methods as well as in the scientific, historical, or scholarly principles underlying such skills and methods." In fact, licensing Acts now require engineers and geoscientists to complete at least four years of formal education and three to four years of relevant work experience before they can practise. This equals the preparation required in medicine and law (two professions that serve as a useful basis for comparison).

Engineering and geoscience also have a "force of organization," in the form of laws and regulations. These have been enacted in every province and territory in Canada (except Prince Edward Island and Yukon, where geoscience is not yet a regulated profession). The Acts, laws, and regulations include Codes of Ethics committing practitioners to "high standards of achievement and conduct" (as discussed in detail in Chapter 11).

Significantly, engineering and geoscience (like medicine and law) are "self-regulating" professions. That is, the government delegates the responsibility for admission, for standards of practice, and for discipline, to the members of the profession. Some differences exist. Unlike medical doctors and lawyers, who are generally self-employed and work with clients on a one-to-one basis, most engineers and geoscientists are employees of large companies, where they work in teams. Moreover, it is a basic fact of life that engineers outnumber every other self-regulating profession (including geoscientists, whose numbers are still rather exclusively small). However, these are minor differences; engineering and geoscience clearly have professional status.

THE TECHNICAL TEAM

Today's complex projects need specialized skills and knowledge, so most engineers and geoscientists work on technical teams. A recent survey of engineers and geoscientists, sponsored by Engineers Canada, showed that the majority (79 percent) worked on teams. Most of the teams (three out of four) were formed on a project-by-project basis. Engineers were the main team members (69 percent), but almost half of the teams (46 percent) included a technician or technologist, and a quarter of the teams (25 percent) included a non-engineering technical person.[17]

This data confirms that, while engineers and geoscientists are the vital link between theory and application, they must work effectively with other team members. Respect for each team member's expertise is essential for a productive work environment. The full technical spectrum includes research scientists, geoscientists, engineers, architects, technologists, technicians, and skilled workers, as described in the next few paragraphs.

- **Research scientist.** Scientists develop ideas that expand the frontiers of knowledge—ideas that may not have practical applications for many years. A doctorate is typically the basic educational requirement, although a master's degree is often acceptable. A scientist is rarely required to supervise other technical personnel except research assistants, and usually is a member of several learned societies in his or her particular field of interest.

Basically, the task of the research scientist is to generate new knowledge, whereas the task of the engineer and geoscientist is to apply that knowledge. The roles of the scientist, the geoscientist, and the engineer overlap, and in some projects, the boundary may be invisible. It is sometimes only the goal of the work that differentiates the two, not the actual duties. Most scientists work in government agencies, universities, or institutes, and their results are usually published in scholarly technical journals. Such new knowledge is occasionally very valuable to industry, and it is often free to anyone who knows where to look for it.

- **Geoscientist.** Geoscience is a fairly new name for the well-established fields of geology, geophysics, and their many related subdisciplines. Geoscientists are mainly concerned with the study, measurement, and analysis of the earth and the many systems, such as petroleum and hydrology, which operate within the earth. The main role of the geoscientist is the practical application of scientific ideas, not the study of natural science. For example, geoscientists analyze seismic, gravitational, magnetic, and other data to discover minerals and fuels, locate stable foundation sites for structures, and identify dangers related to the dynamic movement of our Earth. Geoscientists are licensed (usually under a provincial Geoscience Act) in order to practise. A bachelor's degree in geoscience (typically geology, geophysics, or related disciplines) is the minimum educational requirement. It is important to note that geoscientists and engineers play a key role in any exploitation of the earth (such as resource extraction) that affects life, health, property, or the welfare of the public, because a "qualified person" (as defined in Chapter 2) must approve such activities.

- **Engineer.** Engineers are mainly concerned with the practical application of science. They link theory to practice, so design is a key area of employment—that is, creating plans for devices, systems, and structures for human use. Many engineers are also involved in construction, testing, manufacturing, and a few in the discovery and distribution of natural resources. In these activities, engineers make many decisions that affect life, health, property, or the welfare of the public, so they must be licensed (usually under a provincial Engineering Act) in order to practise. They must have extensive theoretical knowledge, the ability to think creatively, and a knack for obtaining practical results. A bachelor's degree in engineering is the minimum educational requirement.

- **Architect.** Architects are mainly concerned with the planning, aesthetic design, and construction management of buildings, including residences, offices, and institutional and industrial buildings. Architects, geoscientists, and engineers often work together, since each has a specialty that may apply to such buildings. For example, an architect may conduct the aesthetic design and layout plan for a large building; may engage professional geoscientists to assess the foundation design and seismic problems; and may engage professional engineers to design the structural steel and internal heating, ventilation, and air conditioning systems. Architects also make decisions that affect life, health, property, or the welfare of the public, so they must be licensed (usually under a provincial Architects Act) in order to practise. These Acts typically contain several clauses that define the boundary between the responsibilities of engineer, geoscientist, and architect. A bachelor's degree in architecture is the minimum educational requirement.

- **Technologist.** Technologists work most closely with engineers and geoscientists, and often perform key aspects of engineering or geoscience practice, such as design, testing, computing, solving problems, supervising, project management, and so forth, under the direction of a licensed practitioner. The basic educational requirement is usually a diploma from a technology program at a community college, CEGEP (*Collège d'enseignement général et professionel*), or CAAT (college of applied arts and technology), although many technologists have a bachelor's degree (usually in science, mathematics, or technology). Technologists often supervise the work of others and are encouraged to have qualifications that are recognized by a technical society.

In fact, the province of Alberta regulates professional technologists under the Engineering, Geological and Geophysical Professions Act, and technologists are represented on the Association of Professional Engineers, Geologists, and Geophysicists of Alberta (APEGGA) Council, Board of Examiners, and other relevant committees. Alberta has taken the lead in making the boundary between engineers, geoscientists, and technologists less rigid. The technologist designation is a "right to title." That is, the professional technologist designation confers the right to use professional technologist titles, but does not limit the right of unlicensed technologists to practise as technologists. It is a positive step, since more effective use of technologists will likely increase productivity.

Associations of engineering technicians and technologists have been established in all 10 provinces (although not in the territories) to certify the qualifications of technologists. The provincial technician/technologist associations are, in turn, members of an umbrella organization, the Canadian Council of Technicians and Technologists (CCTT), which is a federation of the 10 provincial associations. CCTT was established in 1972 to coordinate activities and facilitate exchanges of information among the provincial associations of technicians and technologists.

In 2007, the 10 provincial associations within CCTT represented more than 49,000 registered technicians and technologists across Canada.[18] However, since certification is voluntary, there were probably many more people actually practising as technicians or technologists. The total number is estimated to be roughly equal to the number of professional engineers and geoscientists practising in Canada (over 160,000).

Certification as a technologist requires an assessment of the candidate's educational background. The candidate also must document at least two years of relevant experience. Certification is voluntary and is not required in order to work as a technician or technologist in Canada; however, only certified technologists are entitled to use the following designations: Certified Engineering Technologist (CET), Applied Science Technologist (AScT), Registered Engineering Technologist (RET), or Technologue Professionnel (TP). Which title is used depends on the province in which certification is granted. In recent years the certification of technologists and technicians has been assisted by provincial Associations of Professional Engineers. In fact, the organization of the provincial technology associations and CCTT closely parallels that of the provincial engineering Associations and Engineers Canada, as described in the next chapter.

- **Technician.** Technicians usually work under the supervision of engineers or technologists in the practical aspects of engineering tests or equipment maintenance. The basic educational requirement is usually a diploma from a program at a community college, CEGEP, or CAAT. This program is usually shorter than for technologists. In most provinces the title Certified Technician (C.Tech.) may be awarded by the provincial associations of technicians and technologists after the applicant completes the appropriate education and acquires two years of appropriate experience. Certification is not essential to work as a technician.
- **Skilled Worker.** Typically, skilled workers apply highly developed manual skills to carry out the designs and plans of others. Master artisans train skilled workers, and the quality of a worker's apprenticeship is more important than the worker's formal education. Each type of trade worker (electrician, plumber, carpenter, welder, pattern maker, bricklayer, machinist, etc.) comes under a different certification procedure, which varies from province to province.

In a large project, each group in this technical spectrum will be required, at different times. Each has a different task and has different skills, knowledge, and training to offer. Knowledge of each group, and mutual respect for their expertise, are the keys to successful projects and a cooperative, productive working environment.

Engineers and geoscientists are usually the key link between scientific theory and practical applications but, as we see in Chapter 2 of this text, legal definitions are required to delineate their roles, and to separate them from natural scientists and architects.

INTRODUCTION TO CASE HISTORIES

As Canadians, we can be proud of many spectacular achievements, from the construction of the transcontinental railway in 1885, to the flight of the Avro Arrow in 1958, to the opening of the Confederation Bridge in 1997. We tend to take success for granted when well-designed structures and devices work properly. In contrast, when structures or projects fail, we focus our attention on the failure. We ask why it happened and how similar failures can be avoided in the future.

When a failure is costly, in lives or in money, an investigation panel or Royal Commission studies the failure impartially and publicly. As a result, we often learn more from failures than from successes, although the lessons are learned at a great cost. Remember that failure itself is not proof of unethical or incompetent practice. Many projects push the limits of knowledge. New projects always involve risk, and even determined, ethical professionals cannot guarantee success every time.

The case histories in this textbook involve engineers and/or geoscientists, and concern ethical aspects such as negligence, incompetence, conflict of interest, or corrupt practices. Many of the cases are fairly well known, occurred in Canada, and are tragic, but some good may result if we can learn from them and avoid similar tragedies in future.

CASE HISTORY 1.1

THE QUEBEC BRIDGE COLLAPSE

The first case history tells the story of the collapse of the Quebec Bridge in 1907, and the negligence that led to that collapse. This case is important, even a century later, because many lessons were learned (as summarized at the end of this chapter) and Canadians were made aware of the need for the professional licensing of engineers.

An Overview of the Project

The Quebec Bridge, which had its official opening in 1919, is the longest cantilever span in the world, with a centre-distance between supports of 549 m (1,800 ft.). The massive size of the Quebec Bridge makes it a very impressive structure. In fact, you must see it in person to fully comprehend its grandeur.

However, the Quebec Bridge is infamous for the many lives lost in the harrowing accidents that occurred during its construction. *The Canadian Encyclopedia Plus* summarizes these tragic losses succinctly:

Québec Bridge Disasters: Construction on the Québec Bridge, 11 km above Québec City, officially began in 1900. On 29 August 1907, when the bridge was nearly finished, the southern cantilever span twisted and fell 46 m into the St. Lawrence River. Seventy-five workmen, many of them Kahnawake (formerly Caughnawaga), were killed in Canada's worst bridge disaster. An inquiry established that the accident had been caused by faulty design and inadequate engineering supervision. Work was resumed, but on 11 September 1916 a new centre span being hoisted into position fell into the river, killing 13 men. The bridge was completed in 1917 and the Prince of Wales (later Edward VIII) officially opened it 22 August 1919.[19]

The residents of Quebec City advocated building a bridge over the St. Lawrence River as early as 1852, and a site had been chosen where the river narrowed just upstream of the city. Designs were prepared, but serious work did not begin until 1900. The success of the cantilevered Forth Bridge, built in 1890 in Scotland, was a factor in the choice of a cantilever design for the Quebec Bridge. The Forth Bridge, the first bridge built entirely of steel, has two spans of 521 m (1,710 ft.) each. At the time, these were the world's longest unsupported (cantilevered) bridge spans, and they would remain so for 27 years, until the Quebec Bridge was successfully completed.

At the time of the 1907 accident, four parties were directly involved in constructing the Quebec Bridge superstructure:

- the Government of Canada, which had provided subsidies and a guarantee of bonds to
- the Quebec Bridge & Railway Company (known simply as the "Quebec Bridge Company"), which had responsibility for the complete structure, and which had contracted with
- the Phoenix Bridge Company in Phoenixville, Pennsylvania, to design and construct the superstructure, and which had subcontracted with the
- the Phoenix Iron Company, to fabricate the steel components.

The Quebec Bridge Company employed a chief engineer, Edward Hoare, on the site, and a consulting engineer, Theodore Cooper of New York, as well as many hundreds of erection and inspection staff. In technical terms, Cooper was highly competent: "In the extent of his experience and in reputation for integrity, professional judgement and acumen, Mr. Cooper had few equals on this continent." Early in the design work, it was decided that Cooper's decisions on technical matters would be final. Cooper insisted on this, so the government gave him full technical authority, in writing, as an order-in-council.[20] Astonishingly, although Cooper was the ultimate technical authority, he visited the Quebec site only while the supporting piers were being built and was never on site thereafter. Furthermore, over the many years that the bridge components were being fabricated, he visited the Phoenix Iron Company shops only three times.[21]

Norman McLure was the Quebec Bridge Company's inspector of erection. Cooper had appointed him, with Hoare's agreement, and McLure received instructions from both of them. He reported to Hoare mainly on "matters regarding monthly estimates, and to Cooper on matters of construction."[22] The Phoenix Bridge Company's chief engineer was Mr. Deans, who was an experienced bridge builder but was more accurately described, after the accident, as its "chief business manager."[23] The design engineer was Mr. Szlapka, a German-educated engineer with 27 years of experience in designing many similar projects. Szlapka was responsible for generating the design details and had the full confidence of Cooper.[24]

A competition for the design was held in 1898. Cooper reviewed the submitted plans and recommended the Phoenix Bridge Company's design, which showed a span of about 488 m (1,600 ft.) between the supporting piers.

The contracts for detailed design and construction were signed, and work began in 1899. Cooper requested further investigation of the riverbed to ascertain the best locations for the supporting piers. After much study, he recommended that the piers be located closer to shore, thus lengthening the unsupported span to 549 m (1,800 ft.).

The First Collapse

In 1907, with the first span of the cantilever now reaching out over the water, it became obvious that some parts of the structure were deforming in unexpected ways. This was communicated to Cooper in New York. H. Petroski summarizes these fatal days concisely in his highly readable book *Engineers of Dreams*:

> The south arm of the Québec Bridge had been cantilevered out about six hundred feet over the St. Lawrence River by early August 1907, when it was discovered that the ends of pieces of steel that had been joined together were bent. Cooper was notified, by letter, by Norman R. McLure, a 1904 Princeton graduate who was "a technical man" in charge of inspecting the bridge work as it proceeded, who suggested some corrective measures. Cooper sent back a telegram rejecting the proposed procedure and asking how the bends had occurred. Over the next three weeks, in a series of letters back and forth among Cooper, chief engineer Deans, and McLure, Cooper repeatedly sought to understand how the steel had gotten bent, and rejected explanation after explanation put forth by his colleagues. Cooper alone seems to have been seriously concerned about the matter until the morning of August 27 when McLure reported that he had become aware of additional bending of other chords in the truss work and, since "it looked like a serious matter," had the bends measured; he explained that erection of additional steel had been suspended until Cooper and the bridge company could evaluate the situation.
>
> Yet, even as McLure went to New York to discuss the matter with Cooper, Hoare, as chief engineer of the Quebec Bridge Company, had authorized resumption of work on the great cantilever. As soon as McLure and Cooper had discussed the bent chords, Cooper wired Phoenixville: "Add no more load to bridge till after due consideration of facts." McLure had reported that work had already been suspended, and so contacting Québec more directly was not believed to be urgent, but when McLure went on to Phoenixville, he found that the construction had in fact been resumed. Some conflicting reports followed, thanks in part to a telegraph strike then in progress, as to whether Cooper's telegram was delivered and read in time for Phoenixville to alert Québec.
>
> In any event, the crucial telegram lay either undelivered or unread as the whistle blew to end the day's work at 5:30 P.M. on August 29, 1907. According to one report, ninety-two men were on the cantilever arm at that time, and when "a grinding sound" was heard, they turned to see what was happening.

"The bridge is falling," came the cry, and the workmen rushed shoreward amid the sound of "snapping girders and cables booming like a crash of artillery." Only a few men reached safety; about seventy-five were crushed, trapped, or drowned in the water, surrounded by twisted steel. The death toll might also have included those on the steamer *Glenmont*, had it not just cleared the bridge when the first steel fell. Boats were lowered at once from the *Glenmont* to look for survivors, but there were none to be found in the water. Because of the depth of the river at the site, which allowed ocean liners to pass, and which had demanded so ambitious a bridge in the first place, the debris sank out of sight, and "a few floating timbers and the broken strands of the bridge toward the . . . shore were the only signs that anything unusual had happened." The crash of the uncompleted bridge "was plainly heard in Québec," and the event literally "shook the whole countryside so that the inhabitants rushed out of their houses, thinking that an earthquake had occurred." In the dark that evening, the groans of a few men trapped under the shoreward steel could be heard, but little could be done to help them until daylight.[25]

Photo 1.3 — The Quebec Bridge (1907 collapse). *The twisted wreckage of the Quebec Bridge (Phoenix design) stretches out toward the south pier, after the collapse on August 29, 1907.*

Source: The Québec Bridge over the St. Lawrence River near the City of Québec: *Report of the Government Board of Engineers,* Department of Railways and Canals Canada, 1919.

The Report of the Royal Commission

Within hours of the accident, a Royal Commission was established to determine the cause. The Commission prepared a thorough report containing lessons—learned at great cost—that have benefited structural engineers and bridge designers in Canada and around the world.[26] As G.H. Duggan later wrote:

> The report of the Royal Commission appointed to investigate the failure of the Phoenix[-designed] Bridge in 1907 is very comprehensive, and goes beyond the mere taking of evidence and the investigation of the faults of the bridge, as the Commission assembled most of the available data on other long span bridges, illustrated their important features, recorded the tests on large size compression members that had any bearing upon the work, and made a number of tests to supply some lacking experimental data of the behaviour of large compression members under stress.[27]

The report concluded that appointing Hoare as chief engineer of the Quebec Bridge Company was a mistake. Although he had a "reputation for integrity, good judgement and devotion to duty," he was not technically competent to control the work. Regarding Deans, chief engineer of the Phoenix Bridge Company, the report concluded that his "actions in the month of August, 1907, and his judgement . . . were lacking in caution, and show a failure to appreciate emergencies that arose." However, the Commission assigned most of the blame for the bridge's collapse to errors in judgment made by Cooper and Szlapka.[28]

Design and Communication Deficiencies

The Commission's report identified several serious deficiencies in the bridge's design and in the construction methods followed. Specifically, the design loads were underestimated, and the engineers failed to investigate, even when the bent members showed that very high compressive stresses existed.

The stresses were originally calculated by Szlapka using an estimate of the total dead weight of the bridge—an estimate made by Cooper at the start of the design process. However, as the detail design progressed and as the precise shapes of the members were determined, the dead weight changed. The stresses should have been recalculated using more accurate estimates of the dead weight. This was not done. It is especially noteworthy that the bridge span had originally been specified as 488 m (1,600 ft.), but Cooper later recommended moving the supporting piers, which increased the span to 549 m (1,800 ft.). When the bridge's span was increased, the dead weight increased significantly, yet this increase was not included in the calculations. This point is explained more clearly by Petroski:

> In short, what Szlapka had done was to let stand an educated guess as to the weight of steel that the finished bridge would contain. Such guesses, guided by experience and judgment, are the only way to begin to design a new structure, for without information on the weight of the structure, the load that the

members themselves must support cannot be fully known. When the loadings are assumed, the sizes of the various parts of the bridge can be calculated, and then their weight can be added up to check the original assumption. For an experienced engineer designing a conventional structure, a final calculation of weights only serves to confirm the educated guess, and so such a calculation may not even be made in any great detail. In the case of a bridge of new and unrealized proportions, however, there is little experience to provide guidance in guessing the weight accurately in the first place; a recalculation, or a series of iterated recalculations, is necessary to gain confidence in the design. . . . According to the findings of the commission, "the failure to make the necessary recomputations can be attributed in part to the pressure of work in the designing offices and to the confidence of Mr. Szlapka in the correctness of his assumed dead load concentrations. Mr. Cooper shared this confidence." Since Cooper was well known to have a "faculty of direct and unsparing criticism," his confidence in Szlapka's design work went unquestioned. . . .

The underestimation of the true weight of the bridge had actually come to Cooper's attention earlier in the design process, but only after considerable material had been fabricated and construction had begun. At this time, a recalculation of the stresses in the bridge led Cooper to consider that the error had meant that some stresses had been underestimated by 7–10 percent. All structures are designed with a certain margin of safety; he felt the error had reduced that margin to a small but acceptable limit, and so the work was allowed to proceed. In fact, some of the effects of the underestimated weights were, in the final analysis, of the order of 20 percent, and this was beyond the margin of error that the structure could tolerate.[29]

Other human failures also contributed to the collapse, and addressing them might have prevented the tragedy or lessened its consequences. Because of advancing age and declining health, Cooper had been unable to visit the construction site during the two final recent years of construction. Also, Szlapka criticized Cooper for making the bottom chords curved "for artistic reasons" and for failing to visit the Phoenixville plant where the bridge parts were being fabricated. The Royal Commission's report commented on Cooper's role and on the design deficiencies and communication problems:

Mr. Cooper states that he greatly desired to build this bridge as his final work, and he gave it careful attention. His professional standing was so high that his appointment left no further anxiety about the outcome in the minds of all most closely concerned. As the event proved, his connection with the work produced in general a false feeling of security. His approval of any plan was considered by everyone to be final, and he has accepted absolute responsibility for the two great engineering changes that were made during the progress of the work—the lengthening of the main span and the changes in the specification and the adopted unit stresses. In considering Mr. Cooper's part in this undertaking, it should be remembered that he was an elderly man, rapidly approaching seventy, and of such infirm health that he was only rarely permitted to leave New York.[30]

Cooper's distance from the construction site and his inability to travel created a communication problem that played a critical role in the days leading up to the disaster. Nevertheless, even today, when cellular telephones (capable of sending photos), fax machines, e-mail, and overnight courier service permit design work to be conducted off-site, it is unimaginable that the key consulting engineer would neglect to ever visit the construction site—especially when that engineer has ultimate technical authority of the sort that Cooper wielded.

Organizational Deficiencies

The Royal Commission also criticized both the Quebec Bridge Company and Cooper for the way in which the project had been organized:

> Mr. Cooper assumed a position of great responsibility, and agreed to accept an inadequate salary for his services. No provision was made by the Quebec Bridge Company for a staff to assist him, nor is there any evidence to show that he asked for the appointment of such a staff. He endeavoured to maintain the necessary assistants out of his own salary, which was itself too small for his personal services, and he did a great deal of detail work which could have been satisfactorily done by a junior. The result of this was that he had no time to investigate the soundness of the data and theories which were being used in the designing, and consequently allowed fundamental errors to pass by him unchallenged. The detection and correction of these fundamental errors is a distinctive duty of the consulting engineer, and we are compelled to recognize that in undertaking to do his work without sufficient staff or sufficient remuneration both he and his employers are to blame, but it lay with himself to demand that these matters be remedied.[31]

This problem persists even today, as shown by the 1988 Burnaby supermarket roof collapse (discussed in Chapter 4). Moreover, in the case of the Quebec Bridge, it seems that this lesson was not fully learned by the government's Board of Engineers. When the bridge reconstruction began, the Board spent more than two years and half a million dollars preparing specifications for the bridge. But then, having expended so much time and money, it expected engineering companies to prepare detailed competitive bids within four months with no remuneration.[32]

Redesign and Reconstruction

In 1908 the Government of Canada, recognizing that the bridge would be a key link in the transcontinental railway, decided that the demolished bridge should be redesigned and reconstructed. The government established a three-person Board of Engineers to prepare plans and specifications and to supervise the reconstruction. The Board's duties and powers were clearly defined.

Having reviewed the earlier plans and the report of the Royal Commission, the Board adopted a modified cantilever structure with a wider base, and with straight lower chords. Removing the twisted steel and debris from the 1907 disaster took two years. After that, new supporting piers were built that went down to bedrock. Under the Board's direction, the super-structure was designed, manufactured, and erected by the St. Lawrence Bridge Company, Ltd., of Montreal. In the new design, the compressive chords were significantly larger than in the original design.

From the published data, it appears that the Phoenix bridge, with less efficient compressive chords, must have been a very slender design, compared to the Firth of Forth Bridge. Cooper designed the steel cross-sectional areas of the original Phoenix design to be slightly larger than those of the Forth Bridge. However, the Phoenix compressive members were rectangular; the Forth Bridge has circular cross-sections.

Circular cross-sections always give a larger resistance against buckling (moment of inertia), but more importantly, circular sections do not require latticework or cross-braces. These secondary members add a great deal of weight. Circular chords were considered for the Quebec Bridge, but rejected as uneconomical. Circular sections could be built easily in Scotland, where shipyards were accustomed to large structures, and had the knowledge and machinery to fabricate curved surfaces. In North America, at that time, facilities for projects of this magnitude were scarce.

The final (St. Lawrence) bridge design was intended to instill confidence in the structure: the massive compressive chords are almost 2.5 times as heavy (per unit length) as those on the Forth Bridge.[33]

The Second Collapse

During the reconstruction, a second disaster occurred. The original (Phoenix) erection plan was to construct the bridge entirely in place by building each cantilever out from the riverbank until the two met at mid-span. For the second (St. Lawrence) design, the erection plan was to build the cantilevers only partway out from the shore. Meanwhile, the central part of the span would be assembled onshore. At the appropriate time, it would be floated out and raised into position. On September 11, 1916, the weather and tides were suitable for floating the middle span to the bridge and raising it into place. All went smoothly, and by mid-morning the span had been lifted about 7 m (23 ft.) above the water. At about 11 a.m., a sharp crack was heard and the centre span slid off its four corner supports into the river. Thirteen men were killed and 14 more were injured.

An investigation conducted by the St. Lawrence Bridge Company and the Board of Engineers found that the accident was unrelated to the design and was caused by a material failure in one of the four bearing castings that supported the central span temporarily while it was being transported and hoisted. The St. Lawrence Bridge Company assumed the responsibility for the

Photo 1.4 — The Quebec Bridge (1916 collapse). *The collapse of the centre span of the Quebec Bridge (St. Lawrence design), photographed at the instant it hit the water on September 11, 1916.*

Source: The Québec Bridge over the St. Lawrence River near the City of Québec: *Report of the Government Board of Engineers*, Department of Railways and Canals Canada, 1919.

failure, a second span was constructed, and the design of the support bearings was changed from a casting to a lead "cushion." The new middle span was successfully lifted into place, over three days, in August 1917. The bridge was opened to traffic in 1918, and a formal ceremony attended by the Prince of Wales was held on August 22, 1919.

Aftermath: Some Hard Lessons Learned

In the decade following the Quebec Bridge disasters, the first Acts to license professional engineers were put into law. The Ritual of the Calling of an Engineer (described in Chapter 5) was instituted, and even today the chain and iron rings used in that ritual are rumoured to be made from the actual steel that claimed the lives of so many men in the cold waters of the St. Lawrence. There are many lessons to be learned from Canada's worst bridge disaster, such as the importance of the following:

- Providing adequate capitalization for large-scale projects.
- Hiring capable and competent professionals.
- Defining clearly the duties, authority, and responsibility of personnel.
- Discussing design decisions and technical problems openly.
- Reviewing details, especially in the iterative task of engineering design.
- Monitoring work on the site adequately.
- Ensuring that communication is rapid and accurate.
- Providing adequate support staff and remuneration for professional people.

Provincial regulation of engineering helps achieve these goals. The professional engineer's stamp on engineering plans and specifications identifies unambiguously who is responsible for the accuracy of the documents and for the computations on which they are based. These lessons were learned at great cost.

Photo 1.5 — The Quebec Bridge (completed structure). *A view of the completed bridge from the north shore, in 1919.*

Source: The Québec Bridge over the St. Lawrence River near the City of Québec: *Report of the Government Board of Engineers,* Department of Railways and Canals Canada, 1919.

Where to Learn More

The two-volume book cited below describes the Quebec Bridge in impressive detail. The book is a classic of project documentation, and is well worth reading, even a century after the disaster, by anyone interested in structural design. It is available in most university libraries. More reading suggestions for Canadian engineering history are listed in Appendix E.

> *The Québec Bridge over the St. Lawrence River near the City of Québec: Report of the Government Board of Engineers,* Department of Railways and Canals Canada, printed by order of the Governor General in Council, 31 May 1919.

DISCUSSION TOPICS AND ASSIGNMENTS

1. The section "An Inspiring Legacy" in this chapter summarizes over 15 Canadian engineering and/or geoscience achievements. Select one of these achievements, and investigate it in more detail, using the Internet or your library. Write a brief description of the project (not more than two or three pages). Be sure to include: the motivation for the project; the name of the key person or group responsible for the project; any major technical or financial problems facing the designers or builders. What was (or is) the major impact of the project on Canada?

2. The Avro Arrow, a supersonic interceptor aircraft designed and built in Canada in the 1950s, is not included in the section "An Inspiring Legacy" in this chapter, but it should be. Although the Arrow was decades ahead of other military interceptor aircraft of the time, it was cancelled in 1958. Write a brief description of the project (not more than two or three pages), using the Internet or your library. Be sure to include the factors listed in the previous question, as well as an explanation of the reason why the Arrow was abandoned, and the effect of the cancellation on the Canadian aircraft industry.

3. The RIM (Research in Motion) Blackberry® wireless communicating device is a Canadian invention, and a phenomenal success story for the 21st century. The Blackberry is used by millions of people and organizations around the world, because it provides secure wireless email and communications. Using the Internet, write a brief description (not more than two or three pages) of this device, how it works, the patent problems that RIM has faced, and RIM's potential for the future.

Additional assignments can be found in Appendix E.

NOTES

[1]   Parks Canada, "Rideau Canal National Historic Site of Canada," available at <www.pc.gc.ca/lhn-nhs/on/rideau/plan/plan3_E.asp> (May 11, 2008).

[2]   Engineering Institute of Canada (EIC), "Engineering Centennial and Achievements," Article 12 in *History and Archives*, available at <www.eic-ici.ca/english/tour/article12.html> (May 11, 2008).

[3]   N.R. Ball, *Mind, Heart, and Vision: Professional Engineering in Canada 1877 to 1987*, National Museums of Science & Technology, Ottawa, 1987.

[4]   *The Canadian Encyclopedia*, available at <www.canadianencyclopedia.ca> (May 11, 2008).

[5]   Engineering Institute of Canada (EIC), "Engineering Centennial and Achievements."

[6]   Intergovernmental Panel on Climate Change (IPCC), *Fourth Assessment Report*, "The Physical Science Basis" (by Working Group 1, February 2007); "Impacts, Adaptation and Vulnerability" (by Working Group 2, April 2007); "Mitigation of Climate Change" (by Working Group 3, May 2007); "Synthesis Report" (by Working Group 4, November 2007). All reports available from the IPCC website: <www.ipcc.ch> (May 11, 2008).

[7]   Engineers Canada, *Welcome page*, Ottawa, March 2006, available at <www.engineerscanada.ca/e/> (May 11, 2008).

[8]   Canadian Council of Professional Geoscientists (CCPG), *Press Release*, March 29, 2007, available at <www.ccpg.ca> (May 8, 2008).

[9]   Canadian Engineering Accreditation Board (CEAB), Engineers Canada, *2007 Accreditation Criteria and Procedures*, Ottawa, 2007, p. 28, available at <www.engineerscanada.ca/e/pu_ab.cfm> (June 15, 2009).

[10]   Canadian Geoscience Standards Board (CGSB), a committee of CCPG, *Geoscience Knowledge and Experience Requirements for Professional Registration in Canada*, May 2008, available at <www.ccpg.ca> (June 15, 2009).

[11]   Task Force on the Future of Engineering, Engineers Canada (formerly Canadian Council of Professional Engineers—CCPE), *The Future of Engineering*, Ottawa, 1988, p. 33.

[12] Engineers Canada, *The Future of Engineering.*
[13] Engineers Canada (formerly Canadian Council of Professional Engineers—CCPE), *National Survey of the Canadian Engineering Profession, 2002*, Ottawa, 2003.
[14] Services Canada, *Job Futures*, National Career and Education Planning Tool, available at <www.jobfutures.ca> (May 7, 2008).
[15] Services Canada, *Job Futures*, National Career and Education Planning Tool.
[16] Webster's Third New International Dictionary, Unabridged © 1993 by Merriam-Webster, Incorporated. Reprinted with permission.
[17] Engineers Canada, *National Survey of the Canadian Engineering Profession, 2002*, p. 10.
[18] Canadian Council of Technicians and Technologists (CCTT), *About Us*, available at <www.cctt.ca/home.asp> (May 11, 2008).
[19] H.A. Halliday, "Québec Bridge Disasters," *The Canadian Encyclopedia Plus*, McClelland & Stewart, Toronto, 1995, available at <www.canadianencyclo-pedia.ca/index.cfm?PgNm=TCE&Params=A1ARTA0006593> (May 11, 2008). Used by permission, McClelland & Stewart, Inc., The Canadian Publishers.
[20] Canada, Royal Commission, *Québec Bridge Inquiry Report*, Sessional Paper No. 154, 7–8 Edward VII, Ottawa, 1908, p. 49.
[21] Ibid., p. 50.
[22] Ibid., p. 50.
[23] Ibid., p. 51.
[24] Ibid., p. 52.
[25] H. Petroski, *Engineers of Dreams: Great Bridge Builders and the Spanning of America*, pp. 104–105. Copyright © 1995 by Henry Petroski. Used by permission of Alfred A. Knopf, a division of Random House, Inc.
[26] Canada, Royal Commission, *Québec Bridge Inquiry Report.*
[27] G.H. Duggan, *The Québec Bridge*, bound monograph, prepared originally as an illustrated lecture for the Canadian Society of Civil Engineers, January 10, 1918.
[28] Canada, Royal Commission, *Québec Bridge Inquiry Report.*
[29] H. Petroski, *Engineers of Dreams: Great Bridge Builders and the Spanning of America*, pp. 108–109. Copyright © 1995 by Henry Petroski. Used by permission of Alfred A. Knopf, a division of Random House, Inc.
[30] Canada, Royal Commission, *Québec Bridge Inquiry Report*, p. 49.
[31] Ibid., p. 50.
[32] Petroski, *Engineers of Dreams*, p. 115.
[33] Duggan, *The Québec Bridge.*

Chapter 2
Regulation of Engineering and Geoscience

A key role of government is to protect the public. Therefore, when the government confers professional status on a recognized profession, the public welfare must be protected by licences and by regulations that

- admit only qualified people to the profession,
- establish standards of professional practice and Codes of Ethics, and
- discipline negligent, incompetent, or unscrupulous practitioners (when necessary).

This chapter describes the licensing process and gives an overview of other basic regulations. Since you are (or likely will be) a licensed member of a "self-regulating" profession, you should read this chapter critically. How would you regulate your profession?

THE EVOLUTION OF ENGINEERING AND GEOSCIENCE LICENSING

Professional Engineering

Efforts to place engineering on the same professional footing as law and medicine began as early as 1887, when the Canadian Society of Civil Engineers (CSCE) held its first general meeting. The campaign to regulate the engineering profession was led by the CSCE (which in 1918 became the Engineering Institute of Canada, or EIC). In the years after Confederation, most of Canada followed the British model: engineers entered the profession after a period of apprenticeship, and few engineers were university graduates. However, from its very start, the CSCE took it upon itself to establish and maintain high standards for admission to the Society, with the goal of improving professional engineering practices. Applicants were required to be at least 30 years of age and to have at least 10 years of experience, which could include an apprenticeship in an engineer's office or a term of instruction in a school of engineering acceptable to the CSCE Council. Each applicant also had to show "responsible charge of work" for at least 5 years as an engineer, designing and directing engineering works.[1]

In spite of this early Canadian initiative, the United States was, in fact, the first country to regulate the practice of engineering. The State of Wyoming enacted a law in 1907, after many instances of gross incompetence were observed in a major irrigation project.[2] In Canada, the deadly collapses that occurred during the construction of the Quebec Bridge emphatically reinforced the need to regulate the profession. (Chapter 1 describes the Quebec Bridge disasters.)

However, it would be many years before Canadian engineers overcame professional rivalries, business competition, class barriers, and other impediments and agreed on proposals to improve professional standards—and, indirectly, the status of engineers. In August 1918, at a general meeting of the CSCE held in Saskatoon, an Alberta engineer named F.H. Peters called on the Society to seek licensing legislation. In his view, engineers had developed the nation's resources but had yet to receive the remuneration and the respect they deserved.[3] At that time, the First World War was drawing to an end, and the flood of returning soldiers—some of whom had been involved in various aspects of military engineering—was dramatically increasing the number of engineers. This was depressing salaries, increasing competition, and quality was at risk.

The CSCE (which had just changed its name to the EIC) drafted a Model Act, and published it in the *EIC Journal*. In September 1919, the *Journal* announced that 77 percent of EIC members had approved the Model Act by mail ballot. By the spring of 1920, all provinces except Ontario, Saskatchewan, and Prince Edward Island had passed licensing laws. In Ontario, a joint advisory committee redrafted the bill, and it became law in 1922. The laws enacted in British Columbia, Manitoba, Quebec, New Brunswick, and Nova Scotia were "closed," which meant that engineers would require a licence either to practise engineering or to use the title of Professional Engineer (P.Eng.). In Alberta and Ontario, the laws were "open," which limited the use of the P.Eng. title, but licensing was voluntary, so unlicensed people could still practise engineering. Alberta amended its Act to close it in 1930; Ontario closed its Act in 1937.[4]

In the years that followed, all of Canada's provinces and territories and all of the American states amended or passed licensing laws to regulate the engineering profession and the title of Professional Engineer. Prince Edward Island, in 1955, was the last province to enact closed legislation. There is a key difference between the Canadian and American engineering laws. In Canada, the engineering profession is "self-regulating": each province or territory has passed an Act to create an Association of professional engineers, which in turn regulates the profession.

The importance of self-regulation cannot be overemphasized. Each Association's governing council must enforce the Act, regulations, and bylaws, but the licensed members of the Association elect (most of) the Council members. This ensures that well-informed engineers are involved in the regulation of the profession. By contrast, in the United States, the profession is not "self-regulating." State governments appoint licensing boards to license engineers, and the governments establish the regulations that engineers must follow. Therefore, politicians typically play a more significant role in establishing and enforcing regulations in the U.S. system than they do in Canada. Another significant difference is that a Canadian licence confers both the right to practise the profession, as well as the right to the title (P.Eng. or P.Geo). In the U.S., licensing confers only the right to use the title. As a result, anyone can practise. For example, this loophole, well known in engineering as the "industrial exemption," permits entire industries to function with unlicensed engineers.

In some countries, engineering is not a licensed profession, and anyone may use the term "engineer." In those countries, the possession of a degree or membership in a technical society may be the only guide to the person's competence. In Britain, for example, licensing is not compulsory and the title of *engineer* often means *mechanic*—the sign "Engineer on Duty" hangs outside many garages. British engineering societies award the title of Chartered Engineer to members who join voluntarily and meet the Society's admission requirements.

Professional Geoscience

The licensing of professional geoscientists has followed roughly the same path as that of professional engineers over the past 80 years. The following quote gives a brief history:

> The engineering professions were regulated in Canada in the early decades of the twentieth century. From the outset, it was recognized that the work of many geoscientists also affected the public welfare through their involvement in oil, gas and ore reserves estimation, exploration and mining activities, and construction of major engineering works such as dams and bridges. More recently, geoscientists have become major players in the broad area of environmental practice.

> Initially, geoscientists whose work impacted the welfare of the public were licensed as engineers, usually as mining engineers. In Alberta, John A. Allan, a prominent geoscientist and founder of the Geology Department at the University of Alberta, took an active role in establishing the Association of Professional Engineers of Alberta (APEA) in the 1920s and became its president in the 1930s. In the 1950s, the discovery of oil and gas in Alberta focussed attention on the geoscience professions, with the result that geologists, and the practice of geology and geophysics were explicitly identified in the Engineering Act in Alberta in 1955. Separate designations for geologists and geophysicists (*P.Geol.* and *P.Geoph.*) were introduced in 1960 and, in 1966, APEA changed its name to become the Association of Professional Engineers, Geologists and Geophysicists of Alberta (APEGGA).

> Following the pattern in Alberta, geoscientists are now licensed in most Canadian provinces and territories by Associations of engineers and geoscientists, established by legislative acts covering the professions of engineering and geoscience.[5]

The Bre-X fraud, exposed in 1997, spurred the regulation of geoscientists in the same way that the collapse of the Quebec Bridge, 90 years earlier, spurred the regulation of engineers. In 2000, the Walkerton tragedy reinforced the Bre-X lesson. Seven people died and over 2,300 became ill from contaminated drinking water in Walkerton, Ontario. A Public Inquiry revealed "omissions or failures to take appropriate action" on the part of Ontario's Ministry of the Environment, which is responsible for monitoring the operation of municipal water systems.[6]

Shortly thereafter, Ontario passed the Professional Geoscientists Act (2000), Quebec passed the Geologists Act (*Loi des géologues*—2001), and Nova Scotia passed the Geoscience Profession Act (2002). However, Prince Edward Island and Yukon do not yet regulate geoscience. (A case history of the Bre-X fraud is located at the end of this chapter.)

Qualified Persons—An Important Role for Engineers and Geoscientists

Some of the regulations made after the Bre-X fraud and the Walkerton tragedy introduced the term "qualified person" (QP). This term is now appearing in legislation and regulations, and usually refers to professional engineers or geoscientists, as appropriate.

For example, the Canadian Securities Administrators (CSA), an umbrella body for provincial securities regulators, issued a document titled *National Instrument 43-101*, which came into effect on February 1, 2001. This document specifies the (mandatory) format for providing oral statements or written disclosures of scientific or technical information to the public concerning mineral projects. The document is extremely specific—it even lists the headings for technical reports. The CSA also states that only a QP can disclose scientific or technical information to the public regarding a mineral project. The CSA defines a QP as an individual who:

a) is an engineer or geoscientist with at least five years of experience in mineral exploration, mine development or operation or mineral project assessment, or any combination of these;

b) has experience relevant to the subject matter of the mineral project and the technical report; and

c) is a member in good standing of a professional association.[7]

If people who are not qualified (under this definition) assist a QP to prepare a technical report, the QP must assume responsibility for the report, and must ensure that all information is correct. A QP must always visit the site on which the report is based.

The restoration of "brownfields" (contaminated properties) also requires the supervision and approval of a QP. Several provinces are developing legislation to monitor the decontamination of brownfields, and some provincial Associations have also issued guidelines defining the responsibilities of professionals undertaking the role of a QP.[8,9]

PROVINCIAL AND TERRITORIAL ACTS

The provincial and territorial governments regulate engineering and geoscience by Acts of provincial legislatures (in the provinces) or legislative councils (in the territories). These Acts are extremely important, since they are the legal basis for all professional regulations.

Eight provinces or territories regulate engineering and geoscience in the same Act, and three provinces regulate engineering and geoscience in separate Acts. Two jurisdictions (PEI and Yukon) regulate engineering, but do not yet regulate geoscience.

Alberta	Engineering, Geological and Geophysical Professions Act
British Columbia	Engineers and Geoscientists Act
Manitoba	Engineering and Geoscientific Professions Act
New Brunswick	Engineering and Geoscience Professions Act
Newfoundland & Labrador	Engineers and Geoscientists Act
Northwest Territories	Engineering and Geoscience Professions Act
Nova Scotia	Engineering Profession Act and Geoscience Profession Act
Nunavut	Engineers and Geoscientists Act
Ontario	Professional Engineers Act and The Professional Geoscientists Act
Prince Edward Island	Engineering Profession Act
Quebec	Engineers Act (*Loi des ingénieurs*) and Geologists Act (*Loi des géologues*)
Saskatchewan	Engineering and Geoscience Professions Act
Yukon	Engineering Profession Act

Contents of the Acts

These Acts are extremely important, as mentioned above, because they establish the legal basis for engineering and geoscience as professions. The Acts define basic terms; create the Association as a legal entity; define the extent of its powers; and set standards for admission, practice, and discipline of professionals. The following clauses are typical in each Act:

- The purpose of the Act (which, in every case, is to protect the public).
- The legal definition of engineering and/or geoscience.
- The authority to establish a provincial (or territorial) Association.
- The purpose (or objects) of the Association.
- Standards for granting licences (or for admission to the Association).

- Procedures for establishing regulations to govern professional practice.
- Procedures for establishing bylaws to govern the Association itself.
- A Code of Ethics to guide the personal actions of the Licensees (or members).
- Disciplinary procedures.

Every Act is available on the Internet, via the Web pages for the provincial and territorial Associations. (Appendix A lists the Web pages, and Appendix B reproduces key excerpts from all Acts, for easy reference.)

Self-Regulation of the Professions

The Acts establish engineering and geoscience as self-regulating professions, like law, medicine, and dentistry. The term "self-regulating" means that the licensed members elect (most of) the Association's Council, who administer the Act. This ensures that the best-informed people are in a position to establish and enforce the standards of practice, Codes of Ethics, and discipline procedures that protect public safety and the environment. The public also benefits because the Associations run without government funding. Licensing fees pay administrative staff, and licensed members serve voluntarily on the admission, discipline, and other committees.

A Note about Nomenclature

- Ontario uses the term "licensed" but other jurisdictions use "registered" to indicate admission to the profession. The terms are equivalent, but this text uses "licensed" to avoid any misunderstanding that membership is voluntary, or that the Associations are trade unions or special interest groups.
- The terms "licensee," "practitioner," and "member" are interchangeable.
- The term "professional Act," "provincial Act," or simply "Act" refers to the relevant Act or law (as listed above) in each province or territory.
- The term "Association" refers to the Association of Professional Engineers and/or Geoscientists (or *Ordre des ingénieurs* or *Ordre des géologues*) created under the Act, in each province or territory.

LEGAL DEFINITIONS OF ENGINEERING AND GEOSCIENCE

Definition of Engineering

The terms *engineer* and *ingenuity* come from the same Latin root, *ingenium*, which means talent, genius, cleverness, or native ability. Ancient Roman armies marched with a complement of engineers to design and build roads, fortifications, and weapons of war. Over the centuries, non-military or *civil* engineers emerged, to design and build the structures that are essential to a modern, peaceful society. Today, the role of the engineer has expanded. Canada still needs engineers for war, regrettably; but all engineers are committed to protect the public, to build a better world, and to generate prosperity.

Engineering now includes hundreds of branches, subdisciplines, and specialties—so many, in fact, that it is sometimes hard to see the boundary between the professions. Consider this simple example: Architects, geoscientists, and engineers may cooperate in the design of a new high-rise building, but play different roles. The architect may design its artistic appearance and room layout, the geoscientist may carry out the soil and site analysis, but the engineer must approve the size of the structural steel. Professionals must know their roles in any project; authority and responsibility must be clearly defined.

NATIONAL DEFINITION

Engineers Canada (formerly called The Canadian Council of Professional Engineers) proposed a national definition for "the practice of professional engineering." The goal is to assist governments to adopt consistent definitions in their Acts, and thus permit engineers to practise across Canada more easily. The Engineers Canada national definition is as follows:

> **The Practice of Professional Engineering:** any act of planning, designing, composing, evaluating, advising, reporting, directing or supervising, or managing any of the foregoing, that requires the application of engineering principles and that concerns the safeguarding of life, health, property, economic interests, the public welfare or the environment. [An additional paragraph exempts natural scientists from this definition.][10]

This is actually a circular definition, since it uses the term "engineering principles," which needs further explanation. The difference between *engineering* principles and *scientific* or *technological* principles hinges on the purpose and depth of the study, as explained in the Canadian Engineering Accreditation Board's criteria for accreditation:

> Engineering science subjects normally have their roots in mathematics and basic sciences, but carry knowledge further toward creative applications. . . . Application to identification and solution of practical engineering problems is stressed.[11]

In other words, engineering involves almost any act that puts science and mathematics to creative, practical use, and that concerns the safeguarding of life, health, property, economic interests, the public welfare, or the environment. This is a very broad definition; many Acts add further clauses to clarify the boundaries with scientists and architects.

LEGAL DEFINITION

The most important definition of engineering is in the Act for your province or territory, because that definition applies to you. Most Acts adopt something similar to the national definition (above), but not identical. A few provinces list the types of machinery or structures (such as railways, bridges, highways, and canals) that are within the engineer's area of practice. This makes the definition very clear and specific, but also very long and difficult to read. As time passes, such lists become obsolete, as old components (such as steam engines) disappear and new areas (such as nanotechnology engineering) emerge. The shorter definitions are easier to read, but are very general (to the point of being vague). The definition in Ontario's Act is probably closest to the national definition:

> **The Practice of Professional Engineering:** Any act of designing, composing, evaluating, advising, reporting, directing or supervising wherein the safeguarding of life, health, property or the public welfare is concerned and that requires the application of engineering principles, but does not include practising as a natural scientist. [An additional paragraph exempts natural scientists.][12]

In Alberta, the definition is similar, but not identical:

> **"Practice of Engineering"** means
>
> (i) reporting on, advising on, evaluating, designing, preparing plans and specifications for or directing the construction, technical inspection, maintenance or operation of any structure, work or process
>
> (A) that is aimed at the discovery, development or utilization of matter, materials or energy or in any other way designed for the use and convenience of humans, and
>
> (B) that requires in that reporting, advising, evaluating, designing, preparation or direction the professional application of the principles of mathematics, chemistry, physics or any related applied subject, or
>
> (ii) teaching engineering at a university.[13]

Note that the Alberta definition does not say that the engineer is responsible for protecting the life, health, safety, and welfare of the public. However, the Alberta Code of Ethics and the Act state this clearly. (The legal definitions for your province or territory are in the Act, found on your Association's website, and excerpts are in Appendix B.)

As a closing comment, we should emphasize that the "factor of safety" (the ratio of load capacity to expected load, on a component, system, or structure) is typically the one calculation that summarizes whether a design properly protects the "life, health, property or the public welfare." The duty of the professional engineer is to ensure that the factor of safety is adequate, and is correctly calculated. Unqualified people cannot assume this responsibility.

Definition of Geoscience

Every province and territory (except Prince Edward Island and Yukon) has an Act that defines the term "professional geoscience" (including geology or geophysics). These definitions draw the boundaries between geoscience and engineering, and between both of these and other professions. The definitions vary slightly among the provinces and territories; however, the Canadian Council of Professional Geoscientists (CCPG) has proposed what it calls the "minimum content" that should be included in a definition of professional geoscience.

The following definition was developed by a task group of the CCPG and has been approved for circulation by the Canadian Geoscience Standards Board and the Board of Directors of CCPG:

> The **"practice of professional geoscience"** means the performing of any activity that requires application of the principles of the geological sciences, and that concerns the safeguarding of public welfare, life, health, property, or economic interests, including, but not limited to:
>
> (a) investigations, interpretations, evaluations, consultations or management aimed at discovery or development of metallic or non-metallic minerals, rocks, nuclear or fossil fuels, precious stones and water resources;
> (b) investigations, interpretations, evaluations, consultations or management relating to geoscientific properties, conditions or processes that may affect the well-being of the general public, including those pertaining to preservation of the natural environment.[14]

The CCPG definition includes an exemption for earth science or earth systems science, which may include activity in fields such as atmospheric science, meteorology, and oceanography; those areas presently fall outside the scope of regulated professional geoscience in all Canadian jurisdictions.[15]

PROVINCIAL AND TERRITORIAL ASSOCIATIONS

Each province or territory has a self-governing Association of Professional Engineers and/or Geoscientists, which administers the Act. (Quebec has an *Ordre*, rather than an Association). In eight jurisdictions, the Associations include engineers, geoscientists, geologists, and geophysicists, but in some jurisdictions, engineers and geoscientists have separate Associations. A list of all Associations and Ordres, their addresses, and websites, is located in Appendix A.

As stated above, the governments delegate the responsibility for implementing each Act to the provincial or territorial Association. Each Association, in turn, has developed regulations, bylaws, and a Code of Ethics (or has assisted in their development, subject to final approval by the government). Variations exist in some provinces, but the following definitions are useful:

- **Regulations** are rules that clarify the Act, and typically include admission requirements, professional conduct, and disciplinary procedures.
- **Bylaws** are rules for running the Association itself, and typically include election procedures, financial matters, committees, and meetings.
- **The Code of Ethics** is a set of rules of personal conduct. Every engineer and geoscientist must be familiar with this code and endeavour to follow it.

The regulations, bylaws, and the Code of Ethics obtain their authority from the Act. As mentioned earlier, the professions are "self-regulating" because members, who are licensed professionals, establish and enforce these rules. In addition, members serve voluntarily on the admission, discipline, and other committees; members elect the majority of the Association's governing council (the government also appoints some councillors); and members usually must confirm (by ballot) any regulations recommended to the government, and bylaws passed by the council. Obviously, for self-regulation to work effectively, members must be willing to serve in these volunteer and elected positions.

ADMISSION TO ENGINEERING

The standards for admission to the engineering profession are very similar, although not identical, across Canada, and generally follow the guideline for engineering admission procedures, published by Engineers Canada.[16] The following conditions must be satisfied to become a licensed professional engineer:

- **Education.** The applicant must have adequate academic qualifications. A university degree from an accredited university engineering program (or the equivalent) is required.
- **Experience.** The applicant must satisfy the experience requirements. Most jurisdictions now require four years of suitable experience (Quebec requires three years of experience). At least one year of experience must be in a Canadian environment.
- **Knowledge of professional practice and ethics.** Typically, every applicant must write and pass the professional practice exam on Canadian law and ethics.
- **Language.** The applicant must be able to communicate effectively, both orally and in writing, in the working language of the province or territory.
- **Character.** The applicant must be of good character, as determined mainly from references. Evidence of criminal conviction, fraud, or false statements on applications may affect admission.
- **Citizenship.** All provinces (except Quebec) require Canadian citizenship or landed immigrant status, but also issue temporary, foreign, or non-resident licences.
- **Residence.** Most provinces are eliminating residency requirements. Such requirements impede mobility and are not indicative of competence, ability, or character. Provinces with this requirement issue temporary licences to non-resident engineers.
- **Age.** The applicant must have reached the legal age of majority, which is 18 years in most of Canada.

Requirements vary slightly across Canada. The following paragraphs give more details, but if you are applying for admission, check your Association's Web page for recent changes.

Academic Requirements

Academic qualifications are the most important requirement for admission. Applicants must submit originals (or certified copies) of all transcripts and other academic documents for evaluation by the Association's Board of Examiners (or Academic Requirements Committee).

CEAB-ACCREDITED DEGREES   Graduates of accredited university engineering programs are exempt from technical exams, but must write the Professional Practice Exam. The Canadian Engineering Accreditation Board (CEAB) publishes a list of accredited programs. A few international degree programs are also CEAB-accredited.[17]

MUTUALLY RECOGNIZED DEGREES   Engineers Canada has negotiated international agreements with about 10 countries. These agreements recognize that Canadian and foreign accreditation procedures (for these countries) are substantially equivalent and satisfy the academic requirements for admission to professional engineering. For example, the Accreditation Board for Engineering and Technology (ABET) accredits university engineering programs in the United States. One agreement recognizes engineering degrees from ABET-accredited programs as equivalent to Canadian degrees. Graduates applying under these agreements must provide complete academic records. However, if the Association's Board of Examiners judges an applicant's academic record to be deficient, the Board may still assign exams. (See the list of countries below, under "Mobility Agreements.")

NON-ACCREDITED DEGREES   Applicants with engineering degrees that lack accreditation are normally required to write confirmatory or technical exams. The Board of Examiners examines each case carefully. The exam program (and its purpose) may vary, as explained below.

Internationally Educated Graduates

Many internationally educated engineering graduates find the engineering licensing process time-consuming and frustrating. In 2003, Engineers Canada began a project called "From Consideration to Integration" (FC2I), to see how the transition into the Canadian work force can be simplified. The Associations must assess the qualifications of all applicants, and they cannot lower standards, but they can provide more guidance (usually through easily accessible websites). Some bridging programs are being established, and the Associations are looking at ways to simplify the licensing process.[18] The process for internationally educated engineering graduates usually follows this path:

APPLICATION   An applicant educated in a foreign college or university must provide the Association with originals (or certified copies) and notarized translations, if needed, of all transcripts and diplomas. The Association's Board of Examiners (or Academic Requirements Committee) assesses those documents.

When an applicant's degree is from a non-accredited program, the Association must obtain and evaluate other evidence in order to justify admission. A request for corroborating evidence of academic ability does not imply any lack of respect for the individual, or the individual's university. The Association evaluates each case carefully, and gives credit for many relevant factors, leading to several possible outcomes. The following three categories are only a rough summary.

QUALIFICATIONS ACCEPTED   Several formal agreements (such as the mobility agreements) recognize a few foreign qualifications and university programs as equivalent to accredited Canadian university programs, and thus grant full exemption from the exam program. However, if an applicant's degree is not accredited, the routine procedure is to assign confirmatory exams. Under certain conditions, the Association may waive these confirmatory exams, as explained below.

CONFIRMATORY EXAMS REQUIRED   Associations usually ask internationally educated applicants to write a set of from three to five confirmatory exams. This decision is required, even when the applicant's academic record from the non-accredited university appears to be equivalent to the Syllabus of Examinations published by Engineers Canada. Confirmatory exams cover the advanced topics in a small portion of the full engineering program. If the applicant obtains good grades on the first two exams, then the Association normally waives the rest of the exams.

In the case of a senior, experienced foreign applicant, the Association may be "looking to exempt" the applicant from exams, but some additional evidence must be provided. Associations normally interview senior applicants (with over 10 years of experience) to evaluate their academic qualifications and engineering achievements, and may then waive some (or all) of a confirmatory exam program.

Moreover, Associations may waive some confirmatory exams if applicants can provide other evidence of engineering competence. For example, applicants who can show that they have completed a master's degree (or even some advanced undergraduate courses) at an accredited Canadian university may receive full or partial exemption from the confirmatory exams.

In Alberta, engineering applicants assigned confirmatory examinations by APEGGA's Board of Examiners have the option of satisfying this requirement by writing the Fundamentals of Engineering (FE) examination, administered by the U.S. National Council of Examiners for Engineering and Surveying (NCEES). The FE Examination is a comprehensive eight-hour technical examination covering basic concepts of engineering (with 120 basic general questions in a four-hour morning session, followed by 60 discipline-specific questions in a four-hour afternoon session). The FE Exam is the first step to Professional Engineering licensing in all U.S. states. APEGGA arranges for the FE exam to be held in Alberta on specific dates. More information about the FE Exam is available from APEGGA or from the NCEES website at <www.ncees.org>.

Applicants whose engineering degrees are judged not to be equivalent to the Syllabus of Examinations published by Engineers Canada may still enter the engineering profession, but will be required to write technical examinations, as described below.

Technical Examinations

Applicants who have non-accredited engineering degrees, or engineering-related degrees (such as the B.Tech. or B.Sc.), or three-year technology diplomas, may enter the engineering profession by writing technical exams, based on the Engineers Canada syllabus.[19]

For each branch of engineering, there are between 14 and 18 three-hour technical examinations. After a detailed evaluation of the applicant's academic transcripts, the Association assigns a subset of technical exams to make up academic deficiencies. Associations normally do not admit applicants to the exam program if the subset exceeds 9 exams. Some Associations permit applicants to take equivalent university courses in lieu of the assigned exams.

The possession of a postgraduate degree may reduce the number of exams assigned, but postgraduate degrees from other disciplines are rarely relevant. Moreover, accredited engineering programs teach the important engineering subjects in their undergraduate courses, and this knowledge is essential for licensing.

Permission to enter the examination system varies slightly across Canada. Engineers Canada recommends that an applicant should have at least 15 years' education, including a bachelor's degree. For example, in Ontario, the examination system is open to those who hold, as a minimum, one of the following: a three-year engineering technologist diploma from a college of applied arts and technology; a technologist-level certificate from the Ontario Association of Certified Engineering Technicians and Technologists (OACETT); or equivalent acceptable education as determined by the Association.

The examination system provides an alternative route into the profession, but it is not an easy route, because applicants must prepare themselves to write and pass the exams. The Associations do not offer classes, laboratories, or correspondence courses.

Professional Practice Examination

Almost all applicants, whatever their academic qualifications, must pass a Professional Practice Examination (PPE) in professional practice, law, contracts, liability, and ethics. British Columbia also requires applicants to attend a law and ethics seminar before writing the exam. The BC law and ethics

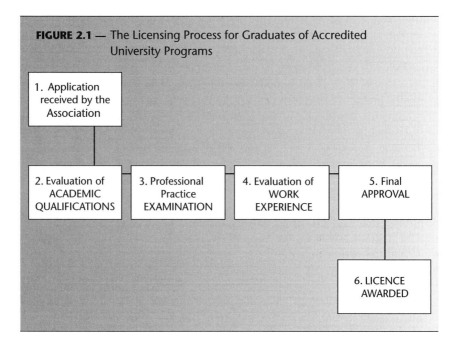

FIGURE 2.1 — The Licensing Process for Graduates of Accredited University Programs

seminar is also available on CD for the convenience of those unable to attend. Applicants transferring their licences from elsewhere in Canada do not need to write the PPE because of mobility agreements negotiated between Canadian Associations, based on equivalent admission standards.

Figure 2.1 shows the typical admission process for graduates of CEAB-accredited university engineering programs. However, it is advisable to start the licensing process as soon as possible. Many Associations provide assistance, internships, mentoring, and advice to help the applicant.

Experience Requirements

Satisfactory work experience is required for obtaining a licence. Applicants must submit a summary of their experience for the Association's Experience Requirements Committee (or Board of Examiners, depending on the province) to assess. Associations generally use the standards for engineering experience developed by Engineers Canada, with minor modifications.[20] The nature, duration, currency, and quality of the applicant's experience are important. As an example, a condensation of the Ontario experience requirements, based on the Engineers Canada guideline, follows.[21]

NATURE OF EXPERIENCE

Engineering experience is normally in the same area as the applicant's academic study. When the experience differs from the academic study—for example, if a mechanical engineering graduate is working in an electrical engineering job—the Association may require additional experience or additional studies before granting a licence. In addition, some jobs may have similar descriptions but, depending on the activities performed, may have different credit as engineering experience. An applicant should not presume that his or her area of employment will automatically be accepted (or rejected) as valid engineering experience.

In particular, an applicant whose employment falls into one of the following categories should consult the provincial Association for more specific advice:

- Teaching (at any level)
- Sales and marketing
- Military service
- Project management
- Operations and maintenance
- Computer engineering

DURATION OF EXPERIENCE

Since 1993, almost every jurisdiction has had a four-year experience requirement. The exception is Quebec, which in 2002 raised its experience requirement from two years to three.

CURRENCY OF EXPERIENCE

The Associations consider it important that the applicant's experience must be recent, because engineering evolves. Of course, theory evolves slowly; procedures and standards change more quickly; and computer applications change very fast. An applicant must show that experience obtained at the start of a career is still relevant.

QUALITY OF EXPERIENCE

Each applicant must prepare an experience résumé and explain how that experience satisfies five quality criteria:

- Application of theory
- Practical experience
- Management of engineering
- Communication skills
- Social implications of engineering

The application of theory is a mandatory requirement, and it must appear in a substantial part of the experience period (though not necessarily all of it). Theory must be supplemented by exposure to, or experience in, the remaining four criteria. Some Associations may place a slightly different emphasis on each of these criteria, so consult your Association's guidelines when you prepare experience documentation.

PRE-GRADUATION EXPERIENCE

Most Associations grant up to 12 months' credit for work experience obtained before university graduation. The pre-graduation experience (preferably obtained after the midpoint of the academic program) must satisfy the quality standards above. Quebec permits the three-year experience requirement to be reduced by as much as eight months if the candidate completes an optional sponsorship program, which involves a series of meetings with senior engineers to discuss specified topics, including skills, responsibility, ethics, and social commitment.

CANADIAN EXPERIENCE

All Associations require all applicants to obtain 12 months' work experience in a Canadian environment. This usually means working in Canada under the direction of a licensed professional engineer. However, except for Quebec, the work may be outside Canada—for example, in a Canadian company with foreign contracts, where the applicant is working with Canadian engineering laws, practices, standards, customs, codes, conditions, and climates. (Quebec requires 12 months of work experience in Canada.)

POSTGRADUATE EXPERIENCE

Most provinces grant 12 months' work experience for completing a postgraduate degree in engineering. In fact, some provinces permit even more credit (up to the total time spent in postgraduate studies), depending on how well the postgraduate experience satisfies the five quality criteria described earlier.

Nonresident, Temporary, and Provisional Licences

Most provinces and territories offer other forms of licence, depending on local needs and practices. For example, Associations typically offer licences (memberships) for residents, and temporary licences for nonresidents. The procedures for obtaining a nonresident or temporary licence vary slightly across Canada. Typically, the applicant must be licensed in another province or territory, qualified to work on a specific project, and familiar with the applicable codes, standards, and laws. Usually the nonresident must be collaborating with a member of the Association on the project specified for the temporary licence. However, this last requirement is waived for highly qualified applicants.

Several provinces recently introduced Provisional Licences for applicants who have satisfied all of the application requirements (including experience), except the requirement that the applicant must obtain 12 months of experience in Canada. Provisional Licences entitle the holder to practise professional engineering, but only under the supervision of a licensed professional engineer. The supervising professional engineer must sign all final drawings, specifications, plans, reports, or other documents. Provisional Licences give holders an officially recognized status during the 12-month Canadian experience period.

ADMISSION TO GEOSCIENCE

The Canadian Geoscience Standards Board (CGSB) has prepared recommended minimum requirements for admission to the practice of professional geoscience, which are very similar to the engineering requirements above. The recommendations are not binding on the provincial and territorial geoscience Associations, which make final decisions on all applicants, and may adopt, modify, or reject the guidelines. A condensed summary of the CGSB requirements follows:

- **Knowledge Qualifications.** Minimum knowledge qualifications are similar to Canadian university degree programs in geoscience. A university honours Bachelor of Science (B.Sc.) Geoscience program in Canada typically requires four years of study. In Quebec, university honours B.Sc. programs are three years, after graduation from a CEGEP. However, the CGSB recommends that no specific university program be designated as a standard. Instead, a number of minimum knowledge requirements must be satisfied, as defined in a CGSB Guideline. Each applicant must submit transcripts and documentation. The Association reviews these documents and compares the courses with the CGSB Guideline.[22]
- **Experience Qualifications.** A minimum of 48 months of cumulative and progressive geoscience work experience is required. At least 12 months of work experience should be in a Canadian work environment. Work must be reasonably current (preferably within the preceding 10 years). Summer work, co-op work, or similar geoscience work experience gained before satisfying the knowledge requirement is acceptable, up to a maximum of 12 months. Associations should evaluate experience in postgraduate study, university research, or teaching of geoscience in the same way as industrial, consulting, or government experience. Graduate degrees should receive no more than 24 months of experience credit.
- **Professional Practice Examination.** A Professional Practice Exam (PPE) is a requirement for registration.
- **Language.** Competency in a language of commerce of the jurisdiction of registration is necessary.
- **Good Character.** Applicants for licensure must be of good character and reputation.
- **No Canadian Citizenship or Residency Requirement.** In compliance with international agreements on the mobility of professionals, residency in Canada and/or Canadian citizenship is not required. Some provinces designate non-citizens as Foreign Licensees.

CANADIAN AND INTERNATIONAL MOBILITY AGREEMENTS

National Mobility Agreements

ENGINEERS   The 12 Associations that license professional engineers have signed an agreement for mutual mobility of their members. Any professional engineer who applies for registration with another Association shall be accepted, if the applicant satisfies the conditions below.

GEOSCIENTISTS   Similarly, the 10 Associations that license professional geoscientists have signed an almost identical agreement for mutual mobility of their members. Any professional geoscientist, geologist, or geophysicist who applies for registration with another Association shall be accepted, if the applicant satisfies the conditions below.

CONDITIONS   The applicant is a professional engineer, geoscientist, geologist, or geophysicist in good standing, and

- Has not been disciplined (either past or pending);
- Provides all of the required documentation;
- Permits the Associations to exchange files;
- Agrees to meet the rules for continuing competence in the new Association;
- Agrees to satisfy any language requirements for the new Association.[23,24]

DUE DILIGENCE   Although admission requirements are similar across Canada, professional registration is a provincial (or territorial) responsibility, so applicants must satisfy the admission requirements set by the provincial (or territorial) Act. Mobility agreements therefore include "notwithstanding" clauses to deal with exceptional cases. The final decision concerning an applicant's admission, or transfer of licence, lies with the Association.

International Mobility Agreements

ENGINEERING   Engineers Canada, on behalf of all Canadian engineering Associations, has agreed with several foreign countries for reciprocal recognition of qualifications. Canadian engineers who wish to work in those countries (or immigrants to Canada from these countries) may be interested in learning more about these agreements from Engineers Canada.[25]

- **United States/Canada:** Agreement with the Accreditation Board for Engineering and Technology (ABET).
- **The Washington Accord:** Agreement with engineering organizations in the United States, Ireland, the United Kingdom, Australia, New Zealand, Hong Kong, and South Africa.
- **NAFTA Agreement:** Agreement with the United States and Mexico in accordance with the North American Free Trade Agreement. [NOTE: This agreement has not been implemented.]

- **France/Canada:** Agreement with France that undergraduate engineering programs are substantially equivalent. The agreement also provides mechanisms for the reciprocity of "ingénieur diplômé" and "P. Eng./ing." designations.

GEOSCIENCE   The Canadian Council of Professional Geoscientists (CCPG), the national organization in Canada for geoscience licensing, and the following organizations have agreed to cooperate, since their objectives concerning the professional practice of the geological sciences are similar. The agreement details are on the CCPG website.[26]

- **The American Institute of Professional Geologists (AIPG)**
- **The Australasian Institute of Mining and Metallurgy (AusIMM)**
- **Australian Institute of Geoscientists (AIG)**
- **European Federation of Geologists (EFG)**
- **The Geological Society of London (GeolSoc)**
- **Institute of Geologists of Ireland (IGI)**
- **National Association of State Boards of Geology (ASBOG)**

LICENSING OF CORPORATIONS

A corporation, as a legal entity, may obtain a licence to practise engineering or geoscience. However, the purpose of licensing is to protect the public against incompetence, negligence, and professional misconduct, and these are qualities of human beings. The simple question arises: Who is being licensed to provide the engineering or geoscience services?

In almost every province and territory, the Act solves this problem by requiring each corporation to obtain a Permit to Practise (also called a Certificate of Authorization). To obtain this permit, the corporation must employ a professional engineer or geoscientist who acts in a supervisory capacity, and who assumes personal responsibility for the services provided by the corporation. The corporation must also obtain liability insurance. In addition, the corporation's engineers and geoscientists must normally participate in a continuing competence program (as discussed in Chapter 3).

A professional engineer or geoscientist working for a corporation that already has a permit to practise (or certificate) does not have to apply for an individual permit and, obviously, a corporation that does not offer such services to the public does not need a permit.

However, some provinces, such as Ontario, require that every entity—be it an individual, a partnership, or a corporation—that offers or provides professional engineering services to the public requires a Certificate of Authorization.[27] An engineer who plans to provide services to the public must discuss this issue with the provincial Association before providing such services. An engineer who "moonlights" at night or works on weekends without a permit (or certificate) may be breaking the law.

CONSULTING ENGINEERS

At present, Ontario is the only jurisdiction in Canada that regulates the designation of Consulting Engineer.[28] To qualify as a consulting engineer, a member must

- have been continuously engaged for at least two years in private practice,
- have at least five years of satisfactory experience since becoming a member, and
- pass (or be exempted from) exams prescribed by the Association Council.

Since applicants for the Consulting Engineer designation must be engaged in private practice, and therefore offering their services to the public, they must also be holders of a Certificate of Authorization in Ontario; or they must be associated with a partnership or corporation that is a holder of a certificate.

Photo 2.1 — The Confederation Bridge. *The Confederation Bridge opened on May 31, 1997, and links Prince Edward Island to New Brunswick. It is 12.9 km (8 miles) long and has a navigable clearance of 60 m (197 ft.) above the water. The central portion (11 km; 7 miles) has 44 spans, typically 250 m (820 ft.) long. The PEI approach bridge has 7 spans (580 m; 1,903 ft.), and the NB approach bridge has 14 spans (1,300 m; 4,265 ft.). The Confederation Bridge is the world's longest bridge over ice-covered waters. It carries two lanes of traffic 24 hours a day, seven days a week.*

Source: CP/Andrew Vaughan.

THE PROFESSIONAL SEAL

When an Association awards a licence, it also includes a "seal" (usually a rubber stamp), which the engineer or geoscientist uses to approve documents. The professional must sign, seal, and date all final drawings, specifications, plans, reports, and similar documents before releasing them for action. Write your signature and the date next to the seal, but do not obscure the seal. When two or more professionals collaborate on different aspects of a project, both seals are applied. Each professional specifies his or her area of responsibility in writing, next to the seal.

The seal has important legal significance—it means that you approved the document. A seal identifies the person responsible, and assures that it was competently prepared. Only *final* documents are sealed. *Preliminary* documents are not sealed; they must be marked "preliminary" or "not for construction." Costly errors can occur when preliminary documents are mistaken for approved final designs.

Only the engineer or geoscientist who prepares or approves a document should seal it. The seal implies an intimate knowledge of and control over the document. Do not use the seal casually.

A professional who knowingly signs or seals a document that has not been personally prepared (or prepared by assistants under supervision), may be charged with professional misconduct, and may also be liable for damages if the misrepresentation results in a loss.

A potential problem arises when a superior asks you to "check," sign, and seal documents prepared by someone else. Such documents should carry the seal of the engineer who prepared them, or who supervised their preparation. If a nonprofessional prepared the documents, then perhaps he or she should have been working under the supervision of an engineer. Many disciplinary cases have arisen because professionals signed and sealed documents prepared by others that later proved to have serious flaws.

The work required to "check" a document is defined in Chapter 6. A proper check might require complete duplication of the analysis. Of course, if you duplicate the analysis, then it is appropriate for you to assume responsibility for it. However, never approve a document until you thoroughly and independently review it. (Chapter 6 gives much more information on seals, and discusses electronic seals.)

THE CODE OF ETHICS

The Code of Ethics defines a high standard of personal professional conduct, required by every provincial and territorial Act. The Code of Ethics incorporates common sense, natural justice, and basic ethical concepts. The code defines, in general terms, the duties of the professional to the public, to the employer or client, to fellow professionals, to the profession, and to oneself. The code protects the public by requiring professional behaviour, and is the basis for disciplining unscrupulous practitioners.

Every professional should read and understand the code, but it is not necessary to memorize it. Most professionals use common sense in their ethical behaviour, follow the code intuitively, and never fear charges of professional misconduct. The Code of Ethics for each Act is in Appendix B. Chapter 11 discusses the ethical basis for the codes, and many cases studies in this textbook cite the codes.

In addition to the Code of Ethics, there is a much older voluntary oath, written by Rudyard Kipling and first used in 1925, called the Obligation of the Engineer. The iron ring, worn on the working hand, easily identifies engineers who have taken this oath. In recent decades, earth science rings, created for geoscientists, are part of a ceremony comparable to that of the iron ring. (These ceremonies are described in more detail in Chapter 5). Wearing an iron ring or an earth science ring does not mean that the wearer received a degree, but indicates that he or she participated in the ring ceremony and has voluntarily sworn to maintain high standards in professional work.

Although they have the same purpose, do not confuse the Obligation and the Code of Ethics. The Obligation is a voluntary commitment to high standards, but the Code of Ethics requires high standards. Infractions of the Code of Ethics are subject to discipline, under the Act.

ENFORCEMENT AND DISCIPLINE

To protect the public welfare, the Associations must ensure that only qualified individuals are practising. In addition, it is sometimes necessary to discipline the few engineers or geoscientists who commit professional misconduct. To enforce the Act, each Association has a staff that receives complaints, prosecutes people practising under false pretences, and administers any complaints against licensed members. These functions are explained in detail in Chapter 4, but are summarized briefly, as follows.

Enforcement

Each Association enforces the Act by prosecuting unlicensed individuals who practise engineering or geoscience, or who use the titles of Professional Engineer, Professional Geoscientist, Professional Geologist, or Professional Geophysicist (or the French equivalents). To use any of these titles, a person must obtain a licence from a provincial or territorial Association of engineers and/or geoscientists (in Quebec, the *Ordre des ingénieurs du Québec* or the *Ordre des géologues du Québec*).

Professional Misconduct and Discipline

Each Act requires the Association to reprimand, suspend, or expel members who are guilty of professional misconduct. Each Act typically defines professional misconduct as negligence, incompetence, or corruption, including a serious infraction of the Association's Code of Ethics. Therefore, it is very important to distinguish between your Association's Code of Ethics (which is enforceable under the Act) and the codes of ethics endorsed by many technical societies, since the societies' codes are actually voluntary guides to conduct.

The disciplinary process has three stages: information gathering or investigation, peer review, and finally, hearings for members charged with misconduct. A Discipline Committee, appointed by council, makes final decisions. (Chapter 4 describes the process in detail.)

ENGINEERS CANADA

Engineers Canada (formerly called the Canadian Council of Professional Engineers or CCPE) is a federation of the 12 Associations that license engineers in each province and territory across Canada. Established in 1936 as a federation of the Associations, Engineers Canada does not have individual members; however, every licensed engineer is indirectly a member.[29]

Engineers Canada coordinates the engineering profession on a national scale by promoting consistency in licensing and regulation. It develops policies, guidelines, and position statements. Although these are not binding, the Associations are encouraged to review and adopt the policies. On behalf of the Associations, Engineers Canada holds the Canadian trademarks for engineering titles such as P.Eng., Professional Engineer, Engineer, Engineering, and Consulting Engineer (and others), as well as the French equivalents. Engineers Canada has two important committees (CEAB and CEQB), as described below.

Canadian Engineering Accreditation Board (CEAB)

The Canadian Engineering Accreditation Board (CEAB) evaluates undergraduate engineering programs offered by Canadian universities. CEAB develops academic criteria, arranges accreditation team visits, compares engineering programs against these criteria, and recommends (for or against) accreditation. Accreditation ensures that university academic standards are adequate for professional engineering registration across Canada. CEAB accreditation is important to universities, and particularly important to Canadian engineering graduates, who qualify academically for licensing as professional engineers.

Canadian Engineering Qualifications Board (CEQB)

The Canadian Engineering Qualifications Board (CEQB) develops national guidelines for professional engineering to promote consistent licensing, registration, and other regulations across Canada. CEQB also maintains the Syllabus of Examinations, which describes technical exam programs for 19 engineering disciplines, as well as basic and complementary studies. The Syllabus is the key document for evaluating qualifications of applicants with non-accredited engineering degrees, and for assigning exam programs.

CANADIAN COUNCIL OF PROFESSIONAL GEOSCIENTISTS (CCPG)

The Canadian Council of Professional Geoscientists (CCPG), incorporated in 1996, serves the same purpose for professional geoscientists as Engineers Canada plays for professional engineers. However, CCPG is more recent, so its structure is evolving. CCPG's mandate is to coordinate the geoscience profession on a national scale by promoting consistency in licensing and regulation. It provides unifying advice to the geoscience licensing Associations. The Canadian Geoscience Standards Board (CGSB), a subcommittee of CCPG, develops national guidelines and examination syllabi for professional registration in the geosciences, assesses geoscience education across Canada, makes this information available to the Associations, and provides advice regarding the national mobility of geoscientists. CCPG and CGSB policies, guidelines, and position statements are not binding on the Associations, but they are encouraged to review and adopt these documents, where appropriate. Several geoscience guidelines are available on the CCPG website.[30]

OVERVIEW OF RELATIONSHIPS

Figure 2.2 illustrates how professional engineers and geoscientists interact with the various organizations mentioned in this chapter. As a professional engineer or geoscientist, you obtained your licence from your provincial Association, and you likely joined at least one technical society (as discussed in Chapter 5). The members of the Association (including you) elect the council, which hires the staff and appoints the committees. The Association's staff members communicate with you regarding admission, professional practice, discipline, professional development, and so forth.

In most provinces, the Association licenses both engineers and geoscientists. All engineering Associations are federated members of Engineers Canada (the national body for engineering), and all geoscience Associations are linked to CCPG (the national body for geoscience).

Although both Engineers Canada and CCPG are important bodies, they affect you only indirectly. They probably assessed or accredited your university program, and they assist the profession across Canada; however, they advise the Associations and rarely interact directly with members.

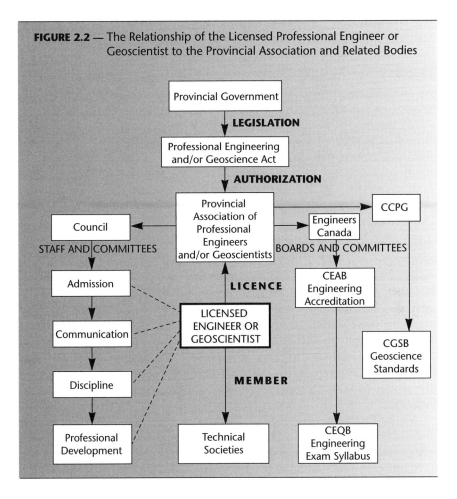

FIGURE 2.2 — The Relationship of the Licensed Professional Engineer or Geoscientist to the Provincial Association and Related Bodies

CASE HISTORY 2.1

THE BRE-X MINING FRAUD

In the spring of 1997, Bre-X Minerals Limited, a mining company based in Calgary, became the focus of a spectacular mining fraud, apparently perpetrated by at least one geologist. The fraud convinced Canadians that geology needed professional regulation. Bre-X claimed to have made a gold strike that was richer than any gold discovery in history, and Bre-X stock prices soared. However, after a mysterious death, the fraud began to unravel. The resulting scandal ruined the reputations of almost everyone involved. More seriously, the fraud caused financial calamity for thousands of investors, some of whom had staked their life savings on the Bre-X geologists' reports. An independent team of investigators from Strathcona Mineral Services later stated: "The magnitude of tampering with core samples . . . is of a scale and over a period of time and with a precision that, to our knowledge, is without precedent in the history of mining anywhere in the world."[31]

Background Information

For our purposes, the story begins in May 1993, when David Walsh, the founder, chairman, and CEO of Bre-X Minerals, announced the discovery of a gold deposit in Busang, Indonesia. One site, drilled previously by an Australian company, was reported to contain an estimated one million ounces of recoverable gold.[32] This modest estimate was to escalate as the months passed, causing a frenzy that pushed Bre-X stock prices from pennies in March 1994 to the equivalent of over $200 per share in September 1996.[33]

Sequence of Events

The roles played by various Bre-X geological staff are not completely clear. John Felderhof, a 1962 geology graduate of Dalhousie University, was the chief geologist[34] of Bre-X Minerals, although he has since declared that his role was that of an administrator or commercial manager. Michael de Guzman, a geologist from the Philippines, was "Bre-X's No. 2 geologist." De Guzman was running four Bre-X camps in Indonesia, so a fellow Filipino geologist, Cesar Puspos, reportedly supervised much of the work at the Busang site.[35]

Walsh was actively involved in raising funds in Calgary. Meanwhile, Felderhof was in Jakarta and de Guzman and Puspos were in Busang. In addition, about 20 others worked as geologists or project managers for Bre-X in Indonesia.

A reputable Australian drilling company was hired to drill core samples to evaluate the gold content of the Busang site. In March 1996, Bre-X reported estimates of 30 million ounces of gold at the Busang site; this soon increased to 70 million ounces, with the potential for 100 million ounces. In early 1997, Felderhof raised the "official" reserve estimate at Busang to 200 million ounces of gold.[36]

However, the golden glow began to tarnish in January 1997, when a storage building containing the core samples at the Busang site burned down, allegedly destroying the records of the drilling results. The unravelling fraud attracted world attention on March 19, 1997, when de Guzman committed suicide by jumping from a helicopter. In his suicide note, he explained that poor health drove him to suicide. A body recovered from the Indonesian jungle was confirmed to be his. At the time, de Guzman had been en route to a meeting with a geological team to discuss discrepancies in the test results. Freeport-McMoRan Copper & Gold, Inc., a company in partnership with Bre-X, had drilled additional test holes next to the Bre-X drill holes and the results were quite different from the glowing results quoted by Bre-X.

Felderhof regularly visited the proposed Busang mine site, and made the public estimates of gold content. His final figure of 200 million ounces was worth about $70 billion (at the 1997 gold price). To develop this immense gold find, Bre-X needed the assistance of a larger company and, under pressure from the Indonesian government, became partners with Freeport-McMoRan to develop the find. Freeport challenged the gold estimates, so Strathcona Mineral Services of Toronto was hired to give an impartial analysis.

Strathcona soon discovered the fraud. The gold, allegedly a type found in local rivers, had been carefully added to the samples to create a false image of the proposed mine's gold content. "The results had to give a very specific three-dimensional picture of a plausible deposit. The whole picture had to make sense. It had to be very well-planned and well-executed."[37] Moreover, a critical breach of accepted practice was discovered. When drill cores are removed from the ground, they are ordinarily sawn in half along the centre-line. One half is tested but the other half is documented and stored, in case it is needed for double-checking. However, Bre-X did not follow this practice. Suspicion for the salting therefore fell on De Guzman, since he was the senior person responsible for the drill core samples at the Busang site, sent thousands of them to a local laboratory to be assayed, and had the knowledge needed to perpetrate the fraud.

Aftermath—Chaos on the Stock Market

After the fraud was discovered, Bre-X hired an investigative team, Forensic Investigative Associates (FIA), to perform an independent audit. The FIA report, published in October 1997, exonerated Bre-X's senior staff and stated that FIA had "reasonable and probable grounds" to conclude that de Guzman and others at the Busang site were responsible for the ore salting.[38]

Many Bre-X employees profited personally by selling shares they had purchased with stock options. The FIA report estimated that de Guzman received $4.5 million in stock sales, Puspos, $2.2 million, and Walsh, about $36 million.[39] Felderhof reportedly sold about $30 million of his shares (later estimated at $84 million).

In March 1997, just before the fraud revelation, the Prospectors and Developers Association of Canada named Felderhof as Prospector of the Year. The award was in honour of the Busang discovery, believed at that time to be the world's largest single gold deposit. A few months later, Felderhof agreed to return the award. He was asked to resign from Bre-X, and he now resides in the Cayman Islands. Bre-X was de-listed from the stock exchange, and its shares are essentially worthless. Many investors are pursuing lawsuits, on the basis that the corporation and the individuals who controlled its geological activities should have shown greater diligence in controlling the assay samples and verifying the gold estimates. The Ontario Securities Commission charged Felderhof for illegal insider trading and for issuing news releases that he should have known were misleading, but the courts dismissed the case in 2007. Bre-X president David Walsh died in 1998.

Estimates of the total loss to investors because of the Bre-X fraud run as high as $6 billion. The Bre-X scandal seriously damaged the Canadian mining industry. Junior mining companies—even those with no links to Bre-X—found it difficult to raise capital.

Comments on the Bre-X Fraud

The Bre-X scandal is a case of skilled geological fraud, apparently perpetrated by de Guzman. If he were still alive, de Guzman would be facing criminal charges for fraud, as well as discipline for professional misconduct. In the face of such deliberate fraud, it may seem trivial to refer to the Code of Ethics. The perpetrator ignored the code, came to a bad end, thousands of people suffered serious financial harm, and an entire industry was held in contempt.

This case also shows that accurate, unambiguous duties and titles are important. The chief geologist was reportedly Felderhof. He certainly seemed to consider himself qualified for this title when he made estimates of the gold content in Busang and when he accepted the Prospector of the Year award, although he later claimed to be merely an administrator. In any case, the chief geologist—whether Felderhof or some other person—had a responsibility to show due diligence in safeguarding the core samples and ensuring that the gold assay was conducted properly; that the gold content, based on the samples, was accurately calculated; and that double-checks were made to confirm the results. As noted above, geologists usually split core samples before testing them, and store half of each sample for further confirmation, if necessary. The Bre-X geologists did not do this. Security was loose, and a single individual controlled the testing of all of the Busang samples. This explains why salting could proceed undetected for months.

Stock promoters who unwittingly encouraged the investment of billions of dollars in a fraud may need to question the standards of due diligence in their own profession as well. The Bre-X case is a forceful reminder that the mining and resource industries must insist on professional people with high ethical standards.

DISCUSSION TOPICS AND ASSIGNMENTS

1. Who should be more concerned about the welfare of the public: the professional person or the average person? Does professional status impose additional responsibilities? Should people in positions of great trust, whose actions could harm the public, obey a higher code of ethics than the average person obeys? If so, what is the best way to impose such requirements?

2. Compare the U.S. and Canadian systems for regulating professions. For example, much information is available about professional engineers in both countries. In the U.S., state boards, appointed by the state governments, regulate professions. Engineers are required to write two sets of exams for admission. Everyone, regardless of education, must write the exams. In Canada, the profession is self-regulated by Associations of engineers and/or geoscientists. Graduates of accredited programs are exempt from technical exams. Statistics show that only about 15 percent of the people practising engineering in the U.S. have licences, whereas about 85 percent of the people practising engineering in Canada have licences. In your answer, respond to the questions: Which system is more effective?

Which is simpler to administer? Which is fairer to the applicant? Which is better for protecting the public? Which criterion of effectiveness, simplicity, fairness, or protection should take precedence?

3. The definition of engineering (or engineering practice) varies across Canada, as described in this chapter. Consult the definition in your Act (on your Association's website or in Appendix B) and review the definitions for at least two other provinces or territories, as well. Do they agree with the national definition offered in this chapter? Select the definition you consider "best" and explain why, briefly.

4. Repeat the above question for geoscience.

Additional assignments can be found in Appendix E.

NOTES

[1]   J.R. Millard, *The Master Spirit of the Age: Canadian Engineers and the Politics of Professionalism, 1887–1922*, University of Toronto Press, Toronto, 1988.

[2]   H.A. MacKenzie [opening address for the debate on the Professional Engineers Act, 1968–69], Ontario, Legislature, Debates.

[3]   Millard, *The Master Spirit of the Age*.

[4]   Ibid.

[5]   W.N. Pearson and G.D. Williams, *Professional Registration of Geoscientists in Canada*, Canadian Council of Professional Geoscientists, 2004, available at <www.ccpg.ca/news/professional_registration_geoscientists.html> (May 10, 2008).

[6]   D.R. O'Connor, *Report of the Walkerton Inquiry: The Events of May 2000 and Related Issues* (Part One: A Summary), published by Ontario Ministry of the Attorney General, Copyright © Queen's Printer for Ontario, 2002.

[7]   Canadian Securities Administrators (CSA), *Standards of Disclosure for Mineral Projects, National Instrument 43-101*. Document NI 43–101 can be found by searching the CSA websites in BC <www.bcsc.bc.ca>; Ontario <www.osc.gov.on.ca>; Quebec <www.lautorite.qc.ca>; or Alberta <www.albertasecurities.com> (May 10, 2008).

[8]   Professional Engineers Ontario (PEO), *Professional Engineers Providing Reports on Mineral Properties*, PEO Guideline, Toronto, September, 2002, available at <www.peo.on.ca> (May 10, 2008).

[9]   *National Instrument 43–101—Standards for the Disclosure of Mineral Projects*. Document NI 43–101 is on the Canadian Council of Professional Geoscientists (CCPG) website at <www.ccpg.ca> (May 10, 2008).

[10] Engineers Canada, *Guideline on the Definition of the Practice of Professional Engineering*, Ottawa, available at <www.engineerscanada.ca/e/pu_guidelines.cfm> (June 15, 2009). Excerpt reproduced with permission of Engineers Canada.

[11] Canadian Engineering Accreditation Board (CEAB), Engineers Canada, *2007 Accreditation Criteria and Procedures*, Ottawa, 2006, p. 12, available at <www.engineerscanada.ca/e/pu_ab.cfm> (June 15, 2009). Excerpt reproduced with permission of Engineers Canada.

[12] *Professional Engineers Act*, Revised Statutes of Ontario, 1990, c. P28, s.1.

[13] *The Engineering, Geological and Geophysical Professions Act*, Revised Statutes of Alberta, 2000, Chapter E-11, Section 1(q).

[14] Canadian Council of Professional Geoscientists (CCPG), *Definition of the Practice of Professional Geoscience*, approved January 11, 1999. Also included in *Geoscience Knowledge and Experience Requirements for Professional Registration in Canada*, published by CCPG, May 2008, available at <www.ccpg.ca> (June 15, 2009). Excerpt reproduced with permission of CCPG.

[15]   Canadian Geoscience Standards Board (CGSB), a committee of CCPG, "Outlines of Required Knowledge," Appendix 3 of *Recommended Minimum Requirements of Geoscience Knowledge and Work Experience for Professional Practice*, available at <www.ccpg.ca> (June 15, 2009).

[16]   Canadian Engineering Qualifications Board (CEQB), Engineers Canada, *Guideline on Admission to the Practice of Engineering in Canada*, Ottawa, 2001, available at <www.engineerscanada.ca/e/pu_guidelines.cfm> (June 15, 2009).

[17]   Engineers Canada, *2007 Accreditation Criteria and Procedures*, p. 27.

[18]   Engineers Canada, *From Consideration to Integration*, Ottawa, available at <fc2i.engineerscanada.ca/e/index.cfm> (June 15, 2009).

[19]   Engineers Canada, *Engineers Canada Examination Syllabus*, Ottawa, available at <www.engineerscanada.ca/e/pu_syllabus.cfm> (June 15, 2009).

[20]   Engineers Canada, "Components of Acceptable Engineering Work Experience," *Interpretive Guide IV*, Ottawa, available at <www.ccpe.ca/e/files/guideline_admission_intguide.pdf> (May 10, 2008).

[21]   Professional Engineers Ontario (PEO), *Guide to Required Experience for Licensing as a Professional Engineer in Ontario*, Toronto, 2002, available at <www.peo.on.ca> (May 10, 2008).

[22]   Canadian Geoscience Standards Board (CGSB), a committee of CCPG, *Geoscience Knowledge and Experience Requirements for Professional Registration in Canada*, May 2008, available at <www.ccpg.ca> (June 15, 2009).

[23]   Engineers Canada, *Agreement on Mobility of Professional Engineers within Canada*, available at <www.engineerscanada.ca> (May 10, 2008).

[24]   Canadian Council of Professional Geoscientists, *Agreement on Mobility of Professional Geoscientists within Canada*, June 2003, available at <www.ccpg.ca> (May 10, 2008).

[25]   Engineers Canada, "International Mobility", available at <www.engineerscanada.ca> (June 15, 2009).

[26]   Canadian Council of Professional Geoscientists (CCPG), *International Cooperation Agreements*, available at <www.ccpg.ca> (May 10, 2008).

[27]   Professional Engineers Ontario (PEO), *Guideline to Professional Practice*, Toronto, 1998, p. 15, available at <www.peo.on.ca> (May 10, 2008).

[28]   Ibid.

[29]   Engineers Canada, *Home Page*, Ottawa, available at <www.engineerscanada.ca> (May 10, 2008).

[30]   Canadian Council of Professional Geoscientists (CCPG), *Home Page*, available at <www.ccpg.ca> (May 10, 2008).

[31]   A. Willis and D. Goold, "Bre-X: The One-Man Scam," *Globe and Mail*, July 22, 1997, A1.

[32]   J. Stackhouse, P. Waldie, and J. McFarland (with files from C. Donnelly), "Bre-X: The Untold Story," *Globe and Mail*, May 3, 1997, B1.

[33]   A. Spaeth, "The Scam of the Century," *Time*, Canadian ed., May 3, 1997, p. 34.

[34]   J. Wells, "The Bre-X Bust," *Maclean's*, April 7, 1997, p. 50.

[35]   Stackhouse et al., "Bre-X: The Untold Story," B1.

[36]   Wells, "The Bre-X Bust," p. 50.

[37]   Spaeth, "The Scam of the Century," p. 34. © 1997 Time Inc.

[38]   P. Waldie, "De Guzman Led Tampering at Gold Site," *The Globe and Mail*, October 8, 1997, A1.

[39]   Ibid.

Chapter 3
Continuing Professional Development

Engineering and geoscience are constantly advancing, so keeping up is a challenge. New techniques, theories, software, and hardware emerge every year. How does a professional stay up-to-date? Engineers Canada published a guideline proposing that licensed members should participate in a wide range of learning activities, and report these activities to their Associations annually. All licensing Associations are now following this guideline (or beginning to follow it). This chapter gives an overview of the need for continuing competence, the guideline requirements, and a summary of typical activities that meet the requirements. Fortunately, most professionals rise to the challenge, easily.

CAREER MOMENTUM VERSUS OBSOLESCENCE

Graduates bring new ideas into the workplace and are enthusiastic to learn more. This positive attitude is the start of a successful career. However, your bachelor's degree is like a radioactive mineral that decays over time. In previous decades, the half-life of a university degree was about 10 years, but in high-tech disciplines, it is even shorter. Even if you were at the top of your graduating class, you will eventually be out of date without professional renewal. Keep your skills up-to-date, because professional development maintains your career momentum. More importantly, most provinces now have (or are moving to) mandatory programs, and by neglecting professional development (or failing to document it) you can risk the loss of your professional licence.

A recent survey asked Canadian engineers and geoscientists to identify the skills needed for a successful career (in addition to their basic professional education).[1] The top three were negotiation skills (identified by 30 percent of the respondents), business skills (27 percent), and personnel management skills (20 percent). Engineers Canada recommends developing similar skills in their Guideline on continuing competence. Obviously, you must develop most of these key skills on your own, because very few of them are in the undergraduate engineering and/or geoscience curriculum (unless your university permitted optional or minor courses in management).

Non-Technical Skills

- **Communication** (written and oral)
- **Interpersonal Skills** (cultural sensitivity, working with subordinates, negotiation, delegation, decision making, etc.);
- **Project Management** (scheduling, estimating, budgeting, quality assurance, etc.)
- **Problem Solving**
- **Management** (recruiting, training, performance evaluation, human rights, motivational methods, mentoring, harassment issues, etc.)
- **Lifelong Learning**
- **Business** (contract negotiation, financial accounting, risk analysis, law, etc.)

Technical Skills

- **Dangerous/hazardous materials management**
- **Environmental regulations**
- **Codes and standards**
- **Regulatory compliance** [such as the Building Code].[2]

RESPONSIBILITY FOR CONTINUING COMPETENCE

Of course, you are responsible for maintaining your competence. The good news is that maintaining competence is a win–win arrangement: it helps your career by keeping you productive and competitive, but it is also a quality assurance measure that protects the public.

PROFESSIONALS ARE RESPONSIBLE Each of the licensing Acts contains a clause (usually in the Code of Ethics) requiring continuing competence. For example, the Engineers Canada Code of Ethics (which is a national model code) states this quite clearly. Professional engineers shall "keep themselves informed in order to maintain their competence, strive to advance the body of knowledge within which they practise and provide opportunities for the professional development of their subordinates."[3]

This means that you must continually assess your competence and keep abreast of new theories, equipment, and methods. However, you cannot do it alone; you need help and guidance.

EMPLOYERS ALSO HAVE A ROLE To retain the top talent, employers must help employees achieve their full potential. Professionals want to work hard, but they also want assignments that challenge them, and opportunities for cutting-edge courses, workshops, and conferences to keep their skills sharp. Employers are accepting this role. A recent survey by Engineers Canada showed that 64 percent of licensed engineers and geoscientists had taken additional training in the past three years to maintain or upgrade their professional competence. Of those taking the extra training, most (four out of five) received some financial support from employers, and half received some time off.[4] Career development, today, supersedes the job security of the past.

UNIVERSITIES AND TECHNICAL SOCIETIES SHOULD LEAD THE WAY Universities have an obvious responsibility to provide relevant courses, seminars, and conferences, but technical societies also play a key role in helping you to maintain your professional competence. Technical societies bring together people with similar professional interests, and create a vast wealth of knowledge, research, codes, standards, and techniques. A practising professional should join at least one society to benefit from the help that these societies distribute so freely. It is simple common sense. (Chapter 5 discusses technical societies in detail.)

COMPETENCE PROGRAM REQUIREMENTS

Before Associations introduced continuing competence requirements, they examined other professions in Canada. Law, medicine, accounting, and architecture have one thing in common: continuing professional development is a standard expectation. Some licensing boards in the United States are also introducing similar requirements.

To encourage Associations to develop consistent professional development programs, Engineers Canada published a guideline in 1996 (reissued in 2004). Engineers Canada envisages four major requirements for complete assurance of continuing competence, summarized as follows:

- **Continuing Professional Development (CPD).** Each Association should have (at a minimum) a voluntary continuing professional development program, which advertises Continuing Professional Development (CPD) activities to help members acquire new knowledge, skills and experience. CPD activities (as defined below) include a wide range of activities, including professional practice, formal courses and informal study.
- **Reporting and recording.** The Associations should make it easy to document one's CPD activities, preferably by online electronic reporting.
- **Compliance statement.** The Associations should ask licensed members to make an annual declaration that they have complied with the CPD requirements.
- **Practice Review.** To check that all members maintain continuing competence, the Associations should audit compliance declarations by selecting a small, random sample of members to undergo a practice review.[5]

Engineers Canada recommends that all four aspects of the continuing competence program should be mandatory and that Associations should sanction members who are unwilling or unable to comply. As of May 2007, all provincial and territorial Associations have adopted the Guideline, or are in various stages of adoption. Some programs are voluntary and some are mandatory. APEGGA (Alberta) introduced its mandatory CPD plan in 1998, and was the first Association to respond fully to the Engineers Canada Guideline, although Quebec has had a voluntary CPD policy since 1982.[6] The programs may be classified as follows:

- **Mandatory CPD Programs:** Alberta (APEGGA), New Brunswick (APEGNS), Newfoundland & Labrador (PEG-NL), Ontario (APGO), Prince Edward Island (APEPEI), Quebec (OGC), Saskatchewan (APEGS).
- **Voluntary CPD Programs:** British Columbia (APEGBC), Manitoba (APEGM), Northwest Territories (NAPEGG), Nova Scotia (APENS), Nova Scotia (APGNS), Ontario (PEO), Quebec (OIQ), Yukon (APEY).

Program requirements vary widely. Some require full reporting, compliance, and possible practice review, whereas most of the voluntary programs have no practice review. Each Association has a professional development web page, brochure, or guideline that defines their program. Visit the website to see the CPD and compliance procedures that apply to you. (Appendix A lists your Association's Web address.)

PROFESSIONAL DEVELOPMENT ACTIVITIES

There are many ways to maintain competence. The following suggestions use the same six classifications as the Alberta program,[7] since several Associations follow the Alberta format.

Professional Practice

Your on-the-job engineering or geoscience experience, in progressively more challenging tasks, is important for maintaining competence. In Alberta, the Association (APEGGA) allows one hour of PDH credit for 15 hours of professional practice, to a maximum of 50 PDH per year. However, some Associations (such as APEGBC) automatically give credit for professional practice, and reduce other CPD requirements accordingly. Check with your Association in case the weighting is different for this activity.

Formal Activity

Some formal activity should be included in your program. Formal courses, workshops, and in-house instruction are excellent, particularly if permanent records show the candidate's evaluation and performance. CPD programs may be provided by a university, college, technical society, or an internal industry educational program (including video or interactive Internet courses). Alberta (APEGGA) credits one PDH for each hour of course attendance to a maximum of 30 PDHs per year. (For courses rated in continuing education units or CEUs, one CEU equals 10 PDHs.)

Informal or Self-Directed Activity

Informal activities may qualify, if they expand your knowledge, skills, or judgment. Examples include: self-directed study; attending conferences and industry trade shows, seminars, technical presentations, talks and half-day workshops; attending meetings of technical, professional, or managerial societies; and structured technical or professional discussion with one's peers. Keep records of such study, attendance, or events. Alberta (APEGGA) credits one PDH for each hour of informal activity, to a maximum of 30 PDHs per year.

Participation

Activities that promote interaction and/or promote discussion of new ideas or technology are beneficial to the profession and the public. For example, mentoring members-in-training; providing technical service to public bodies; serving on committees for technical, professional or managerial societies; providing community service, such as an elected public service at any level; and engaging in active service for charitable, religious, or service organizations, are valid participation. Alberta (APEGGA) credits one PDH for each hour of participation, to a maximum of 20 PDHs per year (but only 10 PDHs may be community service).

Presentations

Technical or professional presentations at conferences, meetings, courses, workshops, or seminars are eligible for credit, whether in-house or sponsored by a technical or professional organization. (Presentations given several times count only once.) Alberta (APEGGA) credits one PDH for each hour of preparation or delivery, to a maximum of 20 PDHs per year.

Contributions to Knowledge

Activities such as writing or co-authoring journal papers, patents, monographs, books, codes, standards, and so forth on engineering or geoscience topics expand the technical knowledge base of the professions. Alberta (APEGGA) allows a maximum credit of 30 PDHs per year, but gives different credit for each type of achievement, using the following classification:

- Developing codes and standards: One hour of committee-work equals one PDH.
- A patent awarded or a paper in a peer-reviewed technical journal: 15 PDHs.
- One thesis at the Masters or Ph.D. level, (after defence and approval): 30 PDHs.
- Publication of a book: 60 PDHs, claimed over two years.
- Article in a non-reviewed journal or company report: 10 PDHs (max: 10 PDHs/year)

- Reviewing articles for publication: One hour equals one PDH. (Max: 10 PDHs/year).
- Editing papers for publication: One hour equals one PDH.[8]

REPORTING AND AUDITING

Reporting and Compliance

Members typically report their CPD activities annually, when licences are renewed. At that time, the member submits a statement listing the activities completed during the previous year. This process requires personal record keeping. In most provinces, members report their CPD activities electronically on the Association website. In some provinces, members simply assure that they have complied with the CPD requirement, and keep their records for possible future review.

Time Commitment

Each Association typically provides both a weighting system for CPD activities and a recommended annual sum required. The requirements vary, so check your Association website (listed in Appendix A). The unit of measure may be either the professional development hour (PDH) or the continuing education unit (CEU). The CEU is the unit commonly used for formal courses, but the PDH is more useful for informal activities. The following equations permit a conversion:

- The PDH is equal to one contact hour of learning.
- The CEU is equal to 10 contact hours of learning.

ALBERTA APEGGA suggests 240 PDHs over three years, with activities in three of the various categories (above) per year. This implies an average of 80 PDH/year. APEGGA reduces the requirements for special cases, and non-practising members are, of course, exempt from the CPD program.[9] Several jurisdictions follow the Alberta standard.

BRITISH COLUMBIA APEGBC has a "recommended but not mandatory" program, and the requirements differ slightly from the Alberta model. Credit is already included for professional practice, so APEGBC asks members to complete 90 PDH on a three-year rolling average, or an average of 30 PDH/year, in at least two of the following four categories:

- Formal Activities (structured courses, training, certification, etc.)
- Informal Activities (on-the-job training, self-directed study, etc.)
- Participation (mentoring, service to technical and professional societies, etc.)
- Presentations and Contributions to Knowledge (writing, presenting, or reviewing codes, standards, research papers, etc.)

Photo 3.1 — The CANDU Reactor. *The CANDU (CANada Deuterium Uranium) nuclear reactor is impressive. In 1987, it was judged one of Canada's top ten engineering design achievements. The newest CANDU installation is the Darlington Nuclear Generating Plant about 70 km east of Toronto. Darlington consists of four CANDU reactors, with a total electric output of 3,524 megawatts, which satisfies about 20 percent of Ontario's electricity requirements. The photo shows the four CANDU reactors (in the four large, windowless buildings behind the cylindrical vacuum building). The long building behind the reactors is the turbine hall, which houses the electrical generators. The heat from the reactors creates steam in the boilers atop the reactors. The steam is piped to the turbine hall, where it drives turbines coupled to the generators. The electricity produced by the generators flows to consumers via the transmission lines.*

Source: Courtesy of Ontario Power Generation.

APEGBC allows a maximum of 20 PDH in any one category in a calendar year. Members can simply confirm their compliance each year, or may use APEGBC's CPD On-line Reporting Centre. Keep supporting documents for four years, for audit purposes. You may transfer surplus PDHs to future years, but must claim them within three years of the activity.[10]

Exemptions and Deferments

Every provincial Association allows special consideration for special cases. For example, retired or non-practising members are exempt from the CPD program, if the member confirms that he or she is not practising. In addition, members on maternity leave, studying full-time, working outside of the province, employed part-time, unemployed, or disabled may request exemption or deferment, as appropriate. A written request for special consideration is necessary.

Practice Review

Engineers Canada recommends random checks to ensure that the self-assessment is fair and honest. In most provinces, the Association selects a small sample of members each year, and verifies that the records submitted are accurate and properly evaluated. This auditing process has generated some dissension, since it requires time and effort by the Association and by the licensed members.

THE EIC CONTINUING EDUCATION PROGRAM

The Engineering Institute of Canada (EIC) is a useful place to search for appropriate CPD courses. The EIC, Canada's oldest technical society, has taken on the task of coordinating CPD activities provided by others. EIC is a member of the International Association for Continuing Education and Training (IACET) in Washington, D.C.

EIC does not offer courses, but it selects them, validates their quality, and lists them on the EIC website.[11] The courses are not (usually) degree credits, but all provide acceptable CEUs. More importantly, EIC verifies that over 30 course providers deliver CPD programs that meet established quality standards. The providers are entitled to use the EIC logo to show the EIC validation. They include:

- **Technical societies that are members of the EIC.** EIC has several constituent members, including the Canadian Society for Civil Engineering, Canadian Geotechnical Society, Canadian Society for Chemical Engineering, Canadian Society for Mechanical Engineering, IEEE Canada, and others (see Chapter 5).
- **Universities and other teaching institutions.** Many universities participate, throughout the year, as courses become available.
- **Industry associations with expertise in specific areas.** Several industry organizations participate, including the Canadian Nuclear Society, Canadian Dam Association, and others.
- **Other organizations.** Many other organizations and institutions link to EIC for specific courses or conferences.

POSTGRADUATE STUDIES

A postgraduate degree is useful in maintaining competence. It will likely advance your career, particularly if your goal is to be a specialist in engineering or geoscience. The degree (or more accurately, what you learn while achieving the degree) may be vital for changing your career direction, or escaping a dead-end job. For example, it is generally easier to enter a specialty, such as biomedical engineering, through a master's program.

A common rule of thumb is that a postgraduate degree extends a professional career by as much as 10 years, over a bachelor's degree. However, advanced degrees can be expensive, even when taken part-time, and a half-hearted interest is risky. To avoid wasting valuable years, you must weigh the costs and benefits, define a clear goal, and be determined to achieve it.

The best time to think about postgraduate study is while you are still in university, because you can get advice from professors. More importantly, you can apply for research grants and assistantships, which may be crucial. In later years, family, job, or financial pressures may take priority over study. Since most readers of this textbook are recent graduates or senior undergraduates, the best time to think about postgraduate studies is probably right now!

Admission Requirements—Master's Degree

Universities typically award engineering and geoscience master's degrees in applied science (MASc), science (MSc), or engineering (MEng). Admission requirements vary, but you must usually have a B average, or better, from an accredited undergraduate program. If you have been out of university for three or more years, you may qualify as a mature student. Admission standards are more flexible for mature students because they are more goal-oriented, more organized, and thus more effective in their studies.

Programs that include a thesis ("research master's") usually have higher admission standards than those that are mainly courses ("course-work master's"). Thesis projects typically explore a part of the supervisor's research plan and may include pay as a research assistant. A master's degree takes a minimum of one academic year, but 16 to 20 months is typical. Most universities hire master's candidates as part-time laboratory tutors or teaching assistants, if needed.

Admission Requirements—Doctoral Degree

The usual engineering or geoscience doctoral degree is the Doctor of Philosophy (Ph.D.).

Doctoral candidates usually must rank in the upper quartile of their undergraduate classes, and have a master's degree in a related area of study. Some candidates enter the doctoral program directly from the bachelor's degree, but this is not common in Canada. However, master's students who show exceptional ability often transfer into doctoral programs.

A thesis is required for a doctorate. Although the supervisor gives guidance, the candidate defines the thesis topic, which must make an original, independent contribution to the discipline. Consider several research projects (and several supervisors) before making a commitment—your doctoral research will shape the rest of your career! The doctorate takes a minimum of three years beyond the bachelor's degree (or two years beyond the master's degree). However, the length is typically a year longer, particularly if the doctoral candidate works as a part-time lecturer or research assistant, or if the research project is especially challenging.

Making a Cost–Benefit Analysis

A postgraduate degree usually justifies a higher salary than a bachelor's degree. More importantly, a postgraduate degree may increase your job satisfaction, extend your professional career, and give you more years of work at peak salary. However, the degree takes time and costs money, so you may not break even for several years. The early years may also require some sacrifice; family time may be limited. You may have to delay a new home or car purchase. Although the challenge of postgraduate study may be attractive, you must weigh future benefits against present costs. Before you make a commitment, decide: Will the effort pay off for you?

Choosing a University

If you want to explore the possibility further, gather information and define your goal. University catalogues are on the Internet. The most important factors are your enthusiasm for your study topic (or research project), the quality of the supervisor who will mentor you, and the computer and laboratory facilities that will be available to you. You should define your research interests, check out the researchers in that area, contact them, and decide for whom you would like to work, especially if the degree requires a thesis. Ask about research grants and assistantships.

CLOSING COMMENTS

Continuing competence programs have grown remarkably in the past decade, but not without some criticism. For example, former Professional Engineers Ontario (PEO) president Richard W. Braddock said (in 2003) that he supports CPD activities, calling them "desirable and necessary," but he has serious concerns about mandatory continuing competence programs.[12]

Associations must seek a balance in mandatory continuing competence programs by recognizing the many forms of CPD and giving credit where it is due. The documentation and verification process must be simple and unintrusive, and must recognize that some forms of achievement, experience, and education may be difficult to document. In some cases, practical experience (even bad experience) may teach more than formal courses.

Provincial and territorial Associations must show initiative in providing CPD activities. APEGGA (Alberta) seems to have been successful in this proactive role and has set a standard for other Associations to follow. Engineering and geoscience advocacy organizations are rare; the Associations, with their established communication links, are in a unique position to encourage and advertise CPD courses and activities. The Associations must assist in organizing these events, even though this task may be at the limit of their regulatory duties.

The engineering and geoscience faculties of our universities and colleges also have an obligation to provide more evening, part-time, and Internet courses to satisfy the need for continuing education. Universities must encourage professional employees to enroll in postgraduate courses on a part-time basis.

DISCUSSION TOPICS AND ASSIGNMENTS

1. Use the Internet to examine the requirements for the continuing competence program (or CPD program) in your province or territory. As an exercise, compare the CPD requirements for your province or territory with the rules for Alberta (Note 7, below) and British Columbia (Note 10, below). (Alberta and British Columbia residents should substitute another province.) What similarities and differences do you observe in these three sets of rules? Which is most demanding, and which is least demanding?

2. Before Engineers Canada developed their model for CPD, some Associations considered the proposal that professionals should write a formal examination every five to ten years to maintain competence. Compare this formal examination proposal with the Engineers Canada CPD model, based on personal assessment. In your opinion, which is easier to administer, which is simpler for the Association member, and which is more likely to protect the public? Which criterion should take precedence? In your answer, compare the personal self-assessment process in the Engineers Canada model with the well-known self-assessment process for income tax. Are they similar? Are they effective? Explain and summarize your answers on one or two pages.

3. Under the Act, your Association must monitor professional competence to protect the public. However, as in all self-regulating professions, the Association's members must approve (by a vote) the monitoring process. Your opinion is therefore important. Discuss your Association's CPD or continuing competence program, as published on their website (listed in Appendix A). In your opinion, is the program marginal, adequate, or excessive? Include your responses to the following questions:

   a. What continuing competence requirements protect the public best? Is continuing professional development (CPD) the best indicator of competence? If not, what is better? Conversely, does the absence of CPD activity indicate incompetence?

   b. Should CPD reporting be mandatory? Reporting and auditing take time and effort, so at what point does the expense (to the professionals and to the Associations) outweigh the benefit to the public?

c. Should performance be self-assessed by the professional, or should employers and clients judge professional performance and report it to the Association? Conversely, should "free market" competition be encouraged, so that competent members will succeed, but incompetent members will fail and be eliminated? How does the Association detect alcoholism or laziness, which might be as harmful as incompetence?

d. Should random practice reviews be standard procedure? Should Associations revoke the licences of members who refuse to engage in (or refuse to document) CPD activity?

Summarize your response on one or two pages. Consider sending it to your Association if your response confirms, or might improve, their present procedures.

Additional assignments can be found in Appendix E.

NOTES

[1]   Engineers Canada (Canadian Council of Professional Engineers), *Final Report, 2002 National Survey of Professional Engineers*, Ottawa, June 23, 2003, p. 12, available at <www.engineerscanada.ca> (May 14, 2008). Report prepared by EKOS Research Associates Inc.

[2]   Engineers Canada, "Suggested Related Skills," Appendix C of *Guideline on Continuing Professional Development and Continuing Competence for Professional Engineers* (G05-2004), Canadian Engineering Qualifications Board, Ottawa, available at <www.engineerscanada.ca/e/pu_guidelines.cfm> (June 15, 2009).

[3]   Engineers Canada, *Guideline on the Code of Ethics* (G03-2001), Canadian Engineering Qualifications Board, Ottawa, available at <www.engineerscanada. ca/e/pu_guidelines.cfm> (June 15, 2009).

[4]   Engineers Canada, *Final Report, 2002 National Survey of Professional Engineers*, p. 11.

[5]   Engineers Canada, *Guideline on Continuing Professional Development*, p. 22.

[6]   *Summary of Canadian Engineering and Geoscience Continuing Professional Development Programs,* prepared June 2005, available from the APEGBC website at <www.apeg.bc.ca/prodev/mandatorycpd/ccprogams.html> (May 14, 2008).

[7]   Association of Professional Engineers, Geologists and Geophysicists of Alberta (APEGGA), *Continuing Professional Development Program, October 2005*, APEGGA Guideline, available at <www.apegga.org/pdf/Guidelines/08.pdf> (May 14, 2008).

[8]   Ibid., pp. 4–7.

[9]   Ibid.

[10]   Association of Professional Engineers and Geoscientists of British Columbia (APEGBC), *Continuing Professional Development Guideline*, Vancouver, available at <www.apeg.bc.ca/prodev> (May 14, 2008).

[11]   Engineering Institute of Canada (EIC), EIC course providers, available at <www.eic-ici.ca> (May 14, 2008).

[12]   Richard W. Braddock, P.Eng., President, PEO, "Is competence an issue?", President's Message, *Engineering Dimensions*, published by Professional Engineers Ontario (PEO), January–February 2003, p. 3, available at <www.peo.on.ca> (May 14, 2008).

Chapter 4
Disciplinary Powers and Procedures

This chapter defines the typical enforcement and disciplinary procedures in the engineering and geoscience licensing Acts. The self-regulating professions protect the public by removing unprofessional practitioners. Knowing the disciplinary process and the basis for typical complaints helps you to avoid such problems in your professional career. Moreover, as a member of a self-regulating profession, you should read the Act critically, because you may be asked to assist in the discipline or enforcement processes.

INTRODUCTION

Most professional engineers and geoscientists are well-educated individuals with high ideals, who want to leave our society better than they found it. Malpractice and corruption are therefore relatively rare in our professions. In fact, the key message of this textbook is that a competent, ethical professional never needs to worry about such problems. However, some disputes, misunderstandings, and infractions do occur, and each Association has an obligation to respond to complaints from the public, and to resolve disputes among licensed professionals.

When complaints are made, the Associations investigate, try to mediate, and where necessary:

- enforce the Act by prosecuting people who practise unlawfully, and
- discipline licensed engineers and geoscientists who are found guilty of professional misconduct, negligence, or incompetence.

These cases (enforcement and discipline) follow different procedures, as explained below.

ENFORCEMENT: UNLICENSED PRACTITIONERS

The best way to protect the public is to allow only educated, experienced, competent professionals to practise. This is the purpose of professional licensing. People who practise engineering or geoscience without a licence, or who falsely claim to be licensed, are breaking the law (the Act). Associations are responsible for enforcing the Act by prosecuting offenders in the law courts. Each Act typically states that it is an offence for an unlicensed person to

- practise professional engineering or professional geoscience, or
- use a term or title to give the belief that the person is licensed, or
- use a seal that leads to the belief that the person is licensed.

Most Associations receive many more complaints about unlicensed practitioners than allegations against licensed members. Associations typically employ a staff member to receive complaints and contact alleged unlicensed offenders. Most unlicensed practitioners are simply unaware that they are contravening the Act, and when informed, they promptly stop the offending behaviour.

However, some offenders ignore the Association's warnings, and persist in practising without a licence. In these cases, Association staff members gather any required evidence and, with the help of a lawyer hired by the Association, prosecute unlicensed offenders in court. A trial judge presides and, if the case is proven, the judge usually fines the unlicensed offender. The typical fine is small. For example, Ontario's fines are limited to $10,000 for the first offence, and $20,000 for a second offence, which are small amounts compared to the cost of the education needed for a licence.

In 2005, a Toronto man was jailed for 30 days and ordered to pay $20,000 costs after he was found in contempt of an Ontario Superior Court order (issued in 1995) to obey the Professional Engineers Act. The man had never been licensed, but was convicted on four separate occasions, from 1993 to 1998, for repeatedly misrepresenting himself as a "structural engineer" or as "an engineer" on projects in the Toronto area.

In enforcement cases, the Association assists the court to ensure that unlicensed persons comply with the law (the Act). Case History 4.1, presented later in this chapter, discusses a well-known instance of enforcement.

DISCIPLINE: LICENSED PRACTITIONERS

Under the Act, each Association must protect the public by responding to complaints about licensed members and, where necessary, taking remedial action. It is therefore important to define what behaviour is subject to this discipline, and what remedial action could result.

The provincial and territorial Acts are very similar (although definitely not identical). They typically specify six causes for disciplinary action:

- Professional misconduct (also called unprofessional conduct)
- Incompetence
- Negligence
- Breach of the Code of Ethics
- Physical or mental incapacity
- Conviction of a serious offence

Each of these terms is defined briefly in the following paragraphs (and key excerpts from the Acts are located in Appendix B).

Professional Misconduct

Professional misconduct (or unprofessional conduct) is the main type of complaint made to Associations. In about half of the Acts, the term is not defined, and this places an additional burden of proof on the Association: The Association must prove both that the alleged misconduct occurred and that it constitutes professional misconduct.

Alberta, Newfoundland, and Prince Edward Island have very general definitions. For example, Alberta's Act defines "any conduct . . . detrimental to the best interests of the public" or that "harms or tends to harm the standing of the profession generally" as unprofessional conduct.[1] Such general clauses are really not specific enough to serve as guidance in individual cases (although the Association's Code of Ethics may give more specific guidance).

At the other extreme, Ontario's definition of professional misconduct includes some very specific acts, such as "signing or sealing a final drawing . . . not actually prepared or checked by the practitioner."[2] Such guidance is clear and unambiguous. However, the regulations cannot define every possible form of professional misconduct, so they contain a general clause stating that professional misconduct includes any act that "would reasonably be regarded" as unprofessional.[3] This circular definition is rather general, so any complaint based on this clause would first have to prove that the person's actions were "unprofessional."

Incompetence

As we would expect, incompetence is defined in several Acts as a lack of knowledge, skill, or judgment that demonstrates the member is unfit to carry out duties as a professional. A licensed professional must practise within the limits of competence, and the provincial Act and the Code of Ethics require you to judge your own competence as you progress to more difficult tasks. When you move to a new job or project, you must seek guidance, training, or experience to ensure that you are competent in the new area. If you do so, the Act gives wide latitude. However, many discipline cases involve practitioners who were judged incompetent in a new expertise, even though they were fully competent in their usual fields of practice.

Negligence

In most Acts, "negligence" means "carelessness," or carrying out work that is below the accepted standard, or a lack of adequate thoroughness. However, negligence can also include a flagrant disregard for public welfare. Negligence is a particularly serious discipline complaint when it involves financial loss or a failure to safeguard life, health, or property.

Breach of the Code of Ethics

In four provinces (Alberta, New Brunswick, Newfoundland, and Nova Scotia), a breach of the Code of Ethics is specifically defined in the Act to be equivalent to professional misconduct (or unprofessional conduct). These codes therefore have the full force of the Act. In other provinces (British Columbia, Manitoba, Prince Edward Island, Quebec, and Saskatchewan) and the territories, where the term "professional misconduct" (or "unprofessional conduct") is undefined or defined in very general terms, it would likely be understood to include the Code of Ethics, thus giving the code enforceability under the respective Act.

However, in Ontario, the Code of Ethics is specifically not enforceable under the Act. Instead, a more detailed definition of professional misconduct (in the Regulations) contains many concepts that are in the code, such as "failure to act to correct or report a situation that the practitioner believes may endanger the safety or the welfare of the public," as well as failure to disclose a conflict of interest, and about 16 additional clauses.[4] In other words, the Ontario Code of Ethics describes the ideal professional conduct, but the definition of professional misconduct identifies the lower limit of acceptable behaviour.

Physical or Mental Incapacity

Most Acts also include a "physical or mental condition" as a definition of incompetence, provided the condition is of a nature and extent that, to protect the interests of the public (or the practitioner), the practitioner should not be allowed to practise.

Conviction of an Offence

The Acts also permit disciplinary action against a practitioner who is found guilty under other laws. In other words, if a practitioner is convicted of an offence, and the nature of the offence affects the person's suitability to practise, then the person can be found guilty of professional misconduct. Proof of the conviction must be provided to the Discipline Committee. This clause is used relatively rarely, since convictions for minor offences (traffic violations, local ordinance violations, and so forth) do not affect one's suitability to practise. However, conviction of a serious offence such as fraud or embezzlement, which involves a betrayal of trust and questionable ethics, could be grounds for such action. In a recent case in Ontario, an engineer's licence was revoked when he was found guilty of engaging in child pornography. The standard of conduct for professional people is higher than that expected of the ordinary person.

AN OVERVIEW OF THE DISCIPLINARY PROCESS

Any member of the public can make a complaint against a licensed engineer or geoscientist, although complaints typically originate from building officials, government inspectors, and other practitioners. A serious complaint usually sets in motion a three-stage process. To ensure complete impartiality, the three stages of the disciplinary process are normally carried out by three different groups of people. No one who participates at an earlier stage is permitted to participate (in a decision-making capacity) in the final hearing and judgment.

The complaints process is very similar across Canada. The explanations of the process, published in Alberta (APEGGA)[5] and in Ontario (PEO)[6] are almost identical. Here's what happens when a member of the public alleges improper conduct on the part of an engineer or geoscientist.

Stage 1: Gathering Evidence

The first stage involves collecting information. When an inquiry or allegation is received, a trained Association staff member discusses it with the person, answers questions, explains the Act, and advises what material is essential to support a formal complaint. In some cases, independent information may be needed, and the Association may obtain it.

If the evidence appears to support a formal complaint, the Association staff member assists the complainant to prepare it in an appropriate format. If the evidence does not support a formal complaint, staff may suggest other means of addressing the concerns. However, individuals have the right to insist that a complaint be pursued to the next step. The signed complaint is sent to the practitioner and a response is requested, within a specified time.

Stage 2: Investigation and Peer Review

The second stage is an investigation and evaluation of the complaint, carried out by an Investigative Committee (also called the Complaints Committee, in some Acts). This Committee is composed of licensed members, including at least one of the members appointed to Council by the government. The Committee reviews the complaint, the response from the practitioner, and whatever other evidence was provided or obtained, and decides whether to

- refer the complaint to the Discipline Committee for a formal hearing, or for a "stipulated order" (if the practitioner asks to plead guilty, usually to a less-serious charge);
- dismiss the complaint if it is clearly frivolous, vexatious, or if there is insufficient evidence to go further;
- send a "Letter of Advice" to warn the practitioner about actions that are short of professional misconduct, but warrant some concern; or
- direct the Association staff to obtain more information, and return the case to the Investigative Committee.

Stage 3: The Discipline Hearing

The third stage is an independent review and evaluation of the complaint by a Discipline Committee. This Committee is also composed of licensed members, including at least one of the members appointed to Council by the government, but it is totally separate from the Investigative Committee. The Discipline Committee may, at its discretion, conduct a formal hearing to hear the evidence and the response from the accused member, and to render a judgment.

Formal hearings are usually held at the Association's offices, and follow procedures similar to a law court, with a court reporter recording a transcript. The Investigative Committee is responsible for prosecuting the complaint in front of the Discipline Committee. Witnesses may be called to testify, and the accused member and the prosecutor (from the Investigative Committee) are entitled to be represented by legal counsel. The Discipline Committee also has independent legal counsel. The Discipline Committee assigns a penalty, as discussed below, if the accused member is found guilty.

The disciplinary process described above also applies to holders of Temporary Licences, Certificates of Authorization, Limited Licences, Permits, and (in Ontario) designation as a Consulting Engineer. That is, other forms of permit or certificate may be revoked or suspended using this process.

Disciplinary Penalties

The severity of the penalty assigned by the Discipline Committee depends, of course, on the circumstances of the case. Each provincial and territorial Act typically permits the following maximum penalties:

- Revoke the licence of the practitioner (or the permit or certificate of authorization, if a corporation).
- Suspend the licence (usually for up to two years).
- Limit the practitioner's professional work, by imposing restrictions on the licence, such as supervision or inspection of work.
- Require the member to be reprimanded, admonished, or counselled, and publish the details of the result, with or without names.
- Require the practitioner to pay the costs of the investigation and hearing.
- Require the practitioner to undertake a course of study or write examinations set by the Association.
- Have any order that revokes or suspends the licence of a practitioner to be published, with or without the reasons for the decision.
- Impose a fine (up to $10,000 in Alberta; up to $5,000 in Ontario).

DISCIPLINE AND ALTERNATE DISPUTE RESOLUTION (ADR)

Fairness

Disciplinary procedures are important. They must be fair, and must be seen to be fair—both by the public and by the practitioner. The Complaints Committee and Discipline Committee are therefore independent, are composed mainly of "peers" (other licensed practitioners), and must include at least one of the members appointed to Council by the government.

The first two stages of the disciplinary process are confidential, although hearings by the Discipline Committee are typically open to the public, and its verdicts may be published (with or without names) unless the practitioner is found not guilty (or if there are compelling arguments for privacy). When publication is appropriate, the case is usually summarized in the formal "gazette" of the Association, and/or in the Association's monthly magazine, and/or on the Association's website.

Some Statistics

Compared to other professions, engineering and geoscience rarely need to apply the full disciplinary process. For example, fairly recent (2002) data from PEO in Ontario (Canada's largest Association) indicate that PEO receives only about 100 complaints per year, even though PEO has over 60,000 members. All of these complaints are investigated, but many complaints are resolved or withdrawn after consultation with PEO staff. Only about 30 to 35 of the complaints are formally referred to the Complaints Committee. The Complaints Committee evaluates each and every complaint and the response from the licensed member, and refers only about 10 to 12 complaints to the Discipline Committee each year for the formal final stage of the process.[7]

Alternate Dispute Resolution (ADR)

Many complaints involve contractual disputes between licensed practitioners. Such cases must be weighed carefully at the early stages of the process, because the discipline process is not an alternative to the civil courts. Typical legal disputes, such as breach of contract, revoking offers, or substandard performance, are not usually professional misconduct or breaches of the Code of Ethics.

In November 2005, the Ontario Association (PEO) introduced a voluntary Alternative Dispute Resolution (ADR) procedure to resolve disputes where professionalism and contractual matters overlap. Allegations of professional misconduct are referred to the complaints process, but contractual disputes that appear to be minor breaches of the Code of Ethics are encouraged to go through the ADR process. The key to the process is confidential mediation:

> Participation in PEO's ADR process is confidential and without prejudice to either party. That is, the Agreement to Mediate confirms that everything said during the mediation is confidential and documents shared during the mediation cannot be disclosed to anyone outside of the mediation, unless that person would otherwise be entitled to the documents. The Agreement to Mediate also stipulates that if the complainant and the PEO member are unable to resolve the issues through mediation, the mediator will not be called as a witness, or have his or her notes introduced at any other proceeding. However, the mediator may disclose to PEO any information disclosed during the mediation that the mediator believes raises or might raise a concern about the public interest. The mediator may adjourn the mediation to advise PEO of information that might give rise to a public interest concern and to obtain guidance on whether the case should be removed from the ADR process.[8]

The ADR process involves a mediated negotiation that may lead to a solution satisfactory to all concerned, whereas civil courts or disciplinary hearings are almost always a win–lose arrangement. If evidence of professional misconduct is found during the mediation, the ADR is stopped, and the option of a formal complaint is considered, so the public interest is always protected.

CASE HISTORY 4.1

MISUSE OF "SOFTWARE ENGINEER" TITLE

In the 1990s, the Microsoft Corporation introduced a technical course called the Microsoft Certified Software Engineer (MCSE). The use of the term "software engineer" caused some concern in Canada. U.S. state laws regulate "Professional Engineer" but do not regulate variations, such as "software engineer." Canadian laws are different; they clearly restrict the term "engineer" to licensed professional engineers. For example, the Ontario Professional Engineers Act says that an unlicensed person who "uses the title 'professional engineer' . . . or an abbreviation or variation thereof as an occupational or business designation" is guilty of an offence.[9] Clearly, the "software engineer" variation is an occupational designation.

More importantly, the MCSE title makes confusion possible and likely. Software Engineering is an accredited program at more than 10 Canadian universities,[10] and a "software engineer" would reasonably be expected to be a graduate of such a program, with three or four years of acceptable experience, licensed annually, and constantly subject to professional ethics, discipline, and continuing competence requirements. The MCSE course requires only about 42 weeks, usually in a community college, and costs about $15,000, according to an unconfirmed report in *ITbusiness*.[11] It appears that an MCSE holder has no continuing competence obligations, although upgrades may be required for new specialties.

Representatives from Engineers Canada and several provincial Associations met with Microsoft Canada in 2001, explained that their title contravened Canadian law, and suggested that MCSE (as an acronym) would be acceptable. Microsoft Canada initially agreed to this limitation. However, after speaking to MCSE holders, lawyers, and others, Microsoft apparently decided that the MCSE designation would be less valuable in Canada if holders could not refer to themselves as engineers. In 2002, Microsoft reversed itself, and issued a press release stating that the estimated 35,000 Canadian MCSE holders should continue to use the full title.

Microsoft's decision was regrettable since, as one Association said: "we would be in contempt of our own legislation if we did not enforce improper use of title." At the time, one provider of the MCSE course said that it was unlikely that the restriction on the MCSE title would hurt his business, but "I think it holds a lot more weight to be an MCSE if it is [called] an engineer. . . ."[12]

Shortly thereafter, the *Ordre des ingénieurs du Québec* (OIQ) filed penal proceedings against Microsoft Canada for knowingly causing a person who is not a member of OIQ to use the title of engineer, thereby committing an offence under Quebec's Professional Code. In April 2004, Judge Claude Millette of the Quebec court agreed, and ruled that Microsoft Canada contravened a provincial professional code by using the word "engineer" in its international software certification program. A very small fine was also levied. Microsoft appealed the decision, but in June 2005, Justice Carol Cohen of the Superior Court of Quebec rejected Microsoft's appeal. The OIQ president, Gaetan Samson, Eng., stated "The OIQ is very satisfied with the Superior Court decision, which confirms that the title *engineer*, alone or with descriptors, is reserved by the Engineers Act exclusively for our members."[13]

No other provincial Association has announced plans to prosecute Microsoft, but under every Act, any MCSE holder who practises engineering, or who implies that he or she is licensed, could be prosecuted. At least one community college has added a note to their MCSE course description explaining that the full title cannot be used, legally, under provincial law.

CASE HISTORY 4.2

THE BURNABY SUPERMARKET ROOF COLLAPSE

The 1988 opening of the new Save-on-Foods supermarket in Burnaby, British Columbia, started well, but then the roof collapsed! Everyone was very lucky, because no one was killed, although a few people suffered injuries. A formal Inquiry into the collapse yielded some surprising revelations, and led to improvements in building design and project management. However, the lessons were very expensive. Surprisingly, the collapse bears some similarity to an infamous earlier disaster.

Description of the Collapse

In April 1988, just before 9 a.m. on a rainy Saturday morning, a gleaming new Save-on-Foods supermarket was opened in the Station Square development in Burnaby. The supermarket had a parking lot built on the roof, and shoppers were asked to park their cars in the lot and descend to the main floor. About 600 shoppers, mainly senior citizens, came to the opening of the supermarket, which was staffed by about 370 employees.

After an opening ceremony attended by the Burnaby mayor and a host of local dignitaries, the shoppers milled about the store, examining the goods for sale. A sudden, sharp crack was heard, and water sprayed from an overhead sprinkler pipe near a column in the store's produce area. Startled shoppers looked up to see that a roof beam had twisted severely, breaking a sprinkler pipe. A cameraman attending the opening ceremony took a photograph of the beam—a photograph that would later assist in the investigation

Photo 4.1 — Roof Collapse in Burnaby, BC. *The Save-on-Foods supermarket in Burnaby, BC, opened on the morning of April 23, 1988. A parking lot on the roof of the supermarket provided extra convenience to shoppers. However, within minutes of opening, a main beam supporting the roof collapsed, dumping 20 automobiles into the produce section of the supermarket. Fortunately, no lives were lost. The subsequent Inquiry discovered basic errors in the design, which led to changes in licensing procedures for engineers.*

Source: Photo by Craig Hodge. Reproduced with permission of The Tri-City News.

of the failure. While the water sprayed out over a nearby cheese counter, the supermarket staff asked shoppers, over a public address system, to leave the store. Although some shoppers were reluctant, those who saw the twisted beam had no hesitation.

At about 9:15 a.m., less than five minutes after the initial bang, a huge section of the roof collapsed, dropping 20 automobiles from the roof parking lot into the produce area of the store. The remaining shoppers panicked, and people fell as they rushed outside. A pileup occurred at the exit doors. About 21 people were injured, mainly from injuries (such as broken bones) sustained in the melee. One employee was pinned under the falling roof beam and suffered a crushed pelvis. The injured were rushed to hospital. Fortunately, no one died.

The Inquiry and Report

The provincial government appointed a commissioner, Daniel J. Closkey, to head an Inquiry into the cause of the collapse and to suggest how similar failures could be avoided in the future. The Commissioner held 10 days of hearings over two months. Forty-seven witnesses testified, and the hearings were broadcast live on cable television. A final report was published in August 1988.[14]

The report's conclusions were surprising. The engineering design firm was experienced and well established, but several basic errors had been made in the design calculations. Moreover, the calculations had been thoroughly reviewed by a second experienced consulting firm. How could basic design errors have been missed by both of these experienced engineering firms? An excellent summary is presented in the 1990 research paper by Vancouver consultant C. Peter Jones and Professor N.D. Nathan.[15] Jones and Nathan were engaged as advisors to Closkey. Based on their investigation, they identified nine errors by the design engineer (in assumptions, decisions, judgments, or miscalculations) that reduced the factor of safety. Although none of these errors, by itself, would have caused the collapse, together they led to a failure that Jones and Nathan called "inevitable." The engineer who reviewed the design evidently made most of the same errors in assumptions, decisions, judgments, and miscalculations, and neglected to discover the fatal flaws. These errors are described, in detail, both in the Commissioner's report and in the paper by Jones and Nathan, and are summarized very briefly as follows:

Errors in dead load estimation. The original design was correctly carried out. The roof consisted of two concrete slabs separated by a layer of insulation and a waterproof barrier. The bottom slab was part of a concrete/sheet-steel roof deck. During the early design stage, the top concrete slab was made thicker but the beam size was not adjusted to resist the added weight. When the concrete slabs were poured, the actual thickness was (inadvertently) even greater than specified. Finally, a concrete walkway on the roof was widened but this extra weight was ignored. These three factors increased the moment in the beam (caused by dead load) by about 55 percent, effectively eliminating the factor of safety.

Error in beam specification. The beam was originally checked for strength and deflection, and the deflection limit was the governing case. However, the design engineer concluded that a greater deflection was acceptable, and a redesign was carried out. The Commissioner found no records or clear recollection of this second design. Nevertheless, subsequent issues of the structural drawings showed the beam incorrectly reduced to a smaller-weight section. The revision was not flagged on the drawing, perhaps explaining why it was overlooked when reviewed later.

Error in live load estimation. Re-evaluating the area of the roof that was supported by the beam also reduced the live load estimation. This was a valid point, and a minuscule reduction of about 1 percent was appropriate. However, an error in judging the "tributary area" resulted in an erroneous live load reduction of 12.6 percent.

Optimistic calculation of bending moment. We usually calculate the bending moment from the centre-line of a structural-steel support. However, in concrete design it is acceptable to calculate a moment from the edge of the column. This optimistic convention was used, slightly reducing the expected bending moment in the beam.

Optimistic tests of beam strength. When the (erroneous) beam section was analyzed with the (erroneous) load data, it was still slightly under-strength. However, at this point the engineer received "mill certificates" from the steel supplier, showing that the steel yield strength was 25 percent greater than the strength used in computations. The beam was therefore judged adequate in strength. However, the Commissioner concluded that this strength estimate, based on only a few test specimens, was unrealistically optimistic, because the test specimens were taken from the web of the beam, which typically has a higher strength than the flanges, and in bending, the flanges are more highly stressed than the web.

Lack of lateral support. To prevent lateral buckling, long, deep beams must be supported laterally, or loads must be reduced. Lateral supports at the failed column were noted on a shop drawing at one point in the design, but the supports were deleted when the engineer was informed of the extra cost. Clearly, the engineer did not evaluate the reduced capacity of the beam caused by the lack of lateral support.

No check of column buckling strength. The Commissioner concluded that there was no evidence that the load-carrying capacity of the beam–column assembly was considered as a possible mode of failure.

The Cause of the Collapse

After hearing all of the testimony, the Commissioner concluded that the roof failure involved two modes. These modes, illustrated in Figure 4.1, are explained as follows:

- **Beam failure.** The primary cause of failure was that the beam was under-strength for the imposed bending loads, and the laterally unsupported lower compression-flange of the beam buckled at the supporting column.
- **Column failure.** In addition, the vertical beam–column assembly likely buckled simultaneously, although this was probably not the primary cause.

Buckling is usually a rapid and catastrophic occurrence, but in this case, the roof did not crash down immediately. When the beam yielded, the movement snapped the sprinkler pipe, and the noise, motion, and water spray alerted the supermarket staff that total collapse was imminent. However, the roof and the automobiles on top of it were supported by an apparent "membrane action" of the roof components, acting in tension. This unexpected support delayed total collapse for about four minutes, giving the shoppers time to leave the danger area before everything came tumbling down.

In their paper, Jones and Nathan also discuss the "fragmentation" of the design process, which created communication problems, leading to inconsistencies in the design. Although one person was responsible for the design,

FIGURE 4.1 — Collapse of the Main Beam in the Save-on-Foods Supermarket

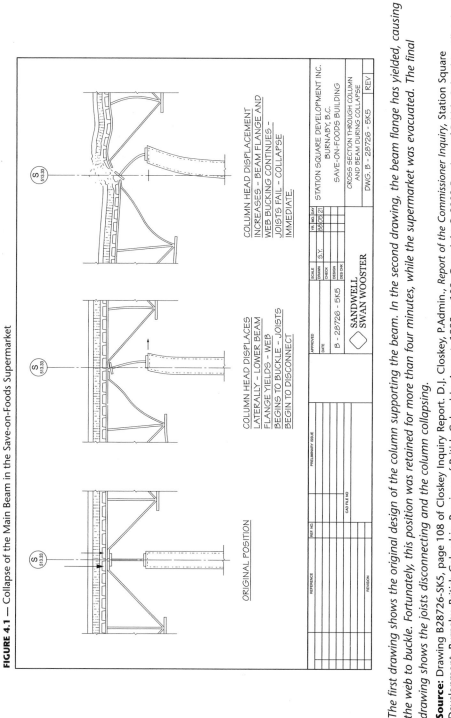

The first drawing shows the original design of the column supporting the beam. In the second drawing, the beam flange has yielded, causing the web to buckle. Fortunately, this position was retained for more than four minutes, while the supermarket was evacuated. The final drawing shows the joists disconnecting and the column collapsing.

Source: Drawing B28726-SK5, page 108 of Closkey Inquiry Report. D.J. Closkey, P.Admin., *Report of the Commissioner Inquiry, Station Square Development, Burnaby, British Columbia*, Province of British Columbia, August, 1988, p.108. Copyright © 2003 Province of British Columbia. All rights reserved. Reprinted with permission of the Province of British Columbia. www.ipp.gov.bc.ca

decisions were actually made by several participating groups. Jones and Nathan conclude with some good advice for design engineers:

> In the design and construction of a structure, hundreds of calculations are made, and hundreds of items of information are communicated from one participant to another. It is certain that many errors will be made, and the process must be designed to eliminate them. In budgeting the manpower for a project, allowances must be made for careful and detailed checking at each step. The engineers checking the design must have nothing else on their minds: they must not be burdened with many other simultaneous responsibilities. They must avoid a mind-set that the design is probably good, particularly if the designer is a respected senior. The checker should cultivate an attitude of mind that "anyone can make mistakes and it is up to me to find them in this design."[16]

Commissioner's Recommendations

The Commissioner made 19 recommendations in his final report, directed at the provincial government, the municipalities, the engineering profession, the Canadian Institute of Steel Construction, the Canadian Standards Association, and the Canadian Sheet Steel Building Institute. All of the six recommendations (5 through 10) directed toward the engineering profession have been implemented.

Recommendations 5 and 6 suggested that companies, partnerships, and other firms be required to register under the Act and should be subject to deregistration for unethical practices. This recommendation was implemented: corporations, partnerships, and other legal entities must now hold a Certificate of Authorization in British Columbia.

Recommendation 7 suggested that APEGBC require structural engineers to satisfy higher qualifications than those required for membership. The status of Structural Engineer of Record (SER) has since been implemented, and requires additional experience and examinations. Only SERs are permitted to approve the design of a building's primary structure. (APEGBC has also established several special interest "divisions," which encourage the exchange of information among engineers in various specialties.)

Recommendation 8 suggested that structural engineers be required to carry a specified minimum of professional liability insurance. This recommendation has been partially implemented: all members, licensees, and certificate holders offering services to the public must notify clients, in writing, whether professional liability insurance is held and applies to the services offered, and must receive the client's acknowledgement before proceeding with the work. (APEGBC has also instituted a secondary liability insurance plan, as explained elsewhere in this text.)

Recommendation 9 suggested that APEGBC establish and enforce a minimum fee schedule. This recommendation resulted from the obvious conclusion that the fees paid to engineers on the Save-on-Foods building design were far too low to permit an unhurried and careful review and analysis of the structure. APEGBC developed a comprehensive fee guideline for engineering services that set minimum recommended fees. This guideline was available from the APEGBC for many years, but is now distributed by the Consulting Engineers of BC (CEBC).[17]

Recommendation 10 suggested that provincial standards of practice be established for building design drawings and calculations. APEGBC now publishes several guidelines for professional practice—including structural, mechanical, electrical, and geotechnical services—with applications to building projects. All of the guidelines are available from the APEGBC website.

In the years since the Burnaby roof collapse, many other provincial Associations have examined their practices and have implemented similar rules, procedures, and guidelines.

Costs of the Roof Collapse

The fortuitous period of about four minutes between the cracking of the roof and its collapse saved the lives of those directly below it. However, the financial costs of the collapse were high. In their paper, Jones and Nathan make an approximate estimate:

- The budget of the commissioner inquiry was $250,000.
- Private legal costs during this period are not known, but 12 legal firms communicated with the Inquiry and many were present at one or more hearings.
- The legal costs to the APEBC [now APEGBC] for the disciplinary inquiry amounted to $80,000. This did not include staff time.
- All defendants had legal counsel.
- All perishable stock in the supermarket was destroyed. All other stock had to be removed. The collapsed area was rebuilt and the entire store upgraded structurally at a cost understood to be approximately $5,000,000. The store finally opened in the fall of 1988.
- Some competitive advantage was possibly lost, as a competing store opened nearby, late in 1988. Six months of sales were lost. Smaller stores in the complex also suffered losses.
- Personal injuries occurred, automobiles were lost, and other damages were alleged.
- The extent of litigation is not known, but total costs are clearly very high.[18]

Some Final Comments about the Collapse

In the preface to his report, the Commissioner made the following statement concerning the public safety roles of engineers, architects, municipalities, and professional Associations:

> Owners are primarily motivated by a satisfactory return on investment. Contractors and suppliers likewise are profit-oriented. The professional

engineers and architects have dual responsibilities. On the one hand, by training and through professional ethics, they have a duty to maintain a high calibre of service to the public. In the context of building design and erection, this translates into the protection of public safety. On the other hand, professional consultants operate businesses in a commercial world. They, too, require a reasonable stream of revenue to survive. In the middle stand the regulators: municipalities, responsible for enforcing building standards, and the professional associations, for maintaining professional standards.[19]

This is the only statement in the Commissioner's report with which I would disagree. The municipalities and professional Associations establish the bylaws, standards, and Codes of Ethics that regulate the profession, but it is the engineer who is "in the middle" between the pressures of profit-making organizations and the escalating demands for safety.

The Save-on-Foods roof collapse bears some similarity to the collapse of the Quebec Bridge more than 90 years earlier (see the case history in Chapter 1). Both collapses were caused by the buckling of a cantilever structure with an undersized cross-section, as a result of a miscalculation of the applied load (specifically, the dead load). In both cases, the design engineer was a respected senior person and the engineering firm was underpaid for the work expected and the responsibility undertaken. These lessons, taught by the 1907 disaster that stimulated the regulation of the engineering profession, are still valid a century later.

DISCUSSION TOPICS AND ASSIGNMENTS

1. Select any three provinces or territories and compare the disciplinary powers awarded to each Association under the engineering or geoscience Act. Which Act provides the most severe fines and penalties? Would you say the disciplinary powers in the Acts are generally similar, or are there serious inconsistencies between them? Point out and discuss these similarities and inconsistencies. The Acts are found on the Association websites. (See list in Appendix A, or excerpts in Appendix B.)

2. In your employment as a professional engineer or geoscientist, you discover that two of your fellow employees who supervise the delivery and storage of materials on the job site (and who are also professional engineers) have been involved in "kickback" schemes with suppliers. The suppliers invoice your employer for materials that have not been delivered, your colleagues validate the invoices, and the suppliers pay them a hidden commission. Obviously, these schemes violate criminal law. In addition, which clauses in your provincial Code of Ethics have your colleagues broken? To what types of disciplinary action have they exposed themselves as a result? Suppose you confront them, and they promise they will discontinue these schemes if you agree not to reveal them. If you agree, would your silence be consistent with the Code of Ethics? Could any disciplinary action be brought against you? Describe the

course of action you should follow. Would your actions be different if your fellow employees were not professional engineers?

3. You receive a registered letter from the Registrar of your provincial Association stating that you are the subject of a formal complaint made by a former client. The letter contains a description of the complaint, which alleges that you are guilty of incompetence because the advice in a report that you wrote was faulty. The client followed your advice and suffered a financial loss. As part of the preliminary investigation conducted by the Association, the Registrar asks you to respond to the complaint. Describe the actions you would take to prepare and protect yourself.

Additional assignments can be found in Appendix E.

NOTES

[1] *Engineering, Geological and Geophysical Professions Act,* Statutes of Alberta, SA 1981, c. E-11.1 (as amended), s. 44(1). Association of Professional Engineers, Geologists and Geophysicists of Alberta (APEGGA). Select links "About APEGGA" and "The EGGP Act" at <www.apegga.org> (May 15, 2008).

[2] Regulation 941/90, s. 72(2)(e), under the *Professional Engineers Act,* Statutes of Ontario, c. P.28, 1990. Select "Complaints" and "Definition of Professional Misconduct" at <www.peo.on.ca> (May 15, 2008).

[3] Regulation 941/90, *Professional Engineers Act,* s. 72(2)(j).

[4] Ibid., s. 72.

[5] Association of Professional Engineers, Geologists and Geophysicists of Alberta (APEGGA), *A Guide to APEGGA'S Discipline Process,* select links "Regulatory" and available at <www.apegga.org/Regulatory/Discipline/toc.html> (June 15, 2009).

[6] Professional Engineers Ontario (PEO), *Making a Complaint: A Public Information Guide,* Brochure, Toronto, select links "Complaints" and "Making a Complaint" at <www.peo.on.ca> (May 15, 2008).

[7] P.J. Greenbaum, "Finding a Place for ADR," *Engineering Dimensions,* Professional Engineers Ontario (PEO), July/Aug 2002, p. 30, available at <www.peo.on.ca> (May 15, 2008).

[8] Professional Engineers Ontario (PEO), *Alternative Dispute Resolution for Complaints,* Brochure, Toronto, available at <www.peo.on.ca> (May 15, 2008).

[9] *Professional Engineers Act,* Registered Statutes of Ontario, R.S.O. 1990, Chapter P.28 Sec. 40 (2). Available from PEO website at <www.peo.on.ca> (May 15, 2008).

[10] Engineers Canada, *2007 Accreditation Criteria and Procedures,* Canadian Engineering Accreditation Board (CEAB), Ottawa, 2007, p. 31, available at <www.engineerscanada.ca> (May 15, 2008).

[11] "Readers weigh in on . . . The engineers' tale," *ITbusiness,* April 23, 2004, available at <www.ITbusiness.ca> (May 15, 2008).

[12] G. Downey, "The engineers strike back," *ITbusiness,* July 31, 2002, available at <www.ITbusiness.ca> (May 15, 2008).

[13] Nicole Axworthy, "Court rejects Microsoft appeal," *Engineering Dimensions,* Professional Engineers Ontario (PEO), September/October 2005, p. 20, available at <www.peo.on.ca> (May 15, 2008).

[14] D.J. Closkey, P.Admin., *Report of the Commissioner Inquiry, Station Square Development,* Burnaby, British Columbia, Province of British Columbia, August 1988.

[15] C.P. Jones and N.D. Nathan, "Supermarket Roof Collapse in Burnaby, British Columbia, Canada," *ASCE Journal of the Performance of Constructed Facilities*, vol. 4, no. 3, August 1990, pp. 142–160.
[16] Ibid., p. 160.
[17] Consulting Engineers of British Columbia (CEBC), *Consulting Engineers Fee Guideline 2008*, available at <www.cebc.org/pulse/cebcFeeGuide08.pdf> (May 15, 2008).
[18] Jones and Nathan, "Supermarket Roof Collapse," p. 158.
[19] Closkey, *Report of the Commissioner Inquiry*, p. x.

Chapter 5
Technical Societies

Technical societies stimulate new research, organize information, and yield a priceless exchange of knowledge. They are therefore extremely useful organizations for professionals. This chapter reviews the role of technical societies, describes the wide range of interests supported by the societies, and explains why every professional engineer and geoscientist should join at least one technical society.

THE ROLE OF TECHNICAL SOCIETIES

For almost two centuries, technical societies have encouraged research, collected and classified new information, and disseminated it to members so that it could be put to use. Technical societies are the most important publishers of new research. They publish journals, conference proceedings, codes, and standards. The world's libraries are bulging with useful publications from these societies, and most of it is available on the Internet. The benefit from this free exchange of information is immense.

Technical societies are equivalent to the *learned societies* that stimulate original thought and discourse in the arts and humanities. However, technical societies should not be confused with the provincial and territorial Associations discussed earlier, which license professional engineers and geoscientists in Canada. Surprisingly, in some countries, the activities overlap. For example, many British engineering societies serve both purposes: they began (and continue) as technical societies, but they now award *Chartered Engineer* status. In the absence of other licensing regulations for engineers, this voluntary certification is highly regarded in Britain.

THE EVOLUTION OF TECHNICAL SOCIETIES

Technical societies originated during the Industrial Revolution when people, eager to reduce physical labour by mechanizing work, came together to discuss ideas and inventions. The first technical society for engineers was the Institute of Civil Engineers, established in Britain in 1818. The Institution of Mechanical Engineers followed, 30 years later. Shortly after that, other societies were established for naval architects and for gas, electrical, municipal, heating, and ventilating engineers.[1] These original 19th-century societies still exist, but they now sponsor many innovative subdisciplines and specialties.

In the United States the first engineering society was the American Society of Civil Engineers, founded in 1852. Many related societies were established in the 1800s—the American Society of Mechanical Engineers (1880), the American Institute of Electrical Engineers (1884), the Geological Society of America (1888), and the American Society of Heating and Ventilating Engineers (1894), to mention only a few.

In Canada, engineers and geoscientists were already at work before the disciplines were invented. The Geological Survey of Canada (GSC), a government institution that dates back to 1842, was one of the most successful scientific societies of its time.[2]

The first engineering society to be formed in Canada was the Engineering Society of the University of Toronto, in 1885. The "society was, indeed, a 'learned society' and published and disseminated technical information . . . in addition to looking after the University undergraduates in engineering."[3] In 1882 the Canadian Institute of Surveys was formed, followed by the Engineering Institute of Canada in 1887 (although the EIC name was not adopted until 1917), the Canadian Institute of Mining and Metallurgy (1898), the Canadian Forestry Association (1900), and many others. These societies are still active today, and are easily contacted on the Internet (see lists later in this chapter).

The growth of societies continued slowly in the early part of the 20th century, but became exponential after the Second World War, as the number of Canada's engineers and geoscientists increased. New societies continue to emerge every year, because they are even more important in the fast-moving world of the 21st century.

CHOOSING A TECHNICAL SOCIETY

Societies often sponsor undergraduate student chapters, so you may already be a member of the principal society in your discipline. The smaller, newer Canadian societies are in the process of building up their reputations and memberships. The older, larger, and better established American and British societies have greater storehouses of technical information and are usually able to offer more services to members. However, the Canadian societies are more effective in dealing with problems that are typically Canadian, and in promoting Canadian interests. Fortunately, the costs of membership are not great, and it is usually possible for a practising professional to join both. Most societies permit membership applications to be submitted electronically. Because of the value and usefulness of societies, membership dues are deductible from personal income (for practising professionals, under Canadian income tax laws).

In addition, many organizations sponsored by government or industry are extremely valuable sources of information, and serve a function similar to technical societies. For example, the Geological Survey of Canada (GSC) is a government organization that collects and disseminates vital geological data, and the American Gear Manufacturers Association (AGMA) is an industry organization that publishes the standards used for most gear strength calculations in North America. A few of these organizations are included in Tables 5.1, 5.2, and 5.3.

Regardless of your discipline, industry, or personal interests, there is a society that can help you professionally, and needs you as a member. Take advantage of this valuable source of useful knowledge. If the lists below do not include a society that specializes in your interests, then a simple Internet search will find it. Every society now has a website.

Canadian Engineering Societies

Several Canadian engineering societies were established in the last few decades, when the Engineering Institute of Canada (EIC), one of the oldest and most prestigious societies, assumed a national role, and several discipline-oriented constituent societies were created. The Engineering Institute of Canada is now a federation of member societies. All of these societies are listed in Table 5.1, along with several related engineering societies.

Agreements signed between EIC and Engineers Canada—which acts on behalf of the engineering licensing Associations when requested—clearly state the roles and duties of the organizations. The licensing Associations are responsible for regulating engineering, and the role of the EIC (and its member societies) is to collect, organize, and disseminate engineering, scientific, and technical information.[4] EIC has taken on this role aggressively, and now provides a vital service by certifying and coordinating courses for professional development.

Canadian Geoscience Societies

The Canadian Federation of Earth Sciences (formerly known as the Canadian Geoscience Council) is a federation of earth science societies. It speaks as a unified voice for earth sciences societies in Canada, and promotes their role in environmental, natural hazard, and climate studies, and in securing, and responsibly developing, Canada's energy, mining, and water resources. The Federation also helps Canada's earth science sectors to share data, knowledge, and new ideas.[5] The federated societies are listed in Table 5.2, along with a few related geoscience societies.

The Canadian Federation of Earth Sciences and the Engineering Institute of Canada have parallel or similar functions. Each of them is a federation of technical societies, separate from the licensing Associations. Their goal is the same: the advancement of their disciplines.

International Societies

The United States hosts the largest and most advanced technical societies. Several American and international technical societies are listed in Table 5.3.

TABLE 5.1 — A Brief List of Canadian Engineering and Related Societies

Acronym	Society	Web address
EIC	**Engineering Institute of Canada**	www.eic-ici.ca
	The Engineering Institute of Canada (EIC) is a federation of member societies, including: CGS, CSCE, CSME, CSChE, CSEM, IEEE-Canada, CNS, CMBES, CDA, MTS, and CSSE (a charitable organization of senior engineers). EIC is also now a leading provider and coordinating body for continued professional development.	
CGS	**Canadian Geotechnical Society** CGS is an active member of both EIC and CFES.	www.cgs.ca
CSCE	**Canadian Society for Civil Engineering**	www.csce.ca
CSME	**Canadian Society for Mechanical Engineering**	www.csme-scgm.ca
CSChE	**Canadian Society for Chemical Engineering**	www.chemeng.ca
CSEM	**Canadian Society for Engineering Management**	www.csem-scgi.org
IEEE—Canada	**Institute of Electrical and Electronic Engineers—Canada**	www.ieee.ca
CNS	**Canadian Nuclear Society**	www.cns-snc.ca
CMBES	**Canadian Medical and Biological Engineering Society**	www.cmbes.ca
CIC	**Chemical Institute of Canada**	www.cheminst.ca
CDA	**Canadian Dam Association**	www.cda.ca
MTS	**Marine Technology Society (Canadian Maritime Section)**	www.mtsociety.org
CSBE	**Canadian Society for Bioengineering**	www.bioeng.ca

Note: All websites are current as of June 14, 2009.

The Institute of Electrical and Electronic Engineers (IEEE) is the largest technical society in the world, with over 370,000 members (including about 80,000 student members), in over 160 countries. In fact, the IEEE includes 39 specialty societies, with interests from aerospace electronics to vehicular technology. The IEEE publishes 132 transactions, journals, and magazines, and arranges or cosponsors 450 technical conferences, worldwide, each year.[6]

The IEEE, like many of the international societies, publishes a Code of Ethics. However, the IEEE does not solicit or otherwise invite complaints concerning the IEEE Code of Ethics, and does not provide advice to individuals.[7]

TABLE 5.2 — A Brief List of Canadian Geoscience and Related Societies

Acronym	Society	Web address
CFES	**Canadian Federation of Earth Sciences** The Canadian Federation of Earth Sciences (CFES) was called the Canadian Geoscience Council (CGC) until 2007. CFES, like EIC, is a federation of Canadian technical societies. CFES includes basic, specialized, and applied earth science societies, and has many links with industry and government.	www.cfes-fcst.ca or www.geoscience.ca
CGS	**Canadian Geotechnical Society** CGS is an active member of both EIC and CFES.	www.cgs.ca
CAG	**Canadian Association of Geographers**	www.cag-acg.ca
CAP	**Canadian Association of Palynologists**	www.scirpus.ca
KEGS	**Canadian Exploration Geophysicists Society**	www.kegsonline.org
CGRG	**Canadian Geomorphology Research Group**	cgrg.geog.uvic.ca/
CGU	**Canadian Geophysical Union**	www.cgu-ugc.ca
CIM	**Canadian Institute of Mining, Metallurgy and Petroleum**	www.cim.org
CMOS	**Canadian Meteorological and Oceanographic Society**	www.cmos.ca
IAH	**Canadian Nat'l Chapter, Int'l Assoc. of Hydrogeologists**	www.iah.ca
CANQUA	**Canadian Quaternary Association**	www.mun.ca/canqua
CSCSOP	**Canadian Society for Coal Science and Organic Petrology**	www.cscop.org
CSEG	**Canadian Society of Exploration Geophysicists**	www.cseg.ca
CSPG	**Canadian Society of Petroleum Geologists**	www.cspg.org
CSSS	**Canadian Society of Soil Science**	www.csss.ca
CWLS	**Canadian Well Logging Society**	www.cwls.org
GAC	**Geological Association of Canada**	www.gac.ca
PDAC	**Prospectors and Developers Association of Canada**	www.pdac.ca
MAC	**Mineralogical Association of Canada**	www.mineralogicalassociation.ca

Note: All websites are current as of June 14, 2009.

That is, these international Codes of Ethics are entirely voluntary. They should not be confused with the Codes of Ethics in Canadian engineering and geoscience licensing Acts, which are enforceable under the Acts, as discussed in earlier chapters.

TABLE 5.3 — A Brief List of International Engineering, Geoscience, and Related Societies

Acronym	Society	Web address
IEEE	**Institute of Electrical and Electronic Engineers**	www.ieee.org
ASCE	**American Society of Civil Engineers**	www.asce.org
AIME	**American Institute of Mining, Metallurgical and Petroleum Engineers**	www.aimeny.org
ASME	**American Society of Mechanical Engineers**	www.asme.org
AIChE	**American Institute of Chemical Engineers**	www.aiche.org
ASABE	**American Society of Agricultural and Biological Engineers**	www.asabe.org
SAE	**Society of Automotive Engineers**	www.sae.org
ASHRAE	**American Society of Heating, Refrigerating and Air-Conditioning Engineers**	www.ashrae.org
AGMA	**American Gear Manufacturers Association**	www.agma.org
ASEE	**American Society for Engineering Education**	www.asee.org
AIPG	**American Institute of Professional Geologists**	www.aipg.org
AAPG	**American Association of Petroleum Geologists**	www.aapg.org
IUGS	**International Union of Geological Sciences**	www.iugs.org
AGI	**American Geological Institute**	www.agiweb.org
AEG	**Association of Environmental and Engineering Geologists**	www.aegweb.org
APGE	**Association of Petroleum Geochemical Explorationists**	www.geotech.org
EFG	**European Federation of Geologists**	www.eurogeologists.eu
GSA	**Geological Society of America**	www.geosociety.org
IAH	**International Association of Hydrogeologists**	www.iah.org
AAG	**Association of Applied Geochemists**	www.appliedgeochemists.org

Note: All websites are current as of June 14, 2009.

CANADIAN STUDENT SOCIETIES

Student Chapters of Technical Societies

Many of the engineering and geoscience societies listed in Tables 5.1 to 5.3 maintain student chapters in Canadian universities. These chapters promote undergraduate technical and social programs, and usually offer free (or inexpensive) subscriptions to society publications, entrance to society meetings, short courses and field trips, as well as opportunities to meet practising professionals in your field. If you are a student, contact your society of interest (see Web addresses in Tables 5.1 to 5.3) and see if it has a student chapter at your university.

Student Programs by Licensing Associations

Recently, several licensing Associations introduced programs to improve communication with the next generation of professional engineers and geoscientists. For example, Alberta (APEGGA) has the APEGGA Student Advantage Program (ASAP), British Columbia has the Membership Advantage Program for Students (MAPS), and Ontario (PEO) has an Internet-based Student Membership Program (SMP). These programs offer undergraduate students the benefits of membership at no cost. The Associations are to be applauded for this initiative, which gives students a smoother transition through the internship (as EIT, MIT, GIT, or similar titles), to licensed professional engineer or geoscientist. Some provinces also have links to faculty members and networks of mentors to assist students. If you are an engineering student, contact your Association (see the Web address in Appendix A) and inquire whether your Association has a student program.

Canadian Federation of Engineering Students

The Canadian Federation of Engineering Students (CFES) traces its history back to a tumultuous inaugural congress of engineering students at McGill University in February 1969, and is organized into four regions: Western Canada, Ontario, Quebec, and the Atlantic region. The Federation's goals are to improve communication between engineering students, and to help engineering students grow culturally, morally, intellectually, academically, and economically. CFES also serves as a liaison between Canadian engineering students and Engineers Canada, as well as the National Council of Deans of Engineering and Applied Science (NCDEAS). CFES organizes many student projects, such as the Canadian Engineering Competition. It also provides interactive online information about job and graduate school opportunities, and publishes *Project Magazine*, a national magazine for engineering students.[8]

CHARITABLE AND HONORARY SOCIETIES

Engineers Without Borders—*Ingénieurs Sans Frontières*

Engineers Without Borders (EWB), established in January 2000, is a registered charity dedicated to international development. The mission of EWB is to promote human development through access to technology. In other words, EWB seeks to narrow the technology gap between the developed world and the developing one by promoting the involvement of engineering students in development issues. EWB focuses on the role of technology in fundamental

Photo 5.1 — The Canadair Waterbomber. *The Canadair CL-215 Waterbomber, designed and built in Canada, is the only aircraft in the world specifically intended to fight forest fires. Canadair designed the CL-215 in the 1960s and it is capable of scooping 5760 kg of water (weighing 12,500 lbs.) from a lake, without stopping, and dumping it on a forest fire. A turboprop version, the CL-415, first flew in 1993.*

Source: AP/Franco Arena.

areas—water, food availability, health, energy, and communications—and tries to address basic problems in developing communities. EWB does not bring technology from the West; rather, it encourages simple technology, developed with local input and innovation. Such solutions are longer lasting and promote sustainability and self-sufficiency. EWB is a Canadian initiative, and EWB chapters have been started at most Canadian universities with engineering programs. The group's newsletters document a startling array of successful projects. The achievements of this group, in the few years since its inauguration, speak well for the initiative and idealism of Canada's engineering students.[9]

Registered Engineers for Disaster Relief (RedR) Canada

Registered Engineers for Disaster Relief (RedR) Canada is a recently established Canadian branch of an international organization that relieves suffering in disasters by selecting and training competent personnel and providing their services to humanitarian agencies. The founding members of RedR Canada are the Association of Consulting Engineers of Canada (ACEC), Engineers Canada, EIC, and the Canadian Academy of Engineering. In January 2001, ACEC signed a memorandum of understanding with RedR International that led to the founding of RedR Canada as a nonprofit organization.

RedR members provide technical assistance to restore roads, bridges, water supplies, and communication systems. They also assist, after a disaster, in managing waste, protecting the environment, and managing financial, material, and human resources. RedR Canada is an independent organization; however, all national offices work together as members of RedR International, based in Geneva.[10]

Canadian Engineering Memorial Foundation (CEMF)

The Canadian Engineering Memorial Foundation was created in 1990 under the stewardship of Engineers Canada, following the events in Montreal at École Polytechnique that resulted in the death of 14 young women, ending their contributions to Canadian society. CEMF is funded entirely through donations from individual engineers, corporations, and the public. Each year the Foundation grants scholarships to outstanding female engineering students at undergraduate and graduate levels.[11]

Canadian Society for Senior Engineers (CSSE)

The Canadian Society for Senior Engineers (CSSE) was originally incorporated as the EIC Life Members Organization in 1967, and is a charitable organization affiliated with the EIC. The primary function of this group of late career and retired engineers is to promote the advancement of science and engineering in Canada and to provide benevolent donations in support of youth, in the advancement of engineering. The CSSE engages in a wide range of charitable works, including research projects and awards for science fairs. It also assists the Tetra Society, which recruits skilled volunteer engineers and technicians to build devices to assist people with disabilities.[12]

Geoscience Charitable Foundations

Several foundations have been established to receive gifts and bequests to support Canadian earth science, including the Canadian Geological Foundation, and several foundations established by earth science specialty groups. The Canadian Federation of Earth Sciences publishes a list of the foundations on its website.[13]

The Canadian Academy of Engineering

The Canadian Academy of Engineering is Canada's highest honorary engineering society. The Academy, located in Ottawa, is an independent, self-governing, nonprofit organization established in 1987. The Fellows of the Academy are professional engineers from all disciplines, elected on the basis of distinguished service and contributions to society, to the country, and to the profession. The Academy presently has about 300 active members, and 90 emeritus members. The Fellows of the Academy are committed to ensuring that Canada's engineering expertise is applied to the benefit of all Canadians. They accomplish this in several ways; for example, by promoting increased awareness of the role of engineering, and giving independent advice on engineering education, research, development, and innovation.[14]

THE IRON RING—A RITUAL FOR ENGINEERS

The Corporation of the Seven Wardens is a little-known group that has performed a vital role in Canadian engineering for many decades. The wardens arrange the Iron Ring ceremonies held on most campuses just before graduation day. This ceremony is a milestone in the engineer's education. The following account of the Iron Ring was written by J.B. Carruthers, P.Eng., and is reproduced with permission:

> Most engineers in Canada wear the Iron Ring and have solemnly obligated themselves to an ethical and diligent professional career through the Ritual of the Calling of an Engineer. This Ritual is the result of efforts by the Corporation of the Seven Wardens, started in 1922 when a group of prominent engineers met in Montreal to discuss a concern for the general guidance and solidarity of the profession. These seven prominent engineers formed the nucleus of an organization whose object would be to bind all members of the engineering profession in Canada more closely together and to imbue them with their responsibility towards society.
>
> They enlisted the services of the late Rudyard Kipling, who developed an appropriate Ritual and the symbolic Iron Ring. Rudyard Kipling outlined the purpose in the following words:
>
>> "The Ritual of the Calling of an Engineer has been instituted with the simple end of directing the young engineer towards a consciousness of his [/her] profession and its significance, and indicating to the older engineer his [/her] responsibilities in receiving, welcoming and supporting the young engineers in their beginnings."

The Ritual has been copyrighted in Canada and the United States, and the Iron Ring has been registered. The Corporation of the Seven Wardens is entrusted with the responsibility of administering and maintaining the Ritual, which it does through a system of separate groups, called Camps, across Canada. There are presently 20 such Camps.

The Corporation of the Seven Wardens is not a "secret society." Its rules of governance, however, do not permit any publicity about its activities and they specify that Ceremonies are not to be held in the presence of the general public.

The original seven senior engineers who met in Montreal in 1922 were, as it happens, all past presidents of the Engineering Institute of Canada. There is, however, no direct connection between the Engineering Institute of Canada and the Corporation of the Seven Wardens.

The wearing of the Iron Ring, or the taking of the obligation, does not imply that an individual has gained legal acceptance or qualification as an engineer. This can only be granted by the provincial bodies so appointed and, as a result, it should also be mentioned that the Corporation of the Seven Wardens has no direct connection with any provincial association or order.

The obligation ceremonies for graduating students are held in cities where Camps are located, and for convenience, in some cases, on the university campus itself. Such ceremonies must not be misconstrued as being an extension of the engineering curriculum. The Iron Ring does not replace the diploma granted by the University or the School of Engineering nor is it an overt sign of having successfully passed the institution's examinations.

The purpose of the Corporation of the Seven Wardens and the Ritual is to provide an opportunity for men and women to obligate themselves to the standard of ethics and diligent practice required by those in our profession. This opportunity is available to any who wish to avail themselves to it, whether they be new graduates or senior engineers. The Ritual of the Calling of an Engineer is attended by all those who wish to be obligated, along with invited senior engineers and, when space permits, immediate family members. A complete explanation of the Ritual, its obligations and history is given to every man and woman before the ceremony so that they may decide in advance whether or not they wish to take part in the spirit intended. A few people, for one reason or another, have chosen to refrain from being obligated, and so cannot rightfully wear the Iron Ring. The Corporation of the Seven Wardens feels that this in no way detracts from their right to practise in the profession and further feels that the obligation should continue to be a matter of personal choice, taken only by those who wish to take part in the serious and sincere manner intended.[15]

THE EARTH SCIENCE RING—A RITUAL FOR GEOSCIENTISTS

The awarding of the Earth Science Ring is a ceremony comparable with the Iron Ring ceremony and, in fact, adopts some of the format and wording of the Iron Ring ceremony. The following description of the Earth Science Ring and award ceremony was written by Philippe Erdmer, P.Geol., and Edward S. Krebes, P.Geoph., and is reproduced with permission:

The Earth science ring ceremony, a ritual of welcome into the profession of newly qualified geologists and geophysicists by senior practising Earth scientists, started in Alberta in 1975. This yearly tradition for the university geoscience graduating classes at Edmonton and Calgary has spread to other provinces and jurisdictions in Canada. The ceremony carries many of the same passages written by Kipling for the Engineers' iron ring ceremony and symbolizes the commitment and responsibility that come with wearing the title of a professional.

Like the engineer's iron ring, the Earth science ring's simplicity and strength bear witness to the calling of the geologist and geophysicist. The ring is made of silver and marked with the crossed hammer of geology and with the seismic trace of geophysics— signifying both the immediate and the remote searching out of Nature's knowledge. Without beginning and without end, it also represents for those who wear it the continuous interplay of ideas and of material realities.

The ceremony includes a charge (speech) by senior Earth scientists and an obligation (pledge) taken by the group of newly graduated geologists and geophysicists. The charge reads in part: "We tell you here that you will encounter no difficulty, doubt, danger, defeat, humiliation or triumph in your career which has not already fallen to the lot of others in your calling. . . ." The obligation includes the words: "I will not pass . . . false information or too casual interpretations in my work as an Earth scientist. My time I will not refuse, my thought I will not grudge; my care I will not deny towards the honour, use, stability and perfection of any project to which I may be called to set my hand . . . . My reputation in my calling I will guard honourably . . . . I will strive my uttermost against professional jealousy and the belittling of my co-workers in any field of their labour."

On a lighter note, following the obligation, new ring bearers are reminded that "From now on, we surrender to you what lies under the earth, and the tools to interpret or misinterpret. Sooner or later, you will drill the holes that bring no return, lose the vein in which lie extra riches and reputation, misinterpret the signal from the depths. This will equally baffle, bewilder and break your heart to your professional and personal education."

Receiving an Earth science ring is neither a prerequisite nor a later condition of professional membership with APEGGA. Although there is no obligation to obtain or wear a ring, it is significant that almost no one in the graduating classes willingly misses the ceremony. In addition, the ceremony is not strictly a graduation event, as it has occasionally included already practising geologists and geophysicists in Alberta who express the wish to receive a ring. Like the iron ring of the obligated engineer, the Earth science ring is a symbol of values that lie at the core of our individual beings and of the trust placed in us by society.[16]

DISCUSSION TOPICS AND ASSIGNMENTS

1. Canadian engineers and geoscientists often debate whether to join the newer Canadian technical societies, or to join the larger, older, foreign-based societies that have well-established journals, committees, and

conferences. Weigh the pros and cons of the two alternatives from your point of view. Does Canada need distinct technical societies? In your answer, discuss the implications for Canadian sovereignty if Canadian societies should be absorbed into the larger, American-dominated societies. Are these societies truly non-political, or do national interests influence their policies and the content of journals and transactions? Is climate change, and its effects on Canada, relevant to a society's role? Are there uniquely Canadian conditions that would justify uniquely Canadian societies? Give examples (for or against). Prepare a two-page summary.

2. Engineering and geoscience students must study hard to succeed. How can they find time for other activities, such as student government, CFES, Association student participation programs, or charitable organizations such as Engineers Without Borders? Discuss whether the CEAB (the Canadian Engineering Accreditation Board) should classify such activities as complementary (non-engineering) studies, which are required for engineering accreditation. How would students earn these credits? Which activities and organizations would be eligible, and how would you weight the various activities?

Additional assignments can be found in Appendix E.

NOTES

[1]   L.C. Sentance, "History and Development of Technical and Professional Societies," *Engineering Digest*, vol. 18, no. 7, July 1972, pp. 73–74.
[2]   R.W. Macqueen (ed.), *Proud Heritage: People and Progress in Early Canadian Geoscience*, Geological Association of Canada, St. John's, NL, 2004.
[3]   L.C. Sentance, "History and Development of Technical and Professional Societies." p. 73.
[4]   "Canadian Engineers Close the Ring," *Engineering Journal*, EIC, vol. 60, no. 1, January 1977, pp. 15–19.
[5]   Canadian Federation of Earth Sciences (CFES) website at <www.geoscience.ca> (May 15, 2008).
[6]   *IEEE Societies, Councils and Technical Communities*, IEEE brochure, available at <www.ieee.org/web/societies/home/index.html> (May 15, 2008).
[7]   J.R. Herkert, "Microethics, Macroethics, and Professional Engineering Societies," *Emerging Technologies and Ethical Issues in Engineering: Papers from a Workshop*, October 14–15, 2003, National Academy of Engineering, pp. 107–114, available at <www.nap.edu/catalog/11083.html> (May 15, 2008).
[8]   Canadian Federation of Engineering Students (CFES) website at <www.cfes.ca> (May 15, 2008).
[9]   Engineers Without Borders (EWB) website at <www.ewb.ca> (May 15, 2008).
[10] Registered Engineers for Disaster Relief (RedR) Canada website at <www.redr.ca> (May 15, 2008).
[11] Canadian Engineering Memorial Foundation (CEMF) website at <www.cemf.ca> (May 15, 2008).
[12] Canadian Society for Senior Engineers, Engineering Institute of Canada, website at <www.eic-ici.ca/english/tour/lmo2.html> (May 15, 2008).

[13] Canadian Federation of Earth Sciences (CFES), available at
      <www.geoscience.ca/found.html> (May 15, 2008).

[14] Canadian Academy of Engineering, "Mission Statement," available at
      <www.acad-eng-gen.ca> (May 15, 2008).

[15] J. B. Carruthers, P.Eng., "The Ritual of the Calling of an Engineer," *la Revue
      Projet / CFES Project Magazine*, April 1985, p. 19. Reproduced with permission.

[16] P. Erdmer, P.Geol., and E. S. Krebes, P.Geoph., "The Earth Science Ring: Made in
      Alberta," APEGGA, available at <www.apegga.org/About/earthr.html> (May 15,
      2008). Reproduced with permission from APEGGA.

Chapter 6

Basic Concepts of Professional Practice

This chapter offers useful advice for graduate engineers and geoscientists entering full-time employment, and answers basic questions, such as: When should I obtain a licence? How do I document my experience? What are reasonable salary expectations? When do I use my professional seal? What working conditions should I expect? This chapter also discusses a key decision that almost all professionals must make: whether to become a specialist in your discipline, or to develop the "people skills" needed for management.

ENTERING THE PROFESSIONAL WORK FORCE

Graduating from university and entering the professional workplace is usually exhilarating: you move to a new location, meet new colleagues, and participate in new projects. You apply your knowledge to real problems and you see your ideas taking shape on the computer screen, on the construction site, in the test laboratory, or on a production line. But, unless you had very good work terms at university, your first professional job may prove to be a challenge.

Your first surprise may be how little supervision you receive, and how much responsibility you get, because you are "a recent graduate and familiar with the theory." You may have to work harder to justify this confidence in your ability. Ask for help if you need it.

A second surprise may be the apparent lack of order and structure. University courses usually have well-defined objectives, but real projects may be "open-ended" and can change drastically at any time. If a crisis arises, you may suddenly be asked to take on a new project. New projects may be chaotic at the start, but they are usually interesting. And anyway, as a professional, it's your job to create order from chaos.

A third surprise may well be the strong emphasis on obtaining useful results. Products must perform as promised; if they don't, the engineer or geoscientist must "find out why and correct it." If you need information, you must be aggressive in getting it. The focus has changed—your goal in university was to get an education, but your employer wants results.

Finally, you may be surprised at the importance of deadlines. Late deliveries cost money—especially when contracts have penalty clauses for lateness, or when "just-in-time" assembly lines have narrow windows for delivering components. Personal time management is much more important in the professional workplace.

Most engineering and geoscience graduates easily overcome these challenges. However, the following hints may help to start your employment on a professional note:

- **Licensing.** As soon as you receive your degree, apply to your Association to begin the licensing process.
- **Experience.** Start to document your experience. It is easy to document experience as you go, but very tricky to remember details later. Follow your Association's criteria.
- **Prepare for advancement.** Eventually you will want to be promoted, so think about your next step. Do you have the knowledge and ambition to be a specialist? Do you have the management skills—especially the "people skills"—to lead the organization? Are you sufficiently self-confident to succeed in private practice or entrepreneurship? These topics are discussed in more detail in the following sections.

APPLYING FOR A LICENCE

As soon as possible, apply to your provincial or territorial Association for a licence to practise (also called membership or registration). Some graduates mistakenly believe they must satisfy the experience requirement first, but this is not true: applicants can apply as soon as they graduate from university. A simple letter, e-mail, fax, or phone call will get the process started. (The Web address of your Association is in Appendix A.)

Every Association has an internship program that you may join as soon as you receive your degree (or otherwise satisfy the educational requirements). When you are accepted into the Association's internship program, you will receive one of the following titles (depending on your province and discipline): Engineering-Internship-Training (EIT), Engineer-in-Training (EIT), Member-in-Training (MIT), Geologist-in-Training (Geol.IT), Geophysicist-in-Training (Geoph.IT), *ingénieur junior* (ing. jr.), Junior Engineer (Jr. Eng.), or Geoscientist-in-Training (GIT). These titles may be used on letters, memos, e-mail, business cards, desk plaques, and so forth. As explained in Chapter 2, graduates must not use titles that would imply that they are licensed, so avoid internal company titles such as Assistant Engineer, Assistant Geologist, Plant Engineer or Sales Engineer. These titles imply that you are licensed.

As an intern, the Association will guide you through the next step, which is documenting your experience for licensing. You will also be invited to attend Association meetings, and you may be able to participate in group insurance, investment plans, and similar benefit programs. Internship simplifies the licensing process, both for CEAB-accredited university graduates, and for foreign-educated applicants, even those with many years of experience.

DOCUMENTING YOUR EXPERIENCE

As soon as you begin working, start documenting your experience. You need four years of documented professional experience to satisfy the licensing requirements (except in Quebec, where the requirement is three years). However, you can shorten the experience needed by as much as one year. As explained in Chapter 2, each Association allows credit for up to one year of pre-graduation experience. Prepare your experience summary in the form of a personal résumé. This format satisfies the Associations, and is also useful for future job applications. Engineers Canada publishes a guideline for documenting experience, which has been adopted by most Associations.[1,2,3,4] The guideline suggests that your experience should satisfy the following five quality criteria:

- **Application of theory.** This is the best form of experience. It includes analysis, design and synthesis, testing methods, and project implementation. Most Associations expect a sizable portion (typically 20 percent) of your experience to be in this category.
- **Practical experience.** Practical experience helps you appreciate the capabilities and limitations of the theory, equipment, systems, procedures, and standards that are typically used in your discipline. For example, you are much more competent if you are personally aware of the capabilities (and limitations) of manufacturing equipment, tolerances, operating procedures, maintenance schedules, equipment reliability, computer software, safety codes, design standards, and so forth, that are commonly used in your discipline. Fortunately, many activities that do not fall under the other headings would likely qualify as practical experience.
- **Management of engineering/geoscience.** Management experience includes planning, scheduling, budgeting, supervision, project control, and risk assessment. New graduates are not usually assigned management duties, so document this experience whenever you have the opportunity.
- **Communication skills.** Professionals must be able to communicate effectively. Your experience résumé should include evidence of effective writing (formal reports, design specifications or standards, contracts, or similar documents), drawings or sketches (where appropriate), and oral presentations to supervisors, management, clients, or the public.
- **Social implications of engineering/geoscience.** This typically includes any experience that heightens the professional's awareness of the responsibility to guard against conditions that are dangerous to life, health, property, or the environment, and to call any such conditions to the attention of those responsible. Most of the topics in this book concern the social implications of technical decisions.

LEVELS OF PROFESSIONAL RESPONSIBILITY

Your future will likely involve many challenging projects, and you will assume greater responsibility as your experience increases. The following list shows the typical levels of engineering and geoscience responsibility. Several Associations publish salary statistics based on these levels.[5]

A word of caution: The average number of years at each level varies slightly, depending on the discipline and the location. Also, managers and specialists have different but equivalent career paths. The higher levels are therefore based on either management responsibility or technical responsibility. Levels often overlap, and some companies may recognize more (or fewer) levels, depending on company size. Most companies (especially in manufacturing) employ more managers than specialists.

LEVEL A—ENTRY LEVEL   A bachelor's degree in engineering, geoscience, or applied science, or its equivalent, is usually required. Recent university graduates—usually with little practical experience—receive on-the-job training in office, plant, field, or laboratory work, or (rarely) in classrooms. Level A employees work under close supervision, preparing simple plans, designs, calculations, costs, and bills of material, in accordance with established codes, standards, or specifications. This stage may last one or two years.

LEVEL B—ENGINEER/GEOSCIENTIST INTERNSHIP   After the first two or three years of work experience, the employee will be assigned tasks of increasing variety, although responsibility is still limited. Typically, the work involves parts of larger projects. Such assignments provide continuing training and development. During this period the employee is usually registered with the provincial Association at the internship level (with the title of EIT, MIT, Geol.IT, Geoph.IT, *ingénieur junior*, or GIT, depending on the province and discipline). Level B employees may give technical advice to technicians or to level A graduates. This stage lasts at least two or three years.

LEVEL C—PROFESSIONAL ENGINEER/GEOSCIENTIST   Level C is the first fully qualified professional level. The engineer or geoscientist carries out responsible and varied assignments in a broad field of engineering or geoscience, and is also expected to understand the effects of decisions on related fields. Combinations of standard methods are used to solve problems, and the Level C professional participates in planning. Typically, this stage requires a minimum of five to six years of related work experience after graduation. Level C professionals make independent analyses and interpret results without supervision, so they must be licensed, of course.

LEVEL D—FIRST SUPERVISORY (OR FIRST SPECIALIST LEVEL)   Job titles at this level have many variations, such as project leader, team leader, lead engineer, site geologist, or engineering/geoscience specialist. This is the first level that involves direct and sustained supervision of other professionals, or the first level of full specialization. This level requires mature knowledge of planning and conducting projects, or of coordinating difficult and responsible assignments. To reach this level, engineers or geoscientists typically require a minimum of seven or eight years of experience in the field of specialization.

LEVEL E—MIDDLE MANAGEMENT (OR SENIOR SPECIALIST LEVEL)   Job titles at this level include chief project engineer, chief project geologist, group head, and senior specialist. This level (in management) involves supervising large groups, containing both professional and nonprofessional staff; alternatively, this level (in specialization) involves authority over a small group of highly qualified professional personnel engaged in complex technical applications. Level E typically requires knowledge of more than one field of engineering or geoscience. The incumbent participates in short- and long-range planning and makes independent decisions on work methods and procedures within a general program. Originality and ingenuity are required for devising practical and economical solutions to problems. The engineer or geoscientist may supervise large groups that include both professional and nonprofessional staff, or may direct a small group of highly qualified professionals in complex technical applications. This level normally requires at least 10 to 12 years of engineering, geoscience, and/or administrative experience.

LEVEL F—SENIOR MANAGEMENT (OR SENIOR CONSULTANT LEVEL)   Job titles at this level include director of engineering or geology, plant manager, and senior consultant. Levels F and F+ may overlap, depending on company size (that is, a chief engineer in a large corporation may have essentially the same duties as the vice president of engineering in a medium-size corporation). The incumbent is usually responsible for an engineering or geoscience administrative function, directing several professional and other groups engaged in interrelated responsibilities; or may be a consultant, recognized as an authority in a field important to the organization. The Level F professional independently conceives programs and problems to be investigated, determines basic operating policies, and devises ways to reach program objectives economically and to overcome problems. The job requires extensive experience, including responsible administrative duties.

LEVEL F+—SENIOR EXECUTIVE LEVEL   Job titles at this level include president; vice president of engineering or geoscience; vice-president of manufacturing; general manager; and partner (in a consulting firm). At this level, the person receives general strategic guidance but conceives independent programs and problems to be investigated. He or she plans or approves projects that require considerable amounts of human and financial resources. This level requires many years of authoritative technical and administrative experience. The incumbent is expected to possess a high degree of originality, skill, and proficiency in the various broad phases of the profession.

SALARY EXPECTATIONS FOR PROFESSIONALS

Salary is not the only motivator for a professional, but it is still important. Several Associations conduct annual surveys of their members' salaries and post summaries on the Association websites. As an example, recent salary data for Alberta engineers and geologists is summarized in Tables 6.1 and 6.2,

TABLE 6.1 — Alberta Employer Salary Survey, 2007: ENGINEERS (All Industries)

Level	Number in Survey	Change in Mean 2006/07	Mean ($)	Lower Decile D1 ($)	Lower Quartile Q1 ($)	Median ($)	Upper Quartile Q3 ($)	Upper Decile D9 ($)
A−	341	5.4%	44,521	37,930	41,089	45,084	48,432	50,448
A	1,024	8.5%	60,138	51,500	55,000	60,000	65,000	69,814
B	1,209	9.2%	69,274	58,000	63,400	69,600	74,796	79,980
C	1,552	9.0%	84,474	70,514	77,100	83,738	90,700	98,400
D	2,106	6.6%	103,828	85,925	94,520	103,360	112,940	122,100
E	1,989	7.6%	127,852	104,544	116,148	129,000	138,170	147,000
F	1,073	5.8%	150,015	117,840	135,346	151,800	163,000	176,190
F+	421	6.9%	183,963	139,260	158,897	175,000	196,400	240,000

TABLE 6.2 — Alberta Employer Salary Survey, 2007: GEOLOGISTS (All Industries)

Level	Number in Survey	Change in Mean 2006/07	Mean ($)	Lower Decile D1 ($)	Lower Quartile Q1 ($)	Median ($)	Upper Quartile Q3 ($)	Upper Decile D9 ($)
A−	12	−2.4%	48,950	46,200	46,200	47,400	49,200	49,200
A	60	9.3%	63,295	53,029	60,538	65,000	67,000	68,602
B	112	7.2%	72,625	65,000	70,000	73,026	77,000	79,000
C	163	8.0%	88,293	78,500	82,745	87,470	92,000	100,100
D	140	6.7%	112,191	96,500	101,000	110,000	120,840	128,935
E	193	5.2%	136,455	117,766	133,920	140,000	144,860	149,730
F	138	7.7%	159,318	148,000	154,000	159,000	163,500	171,983
F+	75	12.3%	191,472	163,765	173,400	183,750	201,000	220,333

Notes for Tables 6.1 and 6.2: The salaries quoted are BASE salaries in effect as of May 31, 2007. The salaries include cost-of-living allowances and bonuses that have a continuing relationship to salary. Commissions, fringe benefits, and profit sharing are not included. Total cash compensation figures are available in the full report: *Value of Professional Services 2007*, available from the Association of Professional Engineers, Geologists and Geophysicists of Alberta (APEGGA). The statistical measures used in compiling the tables are the median, quartiles (Q3, Q1), deciles (D9, D1), and average. The median salary is the salary at which 50 percent of the respondent salaries are higher and 50 percent are lower. The Q3 salary is the salary at which 25 percent of the respondent salaries are higher and 75 percent are lower. The D9 salary has 10 percent of the salaries higher and 90 percent, lower.

Source: Tables 1 and 2 of the "2007 Employer Salary Survey Highlights" from the report *Value of Professional Services 2007*, published by The Association of Professional Engineers, Geologists and Geophysicists of Alberta (APEGGA), available at <www.apegga.org/Members/Publications/ salarysurvey.html> (May 15, 2008). Salary data used with permission of APEGGA.

which are taken from the 2007 Alberta employer salary survey conducted by APEGGA.[6]

The complete salary surveys are usually very comprehensive, and show incomes by responsibility level (defined earlier in this chapter), discipline, year of graduation, type of industry, city or region, and so on. Check your Association's website for the recent salary survey. However, be warned that surveys differ, depending on whether employees or employers provide the data, because employees often report income from several sources. Moreover, salary surveys show past history, and history is not always a good predictor of the future.

In British Columbia, the salary survey by the Association of Professional Engineers and Geoscientists of British Columbia (APEGBC) gathers data from employees, rather than employers, and classifies salaries by "responsibility point levels." APEGBC provides an online Employment Responsibility Evaluation program to calculate the responsibility point levels. The APEGBC point rating is then compared to ratings for similar job descriptions.[7]

In Ontario, the Ontario Society of Professional Engineers (OSPE) has assumed the responsibility for preparing salary surveys from Professional Engineers Ontario (PEO). Salary summaries are available to OSPE members on the OSPE website.[8]

PROFESSIONAL WORKING CONDITIONS

In Canada, professional employees usually receive basic job benefits such as extended health care, disability, and pension plans. In addition, professional employees deserve professional working conditions, challenging technical work, and opportunities for promotion, based on merit. Ideally, the work environment should also include high-quality computer hardware, state-of-the-art software, high-speed Internet connections, friendly colleagues, and clear communication with management. All of these increase productivity and job satisfaction.

Fortunately, most engineers and geoscientists work in these conditions. Engineers Canada surveyed more than 26,000 practising engineers and geoscientists, and the majority agreed, or agreed strongly, that

- they were satisfied with the job (37% agreed; 49% agreed strongly).
- they had freedom to decide on work issues (35% agreed; 47% agreed strongly).
- they were satisfied with their career prospects (37% agreed; 39% agreed strongly).
- they had opportunities for advancement (38% agreed; 23% agreed strongly).[9]

Ideally, working conditions should be specified in the employment contract (which is discussed in Chapter 12). Unfortunately, many professionals

(especially newly hired graduates) do not have personal employment contracts; instead, they are hired on a simple letter of appointment and their
working conditions are set by company policy.

In either case, working conditions should be reviewed on a regular basis.
Enlightened employers conduct a review at least annually, where performance and working conditions are discussed. Failure to review working conditions can lead to declining morale, reduced productivity, and turnover of key
personnel. There must also be some method for negotiating and amending
employment contracts (or company policies). This text cannot offer specific
advice; however, the following two organizations publish advice on professional working conditions.

THE CANADIAN SOCIETY OF PROFESSIONAL ENGINEERS (CSPE)   CSPE is an advocacy group for professional engineers. It is not a union and does not face the
legal and bureaucratic problems typically encountered by unions; it is mod-
elled after the medical associations and bar associations, which work collectively for their respective members, who are also professionals. CSPE has
recently been restructured as a national advocacy group, whose purpose is to
coordinate the activities of provincial advocacy groups. As this text goes to
press (2008), the Ontario Society of Professional Engineers (OSPE) is the only
provincial group that has been created under the CSPE umbrella.[10,11] Both
CSPE and OSPE are proposing extensive advocacy initiatives for their members. However, they have yet to provide the type of extensive information
that is offered by a corresponding U.S. organization, the National Society of
Professional Engineers (NSPE).

THE NSPE GUIDELINES   The National Society of Professional Engineers (NSPE)
has developed professional employment guidelines that may be of interest to
Canadian engineers, since comparable Canadian guidelines do not yet exist.
The NSPE Guidelines are discussed in more detail in a later chapter (and are
included in their entirety in Appendix D). These guidelines contain more
than 60 detailed clauses that discuss general rules for engineering recruitment, employment, professional development, and termination. Most clauses
discuss very specific, practical employment problems and are directed at both
employers and employees.[12]

USING YOUR PROFESSIONAL SEAL

During your career, you will prepare hundreds (maybe thousands) of key documents, such as reports, drawings, plans, maps, specifications, and so forth.
Other people will rely on these documents to make important decisions
affecting the life, health, safety, or welfare of the public. To identify these key
documents, you must stamp them with your professional seal. The seal is usually a rubber stamp, so the terms "seal" and "stamp" are used interchangeably.
The licensing Association sends a seal (a rubber stamp) to each newly licensed

Photo 6.1 — The Avro Arrow Fighter-Interceptor. *The Avro Arrow, a Canadian-designed all-weather fighter–interceptor, was a technological marvel and decades ahead of its time. The Arrow, which first flew in March 1958, could exceed Mach 2 at 50,000 feet in normal flight. On September 28, 1958, the government of Prime Minister John Diefenbaker cancelled the Arrow in a scandalously abrupt manner, on the basis that manned fighters were obsolete and too costly. Avro's 14,000 employees were immediately laid off. Avro was ordered to destroy the two prototypes.*

Source: Photograph courtesy of the Canada Aviation Museum, Ottawa.

engineer or geoscientist, and the rules for using the seal are similar across Canada. The seal remains the property of the Association, and must be returned if the professional retires or resigns (or if the Association asks for it). An Alberta (APEGGA) guideline explains the purpose:

> A professional stamp or seal affixed to a document is intended to indicate that the document has been produced under the supervision and control of a fully qualified professional member of APEGGA, or that it has been thoroughly reviewed by a professional member of APEGGA who accepts responsibility for it. Professional stamps and seals shall be affixed, signed and dated only after the responsible member is satisfied that the document or component, for which he or she is professionally responsible, is complete and correct.[13]

What Does the Seal Represent?

The seal identifies the author of the document, of course, but the seal has a greater significance: It means that the author gave serious thought to the contents, and assures others that they can rely on the facts, decisions, designs, opinions, and judgments in the document. Finally, the seal denotes that the author assumes professional responsibility and accountability if the document should later be found to contain errors.[14] The seal is neither an archaic tradition nor a mindless formality—it is an accepted practice for assuring others that they can rely upon your work.

Which Documents Are Sealed?

The Act typically requires all final drawings, specifications, plans, reports, and similar documents involving professional practice, and issued to the public, to be dated, signed, and sealed by a professional engineer or geoscientist. The use of the seal is not optional, and no fee is charged for sealing documents. Sealing documents is standard professional practice, required under the Act. For example, Ontario's Regulation 941 requires engineers to "sign, date and affix the . . . seal to every final drawing, specification, plan, report or other document prepared or checked" by the engineer, before it is issued to the public.[15]

Documents are sealed before release to the public, but in this case, the "public" means any person or organization, except for the employer. That is, documents intended for strictly internal company circulation need not be sealed (although they may be). However, any document that will be released to clients, customers, government, or anyone outside the company must be sealed.[16]

Which Documents Are Not Sealed?

Only documents related to professional practice must be sealed, so you should not seal a document that has no technical content. Routine memos, letters, and notes are not sealed, although a technical letter containing critical data, specifications, or information should be sealed. Legal and business documents, such as contracts, are not professional practice, so they are not sealed. If the practice is incorporated, it will have a corporate seal. Otherwise, signatures are sufficient for most business purposes.

Only final documents are sealed. Preliminary documents, rough drawings, or draft specifications are not sealed; instead, they are clearly marked "Preliminary," "Not for Construction," "Draft," "For Discussion Only," or something similar to make sure that they are not confused with final documents. Sometimes, to satisfy the requirements of a regulatory agency, a preliminary document may need to be sealed for administrative purposes (for example, to start an approval process). Although the document may be "final" for administrative purposes, it is not final for construction, so it is particularly important that such documents be marked "Preliminary" and/or "Not for Construction." Expensive and dangerous errors can occur when preliminary documents are inadvertently released for fabrication or construction (and such errors have occurred).

Professional seals should not be used in company logos, advertising, letterheads, business cards, or other promotional publications. Also, the seal should not be used on government documents, such as applications for passports or birth certificates, when professional engineers or geoscientists are guarantors.[17]

Document Approval Process

Your seal on a document means you approved it; therefore, applying the seal should be the last step before the document is issued. Your employer likely has a formal document handling system for preparing, approving, identifying, distributing, and filing technical documents. If so, follow that procedure, (even if it seems cumbersome or bureaucratic). It is crucial that everyone use approved, up-to-date drawings and data. However, you should routinely verify your documents before you seal them. The suggestions below apply more to reports than to drawings or other technical documents, but give an idea of the approval process.

- **Scope of work.** Does the document satisfy the contract, or answer the questions, or follow the instructions that initiated it?
- **Accepted practice.** Is the document based on current industry codes, standards, or Association guidelines, or proper theory and correct equations?
- **Accuracy.** Are you confident that the analysis is logical and correct? Were numbers correctly transferred from other documents? Have you followed company policy for double-checking? Do the numbers support the conclusions? Are the conclusions summarized correctly and completely?
- **Completeness.** Is the document complete, with pages numbered in the right order?
- **Format.** Is the document in the accepted company or industry format?

This verification cannot be done in a few minutes. This explains why a professional must supervise the work. You need thorough knowledge of the document before you apply your seal to approve it. If you have not personally prepared the document, or supervised the preparation, you would need to duplicate almost all the work upon which it is based.

Professional and Legal Liability

Perfection is not required in technical documents, but reasonable judgment, based on adequate knowledge and experience, are expected. If you do not have confidence in the document, or have not had time to review it thoroughly, or if it is outside your field of expertise, then you should not seal it. For example, an electrical engineer may be asked to approve the concrete design for a building foundation, because the building is intended to house an electrical transformer. If the electrical engineer has no previous training or experience in reinforced concrete design, then the request is outside his or her field of expertise. A colleague experienced in concrete design must be consulted.

An engineer or geoscientist who signs or seals a document without thorough knowledge of the document may be guilty of professional misconduct. In a case cited recently, the British Columbia Supreme Court ruled that an engineer was liable in a dispute over an improperly designed residence foundation. The court stated: "By affixing his seal to the drawings and by his letter . . . the defendant [engineer] . . . certified that the foundation drawings conformed to all the structural requirements of the 1980 National Building Code."[18] Clearly, the court considered the seal on the drawings a guarantee of their accuracy and conformance with codes. But in a somewhat different type of case, the Supreme Court of Canada ruled that there is some room for error; perfection is not essential: "The seal attests that a qualified engineer prepared the drawing. It is not a guarantee of accuracy. The affixation of the seal, without more, is insufficient to found liability for negligent misrepresentation."[19]

Of course, the best strategy is to avoid these problems. Do not affix your seal to a document unless you are willing to accept responsibility for it based on adequate knowledge of the document and of the project to which it applies, and your reasonable judgment that the document is correct, complete, and within your field of expertise.

Checking Documents

Most Acts allow professionals to sign and seal documents that they have "checked," but "checking" has many meanings. In particular, do not confuse "checking" with "scanning." For example, a colleague may ask you to scan a document for obvious errors. You might do this as a simple courtesy, but a scan is not a check and is not an assumption of responsibility. You would not sign or seal a document that you merely scanned. When the colleague signs, seals, and dates the document, the colleague assumes full responsibility for it, whether or not you scanned it.

As a rule, if anyone asks you to "check and seal" a document that was not prepared under your direct supervision, then you must analyze it sufficiently to ensure that it is correct before assuming responsibility for it. The following two scenarios are commonly encountered:

- **Request from a friend.** Unlicensed practitioners often ask professionals to check and seal documents to avoid the cost of a full technical analysis. For example, a friend who is an engineering or geoscience graduate, but is not licensed, approaches you. She asks you, as a favour, to check, sign, and seal an engineering or geoscience drawing to satisfy municipal bylaws. She assures you that the drawing is completely correct and that your signature and seal are mere formalities.

- **Pressure of work.** You are the designated professional on the certificate of authorization (or permit to practise) for an engineering or geoscience firm. The firm has many projects, and you are unable to monitor all of them adequately. Your employer asks you to sign and seal the final report for a project of which you were previously unaware. The employer says the report is urgent, no further analysis is required, and asks you to sign immediately.

In each of the above situations, you would refuse to sign or seal the document until you had verified the document sufficiently to accept full responsibility for it. Do not be led into a trap by friendships, external pressure, or urgency. For some documents, a proper check would require complete duplication of the analysis. Obviously, if you completely redo the work, it is appropriate for you to assume responsibility for it.

When Professionals Collaborate

If one engineer or geoscientist has prepared a document or drawing and another has approved it, then both seals should be affixed. If this is not possible, or not expected, then only the approving engineer or geoscientist should seal it. This seal indicates that he or she takes responsibility for the document or drawing. Where final drawings cover more than one discipline, it is typically recommended that the drawings be sealed by the approving engineer or geoscientist (typically the project leader) and by the design engineer and/or geoscientist for each discipline. The seals should be "qualified" by an explanatory note that indicates clearly each person's area of responsibility.

Reports and Multiple Drawings

Individual pages of a report (and drawings bound into a report) need not be sealed, provided the report as a whole has been signed, sealed, and dated. The seal is usually placed on the title page near the author's name, or at the end if the report is in the form of a letter.

In large projects, the number of documents that must be sealed may be very large. It is not usually necessary to seal every detail drawing on large structural projects, which may have thousands of details. However, the drawings must be prepared under an engineer's control and supervision, and he or she assumes responsibility for them whether they are sealed or not. For example, the special case of structural steel is described in Ontario's guideline on the use of the seal:

> Generally applicable design details developed by manufacturers or standards organizations, verified by testing and/or approved by government bodies, do not need to be sealed. However, details or subsystem designs produced by manufacturers or contractors for specific projects, or applications that require professional engineering design or judgment, needed for coordination by the design engineer, must be sealed, to ensure there is consistent delineation of design responsibility for all aspects of the work.

For structural steel shop drawings, the building design engineer designs the members and overall stability system and is responsible to indicate member connection forces as required by professional practice standards. Structural steel detailers use this information to produce shop details and connections for the steel members. Many of the connections use standard details from the Canadian Institute of Steel Construction (CISC) handbook, which have been developed over time by qualified engineers.[20]

However, the guideline warns that connections may appear to be similar to standard connections, but not really identical. So, as a rule, all shop drawings should be sealed, or accompanied by a sealed letter to the building design engineer listing the drawings and stating that all detail drawings were prepared and reviewed under the connection design engineer's supervision.

Seal Security

The seal is important; it must be obtained from the Association, must be kept in a secure place, and must not be duplicated or lent to anyone who might use it in an unauthorized way. Unlicensed practitioners occasionally make and use illegal seals, and the Associations prosecute this practice, under the Act.

In the past, the (unsealed) master copies of drawings were usually kept in a secure file, and prints of the master drawings were sealed by hand, as they were issued. This procedure, which is still in use by some professionals, maintains security while keeping drawings up-to-date, since the masters are easily located when revisions are needed.

However, in recent years, Associations have recognized that electronic documents have the same legal status as paper documents, so electronic seals and signatures are now approved for use. The rules for sealing electronic documents are the same as for paper documents, although, of course, the security problems are more difficult. Electronic documents can be much more easily altered or copied. Several companies have developed security methods to protect documents stored as computer files, so if you or your employer are using electronic seals and signatures, make sure that your computer security is adequate to protect them. More information is available on this topic from the Associations. (See Web addresses in Appendix A.)

Failing to Seal Documents

Applying a seal does not increase the author's legal or professional liability, and omitting the seal does not relieve the author of any liability. "The courts assign liability on the basis of the facts, not on whether the document is sealed."[21] However, omitting the seal is a violation of the Act, and although it is not serious, discipline cases often include this charge, because the missing seal may corroborate a pattern of professional misconduct.

Approval authorities (such as municipal building permit officials) routinely reject unsealed documents. Occasionally, however, they may accept an unsealed document if the author is clearly identified as a licensed professional. They may assume that the missing seal is a simple clerical oversight, and process the document as a professional courtesy. However, authorities always give unsealed documents closer scrutiny to find other oversights.

PROMOTION TO SPECIALIST OR MANAGER

A professional engineer or geoscientist may be promoted at any time, depending on ability and promotion opportunities. However, about seven or eight years of experience is usually needed (on average) to supervise other professionals. The title may vary (for example, project engineer, group leader, or supervising geologist) but the position corresponds to Level D in the responsibility levels discussed earlier. The promotion paths diverge at Level D, leading either to specialization or management. This important divergence is sometimes subtle, so be aware of it.

Specialization versus Management

SPECIALIZATION   Specialists are essential. All technical companies need highly qualified experts to give advice to other professionals, solve difficult problems, make discoveries, and create and analyze the new products that generate profits. Employers must develop them through training and experience. To retain these specialists (especially the experts who achieve world-class levels), salaries and benefits must match (or exceed!) those of managers. The typical specialist gives advice and solves problems in an "atmosphere of freethinking creativity."

> **SENIOR ENGINEERING SPECIALIST** Under administrative and/or high technical direction, works as a senior engineer specialist or consultant in a particular field of engineering development or research. Participates in planning, organizes work methods and procedures. Makes independent decisions within own sphere, usually exercising technical authority over a small group of engineer specialists. . . . Gives technological advice and direction to a group of professional specialists. With an appreciation of the necessity of maintaining an atmosphere of freethinking creativity, outlines difficult problems and methods of approach. . . .[22]

Management duties are usually minor, but specialization requires a high level of technical knowledge. An advanced degree is a definite asset to the specialist, especially in high-technology industries. If specialization is your goal, develop your technical problem-solving skills, learn your specialty's codes and standards, join the specialty's technical society, and attend (and contribute to) its conferences. (Technical societies are discussed in Chapter 5.)

MANAGEMENT   Companies also need good managers, and a basic degree in engineering or geoscience is a good preparation, especially for high-tech industries. However, management means working closely with people, so good interpersonal skills are essential. For example, the manager makes key decisions in the "selection, development, rating, discipline and termination of staff."

ENGINEERING/GEOSCIENCE MANAGER Manages a large staff, administers and coordinates several professional, sub-professional and/or mechanical trades functions. . . . Makes decisions regarding the selection, development, rating, discipline and termination of staff. Reviews and evaluates technical work. Selects, schedules and coordinates to attain program objectives. . . .[23]

Managers must have "people" skills. These skills can be developed through personal study and practice. However, in recent years, formal education has become a popular route to management.[24] A master's degree in business administration (MBA) is seen as a fast track to senior management, because an MBA teaches the management and financial skills that engineers and geoscientists usually need. Several universities offer "modular" MBA programs that students can complete while continuing to work.

Choosing a Management Style

A basic study of management skills is interesting and may be useful. Douglas McGregor was the first researcher to explain management theory in a popular way. In his classic book, he described the two extremes of management style: Theory X and Theory Y. These theories define the manager's view of human nature, which is at the root of the management style:

Theory X: Theory X states that work is basically distasteful to most people and that people will avoid it whenever possible. Therefore, employees must be closely monitored and controlled. Furthermore, they must be made to work by threatening or penalties, or by luring them with rewards.

Theory Y: Theory Y states that people are naturally inclined to work and merely need favourable working conditions in order to be productive. Furthermore, psychological factors such as perceived control over one's activities and opportunities for creative work are important for proper motivation. If properly motivated, employees will produce beyond expectations.[25]

McGregor recommended that managers adopt Theory Y. In fact, most managers today would say they try to follow this theory. The "best" management style depends on the situation: a style that works in a software design office may not be effective in managing a police force, a fire fighting squad, or even a construction site. The best management style depends on the personality and maturity of the manager; the type of corporation; the initiative, creativity, education, or skill level of the employees; and their willingness to achieve the corporation's goals.

The spectrum of management styles ranges from the collegial (where the style is based entirely on Theory Y) to the military (based on Theory X). One view of this spectrum is shown in Table 6.3. Most people like the collegial style best and the military style least. Where is your management style (or your future management style) in this spectrum?

TABLE 6.3 — A Comparison of Management Styles

Management Style	Typical Example
Collegial	Manager treats employees as colleagues, and permits them to function independently, within agreed terms of reference. (Based on Theory Y)
Team-Oriented	Manager defines goals, but asks employees to suggest solutions, and guides them to a group decision.
Interactive	Manager presents the problem, obtains ideas and suggestions from employees, and then proposes decision.
Responsive	Manager presents tentative decision to employees, invites questions and discussion. Decision is final only after discussion.
Paternal	Manager presents decision, and explains it to employees, but will change decision only if there are serious objections.
Authoritarian	Manager makes decision and explains it to employees.
Military	Manager instructs employees. (Based on Theory X)

In a professional environment, such as a design office, the manager should adopt a collegial or team-oriented management style. However, an authoritarian style may be needed, occasionally—for example, to insist that safety features be added to a design. A good manager adjusts to the situation to ensure that the goals are achieved.

DISCUSSION TOPICS AND ASSIGNMENTS

1. The Codes of Ethics for provincial Associations all contain the clause that the engineer or geoscientist must consider the health, safety, and welfare of the public to be paramount, yet another clause states that the professional engineer or geoscientist should be loyal to the employer. (The Codes of Ethics are included in Appendix B and are available on the Internet.) In some situations, these two clauses could come into conflict. For example, an employer might ask the engineer or geoscientist to design piping that would discharge untreated wastewater into a public creek. In such situations, what is the proper course of action for the professional? (This type of ethical conflict is discussed in more detail later in this text.)

2. The discussion of management styles in this chapter implies that authoritarian managers are unable to motivate their workers effectively. Discuss this point. Is this necessarily true? Machiavelli would have disagreed. History shows that many authoritarian managers have successfully motivated their workers (or followers) in the past and many will likely do so in the future. While one may disagree with using authority (that is, fear)

to motivate workers, it does sometimes work. Discuss the benefits and disadvantages of an authoritarian management style, and give a few examples (jobs or social situations) where it might be most effective. As a contrast, give a few examples where it might be least effective.

3. The following questions concern management ideas:

a. Two well-known "laws" in management are Parkinson's Law and Peter's Principle. Parkinson's law states: "Work expands to fill the time available." Peter's Principle states: "People are promoted within an organization until they reach their level of incompetence." Using the Internet or your library, find the sources of each of these "laws" and explain them in more detail. Do you agree that they are universally (or even widely) true? Can you find any examples where they may apply in public life, in your personal life, or in your employment?

b. The president of a famous computer manufacturing company is alleged to have said: "We try to promote people without making them into managers." Discuss this concept, briefly. Do you think this is a positive statement about having dual promotion paths for specialists and managers, or is it a negative statement about managers?

Additional assignments can be found in Appendix E.

NOTES

[1]   The Association of Professional Engineers and Geoscientists of British Columbia (APEGBC), "Satisfactory Engineering Experience Guidelines," Appendix to *Application Guide for Membership*, January 2008, p. 21, available at <www.apeg.bc.ca> (May 16, 2008).

[2]   The Association of Professional Engineers, Geologists and Geophysicists of Alberta (APEGGA), *Experience Requirements for Licensure*, 1999 (revised February 2008), p. 6, available at <www.apegga.org/Members/Publications/salarysurvey.html> (May 15, 2008).

[3]   Canadian Engineering Qualifications Board, a committee of Engineers Canada (formerly the Canadian Council of Professional Engineers), *National Guideline for Engineer-in-Training*, Ottawa, available at <www.engineerscanada.ca/e/files/EIT_with.pdf> (June 15, 2009). Reproduced with permission from Engineers Canada.

[4]   Professional Engineers Ontario (PEO), *Guide to the Required Experience for Licensing as a Professional Engineer in Ontario*, January 2002, p. 6, available at <www.peo.on.ca> (June 24, 2007).

[5]   The Association of Professional Engineers, Geologists and Geophysicists of Alberta (APEGGA), "Detailed Job Classification Guide," Appendix A of report *2007 Value of Professional Services*, p. 70, available at <www.apegga.org/Members/Publications/salarysurvey.html> (May 15, 2008).

[6]   The Association of Professional Engineers, Geologists and Geophysicists of Alberta (APEGGA), "2007 Employer Salary Survey Highlights" from the report *2007 Value of Professional Services*, available at <www.apegga.org/Members/Publications/salarysurvey.html> (May 15, 2008).

[7]   Association of Professional Engineers and Geoscientists of British Columbia (APEGBC), *APEGBC 2006 Compensation Survey*, and *APEGBC Online Employment Responsibility Evaluation*, and *Sample Benchmark Employment Descriptions And Corresponding Ratings*, available at <www.apeg.bc.ca> (May 16, 2008).

[8] Ontario Society of Professional Engineers (OSPE), *Member Market Compensation Summary Report*, OSPE Employer Compensation Survey, available at <www.ospe.on.ca> (May 16, 2008).

[9] Engineers Canada, "Executive Summary," *National Survey of the Canadian Engineering Profession*, 2002, Ottawa, p. 9, available at <www.engineerscanada.ca/e/files/surveysummary2002.pdf> (June 15, 2009).

[10] Canadian Society for Professional Engineers (CSPE) website at <www.cspe.ca> (May 16, 2008).

[11] Ontario Society of Professional Engineers (OSPE) website at <www.ospe.on.ca> (May 16, 2008).

[12] The National Society of Professional Engineers (NSPE) website at <www.nspe.org> (May 16, 2008).

[13] The Association of Professional Engineers, Geologists and Geophysicists of Alberta (APEGGA), *Guideline for Ethical Practice*, June 2005, p. 9, available at <www.apegga.org/Members/Publications/guidelines.html> (May 16, 2008).

[14] Professional Engineers Ontario (PEO), *Use of the Professional Engineer's Seal*, July 2005, p. 18, available at <www.peo.on.ca> (May 16, 2008).

[15] *Regulation 941*, Section 53, under the Professional Engineers Act, RSO 1990, c. P. 28, available at <www.e-laws.gov.on.ca/index.html> (May 16, 2008).

[16] Professional Engineers Ontario (PEO), *Use of the Professional Engineer's Seal*, p. 7.

[17] Ibid., p. 8.

[18] Quoted in J.M. MacEwing, "Legal Significance of the Engineer's Seal," *Canadian Consulting Engineer*, July–August 1996, p. 8.

[19] J.M. MacEwing, "Legal Significance of the Engineer's Seal."

[20] Professional Engineers Ontario (PEO), *Use of the Professional Engineer's Seal*, p. 20.

[21] Ibid., p. 19.

[22] Association of Professional Engineers and Geoscientists of British Columbia (APEGBC), *Sample Benchmark Employment Descriptions And Corresponding Ratings*, p. 5, available at <www.apeg.bc.ca> (May 16, 2008).

[23] Ibid., p. 6.

[24] M. Gladwell, "The Talent Myth," *The New Yorker*, July 22, 2002, p. 28.

[25] "Douglas McGregor: Theory X and Theory Y," *Business: The Ultimate Resource*, Perseus Books Group, Cambridge, UK, 2002, pp. 1022–1023.

Chapter 7
Private Practice, Consulting, and Business

Curiously, engineers and geoscientists in "private practice" offer their services to the public. They may adopt the title of Consulting Engineer, Consulting Geoscientist, or simply Consultant. Private practice is entrepreneurial, and while it has more risks than employment, it is usually more rewarding. This chapter is an overview of private practice and consulting, but most topics apply to any technical business. This chapter encourages you to consider making your career in one of these areas.

CONSULTING OPPORTUNITIES

Professionals usually enter private practice and consulting at the peak of a career. The job variety and challenges are very attractive, and clients pay the engineer or geoscientist very well for advice. Moreover, the field is not crowded, but it is very demanding—acquiring knowledge and experience may take years of practice, and consultants must compete for new projects, new clients, or new products. In spite of the attractions of private practice and consulting, only a minority of engineers and geoscientists practise in these areas, as the following statistics show:

- A survey of practising engineers and geoscientists revealed that 82 percent of the 27,108 respondents were permanent employees. Only 9 percent were self-employed, although an additional 3.8 percent had short-term or project contracts.[1]
- In 2008, Canada had about 160,000 professional engineers, but only about 600 consulting engineering firms (including sole proprietorships, partnerships, and corporations). Most member firms employed fewer than 25 people.[2]

These numbers imply that only a small fraction of practitioners are self-employed owners, partners, or "principals" of a private practice. Many readers may see an opportunity here.

CONSULTING ACTIVITIES

The consulting engineer or geoscientist is usually an advisor to a client (an individual or company) who needs advice on a design, development, management, resources, or construction project. The client assumes the risk for the enterprise, but the consultant's advice and skills increase the likelihood of success.

Consulting engineers provide a very broad range of services to the public, as the list of the Association of Canadian Engineering Companies (ACEC) member firms shows. Civil engineering is probably the largest single specialty, but the ACEC members provide advice and assistance for almost any type of project. The tasks most often required are

- **Engineering advice.** Most consulting engineers work on specific projects, in design, development, inspection, testing, quality control, management, and so on. Sometimes a client may require a forensic analysis of a device or structure that has failed.
- **Expert testimony.** An expert witness provides an independent opinion to a court, commission, board, hearing, or similar government or judicial body. An expert witness must be impartial, and must not be a spokesperson for either side (even if only one side is paying for the expert testimony).
- **Feasibility studies.** Consultants are useful in the preliminary stages of a project, when a client requires guidance as to the feasibility, financial justification, duration, and cost of the project. The consultant helps the client decide whether to go ahead with the project.
- **Detail design.** Many consulting engineers are skilled in detail design. This includes preparing drawings, specifications, and contract documents.
- **Specialized design.** Some consultants specialize in custom design and development, such as manufacturing processes, machine design, mining, and similar areas. Consulting engineers may work independently, or may work alongside the client's in-house staff. Consultants may also help to develop inventions or prepare patents.
- **Project management.** Consulting engineers commonly supervise projects, including the design, manufacturing, construction, or assembly phases of a project, or the commissioning (initial start-up) of a large plant.

In summary, engineers and geoscientists in private practice perform any task that requires professional knowledge. They usually do so on behalf of a client who lacks the personnel or the expertise to conduct the work.

ADDITIONAL LICENSING AND INSURANCE REQUIREMENTS

In most provinces, professional engineers or geoscientists may offer their services to the public without additional licensing. However, all provinces and territories (except Quebec) impose rules on partnerships or corporations that offer services to the public. No two Acts are identical on all issues, as shown in the following overview.

Practice by Partnerships or Corporations

All provinces and territories (except Quebec) require firms (companies, partnerships, or corporations) offering services to the public to have an additional licence. The Associations call the licence a "Permit to Practise," a "Certificate of Authorization," or a "Certificate of Compliance." This licence protects the public by requiring firms to identify the professionals responsible for the firm's engineering and/or geoscience work, and by ensuring that they are qualified. The firm must designate the full-time employee(s) of the firm who supervise and accept responsibility for the professional practice. These employees must be licensed and must be qualified to provide the service. The permit (or certificate) lists their names, and holds them responsible for the quality of the professional service.

Most provinces and territories permit licensed individuals (sole proprietors) to offer services to the public, but there are two exceptions: Ontario (PEO and APGO) and Newfoundland and Labrador (PEG-NL) require individuals who offer services to the public to have a Certificate of Authorization (or permit). In Ontario, this certificate requires a minimum of five years of relevant experience.

Designation as a Consulting Engineer

A professional engineer in private practice may use the title Consulting Engineer in most of Canada. However, in Ontario, this title (or any variation with the same meaning) is restricted under the Professional Engineers Act (Ontario). To be a consulting engineer in Ontario, a licensed professional engineer must apply to Professional Engineers Ontario (PEO) and satisfy several requirements:

- **Authorization.** The engineer must obtain a Certificate of Authorization from PEO.
- **Experience.** Consulting engineers must have five years' experience in addition to that required for registration. At least two years of this experience must be in private practice.
- **Liability insurance.** The engineer must give PEO evidence of liability insurance.

Liability Insurance Requirements

Liability—in other words, being sued—is a greater risk for consultants than for employees. As a result, liability insurance (also called "errors and omissions" insurance) is essential for professionals in private practice. The law text by Samuels and Sanders has an extensive section on liability insurance, and states: "Even the most competent professionals get sued, sometimes for no valid reason. The cost of defending a frivolous claim can be enormous."[3] Liability insurance typically includes a duty to defend. That is, if the professional is sued, the insurer pays the legal bills.

In spite of this risk, only a few of the Acts require consultants to maintain professional liability insurance. Several Acts do not mention liability insurance, and some Acts require it for partnerships and corporations, but exempt sole proprietors. The following summary explains this confusing hash of rules. You should check the liability insurance rules for your Association, of course, before committing yourself to private practice.

- **Quebec (OIQ and OGQ):** Liability insurance is mandatory for all members.
- **British Columbia, Manitoba, and Saskatchewan:** Liability insurance is not mandatory, but members, licensees, and certificate holders must notify clients in writing whether professional liability insurance covers the services offered, and receive the client's acknowledgment before proceeding.
- **Ontario (PEO and APGO):** Liability insurance is mandatory for Certificate of Authorization holders (anyone who offers service to the public), unless they notify clients in writing that no professional liability insurance covers the services offered, and receive the client's acknowledgment before proceeding.
- **New Brunswick, Newfoundland and Labrador:** Liability insurance is not mandatory. The Act authorizes bylaws for liability insurance, but the bylaws are silent.
- **Northwest Territories, Nova Scotia (APENS and APGNS), Nunavut, Prince Edward Island, and Yukon:** Liability insurance is voluntary. The Act does not mention it.

Regardless of the Act's requirements for liability insurance in your province or territory, you should consider it essential if you plan to enter private practice.

Secondary Liability Insurance

In every province and territory (except Ontario), professional engineers and geoscientists participate in a mandatory Secondary Liability Insurance Program begun in March 2002.

Engineers Canada administers the insurance plan, which provides members with $100,000 in professional liability coverage, plus an unlimited amount of related legal fees.

The group policy is "secondary" because it is not the main protection against liability claims. Consultants (and design, manufacturing, and similar companies) must maintain separate "primary" liability insurance for professional activities. Secondary insurance typically covers professional engineers and geoscientists who are employees. The secondary plan provides protection for

- claims resulting from actions undertaken as an employee in an engineering position in a firm that does not provide engineering services to the public,

- former employees of consulting firms that no longer carry insurance,
- retirees, for past professional acts,
- gratuitous or incidental advice given to others outside of normal employment,
- small consulting jobs (up to $2,000 in fees) undertaken outside of normal employment.[4]

The secondary insurance coverage is rather narrow, since it does not cover "moonlighting" activities where the fee exceeds $2,000, nor does it cover professionals working for companies without primary protection. Nevertheless, secondary coverage provides an extra layer of protection to the public, so it is worth having if the premiums are reasonable. Group coverage increases participation and reduces premium costs. Information on the insurance agreement, coverage, restrictions, and exclusions is available from your Association.

SELECTING A BUSINESS FORMAT

A consulting firm (or any other business) must have a recognized business structure. The following overview explains the four most common structures, and the law text by Samuels and Sanders discusses the advantages and disadvantages of the structures in more detail.[5] Of course, before you start any business, you should discuss the business structure, legal restrictions, and tax implications with a lawyer.

SOLE PROPRIETORSHIP   A sole proprietorship is a one-person business. It is easily established—the proprietor (owner) simply registers the business name with the provincial or territorial government. The business setup and management are simple, but the proprietor is responsible for all of the business's financial debts and professional problems. For example, if the business should incur large debts, creditors may sue to seize both the proprietor's business and his or her personal property.

GENERAL PARTNERSHIP   A partnership enables professionals to share knowledge, experience, and friendship. Two or more individuals can agree to form a partnership (preferably by a written agreement, but not necessarily). Most often, the partners contribute the capital to start the business and agree to manage it together. They then share the profits (and losses). The partners must register the business with the provincial government (as for sole proprietorships). The partners share the work, but each partner is liable for the business debts or obligations incurred by all the other partners. Therefore, the people involved must have absolute confidence in one another's skills, competence, ethical standards, and personalities before entering a general partnership. Errors, omissions, and fraudulent acts by one partner in the name of the partnership create a liability for the innocent partners. In particular, if a partner does unprofessional work, all partners who authorized, permitted, or even acquiesced in the unprofessional work may be subject to discipline.

LIMITED PARTNERSHIP   Investors form limited partnerships when one (or more) of them does not want to participate in running the business. The general partners carry on the business; the limited partners contribute assets for use in the business (typically money, but sometimes property, such as a building or a patent). The liability of the limited partners is restricted to the value of the assets contributed. However, limited partners cannot intervene in the business's operations. You must register a limited partnership (provincially) to get this liability protection. If a general partner does unprofessional work, only the general partners are required to answer. Limited partners are not involved in the day-to-day running of the partnership, and typically are not at risk.

CORPORATION   A corporation has many of the rights of a real person. For example, a corporation can enter into contracts, own property, and conduct business. To establish a corporation, an individual (or a group of people) must apply to the government; this involves some paperwork and fees. You should first consult a lawyer to help you decide whether to incorporate under federal or provincial law, and for assistance in processing the application. Forming a corporation is usually an effective way to protect personal assets. When a practice is incorporated, creditors (and judgment awards) can seize only the corporation's assets; the personal assets of shareholders are not at risk. A corporation that performs engineering and/or geoscience services must usually obtain a Permit to Practise (or Certificate of Authorization) or must register with the Association. The corporation must designate a licensed professional within the corporation who will take responsibility for the corporation's work. The designated professional must possess the competence and the initiative to ensure that the corporation's work meets professional standards. That person must answer if a charge of professional misconduct should be made against the corporation.

Obviously, each business structure has advantages and disadvantages. When the goal of the professional in private practice is to protect personal assets, the following actions may help:

- Incorporate the practice to limit liability against demands from creditors and against civil judgments such as breach of contract.
- Regardless of the form of business organization, get professional liability insurance.
- Obtain general liability insurance to cover the risk of accidents within the business premises and product liability insurance to cover the risk of damage claims for dangerous products. (Other forms of liability, accident, and corporate insurance are also available.)

Engineers and geoscientists cannot use insurance or incorporation to shield themselves from disciplinary action for negligence, incompetence, or corruption. Negligence, incompetence, and corruption are the result of personal actions, omissions, or traits. However, careful, competent, ethical practice is a good protection for these hazards.

ASSISTANCE FOR CONSULTANTS

Consulting Engineering Organizations

Organizations devoted to assisting consulting engineers exist in every Canadian province and territory except Nunavut. Table 7.1 lists the names, locations, and websites of these consulting organizations.

The consulting engineering organizations in Table 7.1 are voluntary groups that promote the interests of members in several ways.

- **Technical and business information.** These organizations publish a wealth of information on key topics—how to operate a professional practice, write contracts and agreements, and so forth. They provide standard contracts at a nominal charge, usually through the Internet.

TABLE 7.1 — Consulting Engineering Organizations

Association of Canadian Engineering Companies (ACEC)
(formerly the Association of Consulting Engineers of Canada),
Ottawa, ON <www.acec.ca>

International Federation of Consulting Engineers (FIDIC),
Geneva, Switzerland <www1.fidic.org>

Consulting Engineers of Alberta,
Edmonton, AB <www.cea.ca>

Consulting Engineers of British Columbia (CEBC),
Vancouver, BC <www.cebc.org>

Consulting Engineers of Manitoba,
Winnipeg, MB <www.cemanitoba.com>

Consulting Engineers of New Brunswick,
Fredericton, NB <www.cenb.nb.ca>

Consulting Engineers of Newfoundland and Labrador
Paradise, NL (Contact via ACEC website <www.acec.ca>)

Consulting Engineers of Northwest Territories,
Yellowknife, NT (Contact via ACEC website <www.acec.ca>)

Consulting Engineers of Nova Scotia,
Halifax, NS <www.cens.org>

Consulting Engineers of Ontario,
Toronto, ON <www.ceo.on.ca>

Consulting Engineers of Prince Edward Island
Charlottetown, PEI (Contact via ACEC website <www.acec.ca>)

Association des Ingénieurs-Conseils du Québec,
Montreal, PQ <www.aicq.qc.ca>

Consulting Engineers of Saskatchewan,
Regina, SK <www.ces.sk.ca>

Consulting Engineers of Yukon,
Whitehorse, YT <www.cey.ca>

Note: Websites are valid as of June 15, 2009.

- **Communication and representation.** The organizations inform members of current issues that affect the profession. They also represent members before municipal, regional, and provincial governments when requested, or when an issue affects consulting engineers as a group.

Association of Canadian Engineering Companies (ACEC)

Membership in one of the provincial consulting engineering organizations automatically includes membership in the Association of Canadian Engineering Companies (ACEC—formerly known as the Association of Consulting Engineers of Canada), founded in 1925. ACEC comprises about 600 independent consulting engineering firms, who belong to 12 provincial and territorial member organizations, all of which can be contacted through the ACEC website.[6] ACEC promotes good business relations between its member firms and their clients. It also fosters the exchange of professional, management, and business information. ACEC safeguards the interests of consulting engineers when necessary, raises the high professional standards in consulting, and provides liaison with the federal government. The American equivalent of ACEC is the American Consulting Engineers Council (also known as ACEC).

International Federation of Consulting Engineers (FIDIC)

Both North American ACECs belong to the International Federation of Consulting Engineers/Fédération Internationale des Ingénieurs-Conseils (FIDIC), founded in 1913, an umbrella group for more than 50 national consulting engineers' associations. The member associations must comply with FIDIC's code as it relates to professional status, independence, and competence. FIDIC publishes an international directory and works on behalf of consulting engineers at the international level. FIDIC also publishes a wide range of documents to help members draft contracts and agreements, manage projects, and operate consulting firms.[7]

Professional Service Contracts

Standard contract forms are available from the provincial Associations and from consultant organizations such as the Association of Canadian Engineering Companies (ACEC). In addition, the concise law text by Samuels and Sanders (cited earlier) has several chapters on contract law, explaining how to draw up good contracts and what problems and pitfalls to avoid. In this section, we merely want to describe the contracting process for a consultant offering professional services to a client, and the typical evaluation procedure that clients are increasingly following.

Procuring professional services is not the same as purchasing materials. Materials have established standards, so the purchaser's goal is simply to obtain the lowest price. However, it is risky (although not illegal) to select a consultant based on the lowest bid. However, ACEC recommends the Quality-Based Selection (QBS) process, which separates the evaluation of the consultant's qualifications from the fee negotiation. The client negotiates the fee after selecting the consultant.

This process is logical, because fees cannot be set without defining the scope of the project, and this usually requires the client and consultant to work together. In many projects, the most significant cost savings are in the early stages of the project, and employing the best-qualified consultant will usually generate more savings. Penny-pinching at the design stage usually results in higher capital, operating, and maintenance costs.

Quality-Based Selection (QBS)

Guidelines for the contracting process are widely available (from ACEC, FIDIC, and the provincial consulting organizations in Table 7.1). All of the consulting organizations recommend the Quality-Based Selection (QBS) process, summarized below. The process is written for consulting engineers, but applies equally to geoscientists.

THE QBS PROCESS   The basic steps in the QBS selection procedure, as described by the Consulting Engineers of Ontario, are

- Identify a short list of firms with relevant qualifications.
- Request detailed proposals.
- Select the best-qualified and available firm through a detailed process of interviews, site visits, client references, etc., as required.
- Negotiate an appropriate fee based on a mutually agreed scope of services and execute an agreement.
- Notify unsuccessful consultants.[8]

QBS CRITERIA AND EVALUATION   The client should use appropriate criteria, such as those listed below, to evaluate and select the consultant (or the project team). In addition, to ensure a fair and objective evaluation, the Consulting Engineers of Ontario recommend that each of the above criteria (or qualification factors) be given a relative weighting (in the ranges shown), and that each applicant be rated and assigned a mark for each criterion. The sum of the marks gives a relative rating, useful for comparing applicants. (Weighting is a standard method of evaluating alternatives.)[9]

EXPERIENCE AND QUALIFICATIONS OF THE PROJECT TEAM

Project manager/engineer and senior designers	(15–25)
Technical support staff	(5–15)
Sub-Consultants	(0–15)

MANAGEMENT QUALIFICATIONS

Experience on similar projects	(10–20)
Availability of key staff	(5–15)
Stability and reputation of the firm	(0–10)
Multidisciplinary/Specialty capabilities	(0–10)
Quality assurance systems	(0–10)
Local office	(0–10)
Insurance coverage	(0–5)

PROJECT IMPLEMENTATION

Approach and methodology	(10–25)
Scheduling of key activities and resources	(5–15)
Project quality assurance program	(0–10)

QBS BENEFITS   According to the Consulting Engineers of Ontario, the benefits of QBS are

- Within the process, the engineer and client work together in a professional relationship to define and deliver a project.
- Selection focuses on value to the client.
- The life-cycle costs of the asset can be optimized.
- With a detailed understanding and agreement on the required project scope, the engineer is in a much better position to determine resource requirements, cost estimates and fees.[10]

The Consulting Engineers of British Columbia (CEBC) emphasize the benefits of engaging the engineering consultant early in the project:

> Studies have shown that engineering typically represents 1.5% of the total cost of a project, while construction costs represent 16.5% of the total cost, and operations are 82% of the total cost. By hiring a consulting engineer at the beginning of the project, good design can cut 10–15% of construction costs—and more in life cycle costs. It is through engineering services that a client has the best opportunity to manage and potentially reduce the remaining 98.5% of the project's life cycle costs.[11]

COMPENSATION FOR CONSULTANTS

In the past, professional fees were often set as a percentage of the project costs. This procedure is appropriate in some cases (mainly civil engineering); the main drawback is that it penalizes the professional for creating an economical design. A good compensation process should reward efficiency and innovation. Four common methods for calculating consulting fees are found below, adapted from the FIDIC Directory (with permission). The Directory also explains variations of these procedures.[12]

PER DIEM   Per diem payments are simply daily rates. They are the usual payment method when the scope of work cannot be accurately determined.

Studies, investigations, field services, and report preparation are in this category. For example, some consulting companies specialize in short-term overload assistance. When requested, their engineers or geoscientists work side-by-side with the client's personnel. This allows the client to absorb peak workloads without hiring and training employees who may be surplus when the peak passes. Per diem payment is the best compensation method in this case. Direct out-of-pocket expenses, such as travel costs, are added to the per diem payments.

PAYROLL COSTS TIMES A MULTIPLIER Payroll costs, multiplied by a factor to cover overhead and profit, are most often used for site investigations, preliminary design, process studies, plant layout, and detailed design. The multiplier is usually in a range from two to three. Under this method, the client essentially pays the personnel costs as they arise, including a sufficient amount to cover overhead and profit. Direct out-of-pocket expenses are also reimbursed in addition to the payroll costs.

LUMP SUM As the name suggests, the consultant determines, in advance, a "lump sum" fee that includes all costs, overhead, and profit. Many clients prefer this method of compensation because it tells them in advance how much they will be paying for professional services. When the services can be determined accurately, this is a simple approach. The disadvantage is that the consultant may underestimate time or costs, and incur a serious loss.

FEE AS A PERCENTAGE OF ESTIMATED OR ACTUAL COSTS OF CONSTRUCTION This method bases fees on total costs, and is common for consulting services such as preparing drawings, specifications, and construction contract documents. However, this method is becoming less and less popular because of the difficulty in relating design costs to new and rapidly changing construction technologies, and because of the unpredictable market conditions in the construction industry.

STARTING A PRIVATE PRACTICE OR BUSINESS

Starting a Private Practice

Most professionals become "principals" (partners or shareholders) after many years of diligent service to the firm. This route to ownership carries the least risk, but takes a long time. Some professionals may work for an older colleague, and may get an opportunity to buy out the practice or business when the colleague retires. However, most entrepreneurial engineers and geoscientists want to move into senior roles as soon as they can.

Private practice (or consulting) is an attractive option, especially for an engineer or geoscientist who has mastered the technical knowledge, has good self-discipline, and wants a challenge. However, even a well-prepared professional may need two or three years to become fully established, and those early years will test the new consultant's determination. Extra hours of work are essential, of course. In addition, most engineers and geoscientists need better business management skills. Working overtime cannot offset poor business sense or inadequate financial reserves.

To succeed in private practice, you must have an entrepreneurial spirit! Private practice makes you both boss and employee, and it is impossible to be the boss if you object to working long hours or if you cannot cope with stress and uncertainty. To succeed, you need a special set of skills and attitudes. The section below lists the entrepreneurial attitudes, technical skills, and personal characteristics needed.

Starting a Technical Business

A recent survey shows that young people today are more likely to start a business than previous generations were.[13] The present generation is less concerned about security; they know that staying with one company for a lifetime is now obsolete and they know they must adapt to constant change, and triumph over it. Most university students held serious jobs during their university years, and they know far more about the business world than their parents did at the same age. In fact, many university students are already entrepreneurs, some with companies created in high school. Most of today's graduates are well prepared to be the business leaders of tomorrow.

Nevertheless, a leap directly into business could result in disaster. The best advice for graduates is to get some professional experience before you start your own business. Also, before you strike out on your own with a new enterprise, evaluate your skills and your opportunities carefully. The following section may help you do this.

EVALUATING YOUR POTENTIAL AS A CONSULTANT

Regardless of your enthusiasm, private practice is not for everyone. The following checklist of skills, traits, and talents, assembled from various sources, applies to engineers or geoscientists considering a leap into private practice.[14,15] (The successful business entrepreneur would likely show similar skills and personality traits.) Do you have these skills, traits, and talents?

- **Education and licensing.** Obviously, the first two requirements before you can start any engineering or geoscience enterprise are a degree and a licence. The rare exception is entering a consultancy or business with a person who is already licensed.
- **Adequate experience and technical knowledge.** You should know the practice or the business very well. Leaving a paying job to start a totally new and unknown business is very risky. (For example, if a construction engineer wants to be a machine design consultant, some experience in machine design is obviously essential.) If you do not have adequate knowledge, get it first.

Photo 7.1 — Entrepreneurial Education. *This photo shows the Schlegel Centre for Entrepreneurship, established in 2002 at Wilfrid Laurier University, in Waterloo, Ontario. Stantec, a large Canadian consulting firm with offices across North America, designed the structural engineering for the Schlegel Centre, a four-storey, 58,000-square-foot cast-in-place concrete flat slab structure connected to existing buildings. In September 2007, the university began to offer a one-year MBA program, with an innovation and entrepreneurship option, at the Schlegel Centre.*

Source: Photo courtesy of Wilfrid Laurier University. Used with permission.

- **A network of contacts.** A network of people who may become potential customers, suppliers, or investors is extremely valuable. Government contacts may also be useful for advice on new contracts, standards, or regulations. You particularly need these people during the start-up of your business—the critical phase.
- **Determination.** Determination is perhaps the most important personal trait you need, because consultants and entrepreneurs face many obstacles before their efforts yield success.
- **Confidence and independence.** Entrepreneurship is often a lonely business, so you must like being your own boss. You must also be able to learn from your mistakes.
- **Business skills.** Running a consultancy or a business requires discipline and good management skills. Can you manage budgets, business operations, and employees? A basic knowledge of accounting is essential. If terms such as "cash flow," "balance sheet," and "profit and loss statement" are not familiar, then you will need to upgrade your business skills.
- **People skills.** You may have innovative ideas, but you still have to sell your ideas to people. You are well prepared if you can say, honestly, that you have a positive personality, enjoy working with people, and can communicate well.
- **Good health.** Most successful entrepreneurs enjoy getting up early in the morning to attack the day's problems. Good health is important, especially in the early years, when the stress and physical demands are greatest.
- **Intelligence.** Regardless of their education, successful consultants and entrepreneurs are intelligent, think quickly, and enjoy working with new ideas.

GETTING YOUR ENTERPRISE STARTED

Obviously, the above traits will not guarantee success. You also need good planning and enough money to survive until the business starts to pay for itself. Before you risk your time and money, you must conduct a market survey—that is, you must gather as much information as possible about the potential marketplace. When you have solid market research showing that a need exists for your service, product, or idea, you must then prepare a business plan.

A business plan describes what you learned from the market research; it also defines your business objectives and outlines a strategy for reaching those objectives. Your business plan should describe every aspect of your proposal, including business structure, manufacturing, advertising, and marketing, and whatever other topics are appropriate.[16] A good business plan is the key to raising money—the essential ingredient. In fact, the main obstacle to entrepreneurs is the shortage of investors willing to gamble on a new venture— especially in Canada. The details of starting a business are beyond the scope of this text, but further advice and references are found in Appendix G.

DISCUSSION TOPICS AND ASSIGNMENTS

1. Consider the list of nine personal characteristics explained in this chapter under "Evaluating Your Potential as a Consultant." Rate your own ability under these nine headings, on a scale of one to ten, where:

   Zero means you have serious doubts about your ability in the area.

   Five means you have reasonable confidence in your ability in the area.

   Ten means you have absolute confidence in your ability in the area.

   Sum your scores to get a rating out of 90. Give yourself an additional 5 points for each further qualification listed below.

   - You have adequate, current experience in computer software related to your field.
   - You own computer hardware that runs the software mentioned above.
   - You have published three or more technical papers.
   - You have already been involved in consulting.
   - You have a master's degree.
   - You have contacts in three or more companies that might need your services.
   - You have experience in making presentations and writing technical proposals.
   - You enjoy making important (and expensive) decisions under pressure.

   Total your points; they should not exceed 130. If your total is 100 points or more and you have been scrupulously honest in your personal assessment, then you are probably ready to move into private practice or business. If your total is less than 80 points, then you probably need more experience, education, or determination.

2. Assume that you have carried out the quiz in Question 1 and have evaluated your qualifications for entering private practice. The quiz should give you a score between zero and 130.

   a. If your score was higher than 100, prepare a business plan for establishing your private practice or business. Discuss the completed plan with your professor (or your banker).

   b. If your score was lower than 100, consider whether other career paths (such as management or specialization) are better for you. If private practice is still your objective, make a list of steps that would enable you to move into private practice. Do you need to improve your qualifications, or do you simply need to obtain more experience in your field?

3. Imagine that you have decided to enter private practice and that you are trying to become better known so that you can attract more clients and contracts. Advertising is a sensitive issue, since it must be consistent with the Code of Ethics. Read the sections in this text on professional

advertising (consult the Index). Then devise at least five methods for becoming better known as a competent and ethical professional that are clearly consistent with your provincial or territorial Code of Ethics.

Additional assignments can be found in Appendix E.

NOTES

[1]   Engineers Canada (formerly the Canadian Council of Professional Engineers— CCPE), *National Survey of the Canadian Engineering Profession*, 2002, Ottawa, available at <www.engineerscanada.ca> (May 17, 2008).

[2]   Association of Canadian Engineering Companies (ACEC) website at <www.acec.ca> (May 17, 2008).

[3]   B.M. Samuels and D.R. Sanders, *Practical Law of Engineering, Architecture and Geoscience*, Canadian Edition, Pearson Prentice Hall, Upper Saddle River, NJ, 2007, p. 174.

[4]   The Association of Professional Engineers, Geologists and Geophysicists of Alberta (APEGGA), "National Secondary Professional Liability Insurance Program," March 2007, available at <www.apegga.com/members/Benefits/pdf/QuestionsAnswers.pdf> (May 17, 2008).

[5]   Samuels and Sanders, *Practical Law of Engineering, Architecture & Geoscience*, p. 39.

[6]   Association of Canadian Engineering Companies (formerly the Association of Consulting Engineers of Canada—ACEC), Ottawa, available at <www.acec.ca> (May 17, 2008).

[7]   International Federation of Consulting Engineers (FIDIC) website at <www1.fidic.org> (May 17, 2008).

[8]   Consulting Engineers of Ontario (CEO), *Guidelines for the Selection of Consulting Engineers*, p. 3, available at <www.ceo.on.ca> (May 17, 2008).

[9]   Ibid., p. 7.

[10]   Ibid., p. 4.

[11]   Consulting Engineers of British Columbia (CEBC), "Selecting a Consulting Engineer," available at <www.cebc.org> (May 18, 2008).

[12]   International Federation of Consulting Engineers (FIDIC), *FIDIC International Directory of Consulting Engineers, 1997–1998*, Lausanne, Switzerland. Excerpts adapted with permission of RhysJones Publishing Limited.

[13]   "Know future," *The Economist,* December 21, 2000.

[14]   "Assessing Your Entrepreneurial Profile," *Business: The Ultimate Resource,* Perseus Books Group, Cambridge, UK, 2002, p. 818.

[15]   Consulting Engineers Ontario (CEO), *Guide to Engineering Consultancy Practice*, 1994.

[16]   "Writing a Business Plan," *Business: The Ultimate Resource*, Perseus Books Group, Cambridge, UK, 2002, p. 486.

Chapter 8

Hazards, Liability, Standards, and Safety

Engineering and geoscience projects are sometimes hazardous to workers, to the public, or to the environment. If a hazardous project causes damage or loss, the engineer or geoscientist responsible may be held liable. Fortunately, safe design and operation are easily achieved by taking these simple steps:

- For safe design, follow design codes and standards, be aware of safety regulations, and make formal hazard analyses. Where hazards cannot be eliminated, post prominent warnings.
- For workplace safety in factories, construction sites, well sites, or process plants, learn and follow Occupational Health and Safety regulations.

This chapter explains basic concepts of hazards and liability. The importance of design standards are reviewed first, then the importance of government regulations—particularly the Occupational Health and Safety laws, which are vitally important for workplace safety. The chapter concludes with a description of the Westray mining disaster, which led to a clause in Canada's Criminal Code that imposes penalties for workplace safety violations that result in injury or death. Some readers may find hazards and liability to be unsettling or negative topics, but if this chapter helps you to avoid damage, deaths, or lawsuits, you will be very glad you read it.

SOURCES OF PROFESSIONAL LIABILITY

To discuss liability, we must examine the case where a professional engineer or geoscientist gives a client negligent or incompetent advice. (Although this is unfortunate, it does happen.) If the client follows the advice and suffers a loss or damage, then the professional (and/or their employer or insurance company) may be sued, and ordered to pay damages. Similarly, a professional who designs an unsafe product may also be sued to pay damages. Such lawsuits may be based on several legal sources: contract law, tort law, or consumer legislation. These laws are discussed in detail in the law textbooks[1,2] and are summarized briefly as follows:

- **Contract law.** Properly negotiated contracts usually run smoothly, but disagreements, misinterpretations, and breaches may sometimes occur, and if they cannot be solved by negotiation, any party to the contract may sue for damages. The contract document is then examined in a court of law, and the judge decides whether the contract has been fulfilled or not, and what damages should be paid. This explanation is a gross oversimplification, of course, but it stresses the importance of diligence in negotiating and fulfilling contracts. For example, it is usually beneficial for a contract to include bonuses for good results. Conversely, it is also beneficial to consider how a contract could go wrong, to predict the damages that could result, to specify reasonable payments for damages, and/or to set sensible limits to liability. In addition, a professional must work only in the fields where he or she is competent. When a contract is outside the professional's field of competence, he or she must get help with the work, or decline the contract. (This rule is stated in every Association's Code of Ethics.) Contracts are essential, so typical contract problems are discussed at length in the law books cited above.

- **Tort law.** Some readers may be surprised to learn that, even in the absence of a contract, a professional may be sued for negligence. Tort law, which is independent of contract or criminal law, entitles a person who has suffered a loss, which is a direct result of someone's negligence, to seek damages from the negligent person.[3] This important concept is explained later in this chapter.

- **Consumer legislation.** If a defective product is manufactured and sold, the designer, manufacturer, dealer, or seller may be liable for damages under provincial legislation, such as the Sale of Goods Act or the Consumers Protection Act. Canadian and American consumer laws differ on a basic premise. Under the "strict liability" concept in American law, the manufacturer is presumed to be at fault unless it can be proved otherwise. In Canada, the manufacturer is not at fault unless the injured party can prove negligence in the design or manufacture of the product.

In addition to the legal liability noted above, negligence, incompetence, or indifference to public safety might also result in disciplinary action by the licensing Association. As explained in Chapter 11, the Association's Code of Ethics obliges professional engineers and geoscientists to protect the health, safety, and welfare of the public.

PROFESSIONAL LIABILITY—TORT LAW

The word *tort* means *injury* or *damage*. If an injury or damage is caused by wrongful behaviour or by defective merchandise, the plaintiff (the injured party) may sue the defendant (the accused person). The Canadian law of tort is basically fair, because a lawsuit cannot be based on bad luck or a truly random accident. Instead, the defendant's conduct is the key factor. The defendant's conduct may be classified as *intentional, negligent,* or *accidental.*[4]

In general, torts must be intentional or negligent to result in liability. Truly random or accidental events may be tragic, but they cannot be the basis for a tort lawsuit. In other words: "There is no liability without fault."[5] To succeed in a tort action, the plaintiff must prove that

- The defendant owed the plaintiff a duty of care, and
- The defendant breached that duty, and
- The plaintiff suffered loss or damage, and
- The breach was the proximate [significant] cause of the plaintiff's loss.[6]

If any of these criteria are absent, the tort lawsuit will fail. For example, a negligent act, by itself, is not a basis for a claim under tort law—the plaintiff must have suffered some damage.

Defining the Duty of Care

The term *duty of care* in the preceding discussion may require a clearer definition. Certain actions create a duty of care between people, even in the absence of a contract, and even if the people have never met. The most common example is on our highways: all drivers have a duty of care to avoid injury or damage to other drivers. A negligent driver who causes an automobile accident has breached the duty of care, and is liable for the resulting damage. A duty of care exists when an action satisfies the following two conditions:

- A reasonably foreseeable risk of injury or damage to others exists, because of the action.
- Someone is close enough to be affected by the action.[7]

In other words, a duty of care is not owed to everyone, but only to those who are likely to be injured by a potentially dangerous act. However, the scope of the duty of care has expanded over time. For example, legal precedents have extended the duty of care to include people who were not very close during a dangerous act, but who suffered damage later (or the danger did not exist or was not apparent until later). For example, poor building design, poor-quality construction, defective products, environmental pollution, and so on, may be considered dangerous, even if they do not cause damage until many years later.

Other professionals have observed a similar expansion of the duty of care. For example, a psychiatrist who fails to warn others of a dangerous psychiatric patient may be liable for damage caused by the patient. Similarly, a physician may be liable if he or she negligently allows an injured or incapacitated patient to drive an automobile. All provinces now have laws requiring medical doctors to report such patients to the vehicle licence bureau.[8]

Over the years, the law of tort has repeatedly found that engineers and geoscientists have a duty of care for their actions and decisions. When engineers design a device or structure, or when geoscientists plan a mine or drill for oil, they have a duty of care to anyone who may suffer from the harmful effects of these activities, even if the harm occurs years later. The following example demonstrates the duty of care expected from a professional person:

> A house in England was built on a garbage dump, and should have had deep foundations to avoid settling. The foundations were supposed to be inspected by the municipal inspector, but the inspector approved the foundations without inspecting them. Over a period of time, the foundations settled, and could not carry the weight of the building, which collapsed. The municipal authority who employed the inspector was held liable to a later purchaser of the house. In his 1972 judgement, the judge said: ". . . in the case of a professional man who gives advice on the safety of buildings, or machines, or material, his duty is to all those who may suffer injury, in case his advice is bad." (*Dutton v. Bognor-Regis*)[9]

Defining the Standard of Care

If it is determined that a duty of care existed, the next question in a tort case involves determining what *standard of care* was owed. In other words, was the person negligent? The courts will apply a *reasonable person* test, and ask: "What would a reasonable person do, under the circumstances?" In the engineering and/or geoscience licensing Acts, the term *reasonable* appears often in the definition of *negligence*. For example, Ontario's Professional Engineering Act defines negligence as an act or omission that "constitutes a failure to maintain the standards that a reasonable and prudent practitioner would maintain, in the circumstances."[10] Perfection is not required; however, professionals are expected to use reasonable care, established practices, and well-tested principles.

The best protection against negligence is careful, thorough, accurate work. Professionals can also obtain liability insurance to protect against the costs of negligent conduct. This insurance, typically called "errors and omissions" insurance, is a wise investment for engineers in private practice; in fact, in some provinces, it is compulsory. Professional employees are typically covered by the employer's insurance. Of course, regardless of the insurance, a negligent professional is always subject to disciplinary action by the licensing Association (as discussed in Chapter 4).

Proving Negligence

Canadian tort lawsuits require the injured party to provide evidence of negligence. However, courts may accept circumstantial evidence that negligence must have occurred, even when the exact cause cannot be defined precisely. To be more precise, the courts will usually assume that negligence has been proved when

- whatever caused the harm was under the sole control of the defendant, *and*
- the event that caused the harm ordinarily would not occur without negligence.

For example, when a surgeon leaves a sponge inside a patient during a surgical operation, there is no need to discover precisely how the sponge was left there; the surgeon was in control of the surgery, and had the surgeon not been negligent, the sponge would not have been left behind. This type of circumstantial evidence is described by the Latin phrase *res ipsa loquitur* ("the thing speaks for itself").[11]

The precedent for *res ipsa loquitur* dates back to 1863 and the lawsuit *Byrne v. Boadle*. A barrel of flour fell from a warehouse window above a shop, striking and injuring a passerby, who sued for damages. Although no evidence was produced to explain how the barrel came to fall from the window, the judge (on appeal) stated: "A barrel could not roll out of a warehouse without some negligence, and to say that a plaintiff who is injured by it must call witnesses from the warehouse to prove negligence seems to be preposterous." The judge ruled that the barrel was "in the custody" of the warehouse owner (or his employees) and that he was therefore responsible for the control of it.[12]

Relying on circumstantial evidence (using the logic of *res ipsa loquitur*) affects design engineers, because it has been applied to defective products. When an injury has resulted from a product, and when the cause is so obvious that "the thing speaks for itself," the designers must prove that negligence did not occur. In these (*res ipsa loquitur*) cases, Canadian tort law, based on negligence, becomes very similar to the strict liability in American courts.

In summary, tort law requires engineers and geoscientists (and all professionals) to avoid negligence and incompetence, and to eliminate hazards that could endanger others. Failure to do so may create liability for any resulting damage or loss.

PRODUCTS LIABILITY

We are all consumers, and we expect manufacturers to make quality goods. When a consumer purchases a defective product from a manufacturer (either directly, or through the manufacturer's agent, distributor, or retailer), the contract (either written or implicit) is the usual basis for demanding reparation. A claim for damages may be made for three reasons:

- defective manufacturing, *or*
- negligence in design, *or*
- failure to warn of dangers associated with the product.[13]

Contract Conditions and Warranties

The contract clauses must be examined closely. Contract clauses are typically divided into conditions and warranties. Usually these terms are identified in the contract.

- **Conditions.** Conditions are key clauses that must be satisfied, or the contract may be terminated. Obviously, conditions are important, since they have the potential to end the contract.[14]

- **Warranties.** Warranties are clauses that permit the consumer to demand repairs, replacement, or damages. However, a warranty clause does not permit a contract to be terminated. (The term "warranty" is usually applied to goods and products, whereas the term "guarantee" is usually applied to services or agreements.) Warranties are promises that a manufacturer makes about a product. If the product fails to meet the terms in the warranty, the manufacturer may be liable for any resulting damage.

Government Acts Regulating Products

SALE OF GOODS ACT   Every province and territory has a law—typically called a Sale of Goods Act—that may be invoked for those times when a sales contract lacks specific wording. The Sale of Goods Act defines certain conditions and warranties in order to protect the general public. For example, typical sale conditions require products to have a basic quality (or *merchantability*), as well as fitness for use. These terms imply that the product must not be defective and must be usable as intended. Thus, a lawn mower must be able to cut grass, and a refrigerator must be able to keep food cold. Regulations for the sale of goods typically apply to product quality. Recently, however, the courts have been interpreting Sale of Goods Acts to encompass safety as well.[15]

CONSUMER PROTECTION ACT   In addition, every province now has a Consumer Protection Act that imposes further provisions on consumer sales, or that prevents certain basic rights from being waived. However, this legislation is intended to protect individuals; typically, it does not apply to goods purchased for commercial use or resale.

HAZARDOUS PRODUCTS ACT   The federal Hazardous Products Act (also mentioned below under workplace hazardous materials) sets standards of safety across Canada for a wide variety of consumer products, from hockey helmets to kettles. In addition, the federal Motor Vehicle Safety Act (R.S.C. 1985, c. M-10) sets safety standards for automobile manufacturers.[16]

U.S. Products Law—Strict Liability

The concept of strict liability applies mainly in the United States. However, Canadian engineers and manufacturers must be aware of it, because the North American Free Trade Agreement (NAFTA) now permits a freer flow of products across the border into the United States. Moreover, since Canadian and American laws are both based on the British concept of established precedents, legal decisions made in one country may, over time, be applied in the other country.

Strict liability covers product defects and consumer safety. The focus is on the product itself, and no questions of negligence arise. D.L. Marston, in his text *Law for Professional Engineers*, states:

> In products liability cases in the United States, a manufacturer may be strictly liable for any damage that results from the use of his product, even though the manufacturer was not negligent in producing it. Canadian products-liability law has not yet adopted this "strict liability" concept, but the law appears to be developing in that direction.[17]

In both Canada and the United States, the requirements for care were set by a 1932 British case, *Donoghue v. Stevenson*. The judgment in this case stated in part:

> A manufacturer of products which he sells in such a form as to show that he intends them to reach the ultimate customer in the form in which they left him, and with no reasonable possibility of intermediate examination, and with the knowledge that the absence of reasonable care in the preparation or putting up of the products will result in injury to the consumer's life or property, owes a duty of care to the consumer to take that reasonable care.[18]

In the United States, the American Law Institute has published the following two-part rule for products liability, which contains the idea from the British precedent, but applies it more strictly:

1. One who sells any product in a defective condition, unreasonably dangerous to the user or consumer or to his property, is subject to liability for physical harm thereby caused to the ultimate user or consumer, or to his property, if
   (a) the seller is engaged in the business of selling such a product, and
   (b) it is expected to and does reach the user or consumer without substantial change in the condition in which it is sold.
2. The rule stated in Subsection (1) applies although
   (a) the seller has exercised all possible care in the preparation and sale of his product, and
   (b) the user or consumer has not bought the product from or entered into any contractual relation with the seller.[19]

The key difference between Canadian and American law is that the American definition specifically states that the rule applies even when the seller has taken all possible care. In other words, it applies even when no negligence can be shown. Instead of the plaintiff being required to prove negligence, the defendants must prove their product's safety and reliability. Moreover, the American law applies to sellers; as a result, everyone in the design/manufacturing/sales chain is included.

Canadian Products Law—Risk–Utility Analysis

The duty to design safe products is well established in Canada, although Canadian liability law differs somewhat from American law. American law focuses on the *reasonable expectations* of the consumer. Osborne makes the following comments regarding this test, and illustrates the *risk–utility* approach that is more commonly followed in Canada:

> [In the United States:] If the product is not as safe as a consumer might reasonably expect, the design is defective. This test has, however, proved to be very difficult to apply in a fair and predictable manner. The reasonable expectations of consumers in respect to some products may be unreasonably high and in respect of others it may be unduly low. The test also fails to consider if there is an alternative and safer design available. [A different] test, which is favoured in Canada, is a risk/utility analysis that seeks to determine if the utility of the product's design outweighs the foreseeable risks of the design.
>
> The risk/utility test was applied in *Rentway Canada Ltd. v. Laidlaw Transport Ltd.* [1989]. The case involved a head-on collision between two trucks, when both of the headlights of one of the trucks failed. The defendant had designed the lighting system of that vehicle. Flying rubber from a tread separation of a tire knocked out one headlight and, because both headlights were on the same circuit, the other one also failed. The plaintiff claimed that the two headlights should have been on independent circuits, in which case, one headlight would have remained on. The trial judge assessed the safety of the design on a risk-benefit analysis and decided that the design was defective. Consideration was given to the degree of danger arising from the design, the availability of a safer design, and the functionality, the costs, and the risks of that alternative design. The ultimate question was whether or not, in the light of these factors, the product was reasonably safe. The Court held that the danger of having both headlights on a single circuit and the availability of a functional and affordable alternative design outweighed the utility of the single circuit system used by the defendant.[20]

Generally, most lawsuits involving product liability are brought against the manufacturers and sellers of the products, and are usually based on breach of warranty or on strict liability. A lawsuit is usually brought against the design engineer only in the case of alleged negligence. However, some of these lawsuits have been enormously costly, so safety is a good investment. When the designer makes a product safer, this protects the public from harm and simultaneously protects the manufacturer from financial loss.

DESIGNING FOR SAFETY

Obtaining Codes, Standards, and Regulations

Many textbooks describe the steps in the design sequence, and suggest ways to stimulate creative thinking, such as brainstorming. However, very few design texts emphasize the critical importance of codes, standards, and safety regulations, perhaps because these are constantly changing and evolving. Regrettably, some designers may not be aware that such information is now instantly available through the Internet.

Photo 8.1 — The Transcona Silo Foundation Failure. *On October 18, 1913, the foundation for CP Rail's huge grain elevator in North Transcona, Manitoba, tilted almost 27 degrees from the vertical during its initial loading. The structure was eventually righted and still exists today. However, the accident taught a valuable lesson about bearing capacity and the effects of uneven loading.*

Source: Archives of Manitoba, Foote 1801 N2793.

As a first step in any project, designers should search the Internet for relevant codes, standards, and safety regulations. Do not rely entirely on well-known company standards and textbook references. Use the Internet to get state-of-the-art guidance. A general search using a few relevant keywords will find up-to-date information, almost instantly (usually on the websites of technical societies or standards organizations). When a document is not free, you will usually be able to purchase it by credit card and download it immediately.

Codes and standards may not seem important until accidents occur and lawsuits result. The courts usually see design codes as a "minimum mandatory standard."[21] When a designer deviates from accepted design codes, the deviation must be explained by a written analysis or by convincing design calculations. As Samuels and Sanders note, codes are intended to guide the designer, so any design that departs substantially from them is obviously "a faulty design, unless it can be demonstrated that it conforms to accepted engineering practice by rational analysis."[22]

Codes change over time, so designers must use up-to-date information, of course. But sometimes (in very rare cases), a designer must deviate from a code, such as in the following instances:

- **New information.** Recent failures, accidents, or new research studies may prove that a design code is erroneous, so designers must supplement it with further analysis, showing that their designs are safe.
- **Low industry standards.** Some industries set very low internal standards, and designers must routinely exceed the industry code. It is unprofessional for a designer to follow a code that is widely known to be deficient.
- **No code exists.** Finally, some "cutting edge" designs may be so new that no relevant code or standard may exist. In these cases, the designer must use good engineering design principles and follow the Association's Code of Ethics, which puts public health, safety, and welfare ahead of personal profit.

Including Safety in the Design Process

Designers are happy when they find feasible solutions, but the job is not over. The designer is routinely expected to optimize the design, to reduce manufacturing costs, and to eliminate unnecessary parts (also called "value engineering"). In addition, the following four steps for reducing hazards should also be routine:

ELIMINATE KNOWN HAZARDS   The Code of Ethics requires professionals to protect the public, so the first step is to eliminate obvious hazards. For example, in building design, a high walkway is obviously dangerous and the designer should include railings for safety. Designers also have an obligation to remove concealed hazards (that is, hazards that are not obvious to the user). When such a hazard cannot be eliminated, the designer must shield the user from it, if possible, and warn the user about it. For example, a lawn mower may have a hazardous rotating blade that cannot be seen by the user. The blade is needed to cut the grass, so it cannot be eliminated. Therefore, users must be shielded from the blade, and they must be clearly warned that contact with the blade may be lethal. Another example of a concealed hazard is a random flaw in a key component. For example, aircraft have some parts that, if they fail, will cause the aircraft to crash. The designer must specify that such critical parts must be tested to detect flaws and inspected regularly to ensure that old, worn, or damaged parts are replaced.

FOLLOW ESTABLISHED DESIGN STANDARDS   The designer must know and follow (or exceed) the accepted standards—whether they are required by law, or are industry standards, or are simply company design guides—unless there is a convincing analysis to justify deviating from them. As mentioned above, design standards are now easier to obtain than ever before.

FOLLOW LAWS AND REGULATIONS   Everyone must follow provincial and federal safety laws and regulations. For example, every province has Occupational Health and Safety laws to protect workers, and environmental laws to protect the environment. The designer must know (and follow) these laws and regulations. (These are now very easy to find on the Internet.)

FOLLOW GOOD ENGINEERING PRACTICE   In the absence of design standards or government regulations, the designer must simply use good engineering practice. It is reasonable to expect a designer to examine older designs (particularly if they have failed) to see what lessons can be learned, to conduct a methodical hazard analysis of a new design to find unexpected or potential dangers, and (for a complex system) to conduct a failure analysis. These techniques are described below.

HAZARD REDUCTION

Formal Hazard Reduction Methods

HAZARD ANALYSIS   A *hazard analysis* should be conducted for every design, and the designer should keep a permanent record of the analysis. (In simple designs, the analysis may be very brief.) The hazard analysis is a systematic review of the design, in which the designer ensures that the following four questions are asked and answered:

- **Identify:** Have we identified all hazards (obvious or hidden)?
- **Eliminate:** Have we eliminated the hazards (wherever possible)?
- **Shield:** Have we shielded users from hazards, where they cannot be eliminated?
- **Warn, Remedy, Recall:** Have we provided remedial action where shielding is not possible (by warning users, by recalling products for repair, by providing escape routes, and so forth)?

The analysis should consider both the hazards of normal operation, and hazards that might arise from abuse or misuse.

FAILURE ANALYSIS   A *failure analysis* examines the consequences if a single component of a large system should randomly fail. Failure analysis is intended to find which component failures could lead to a disastrous failure of the whole system. Formal failure analysis techniques are complex, so they are typically applied only to very large systems, such as electrical power plants, aircraft, or computer control systems. The two best-known methods are failure modes and effects analysis (FMEA) and fault tree analysis (FTA). In some cases, the risks (or probabilities of failure) can be estimated mathematically using reliability theory, and the design can be changed until the probability of failure is reduced to an acceptable value. Unfortunately, such complex computations are not feasible in most design situations, and the engineer's judgment must suffice. When in doubt, the engineer's bias should always lean toward increasing safety. A more detailed discussion of hazard analysis, FMEA, and FTA is included in Appendix G.

Designer's Checklist

Hazards exist in normal life, and flaws exist in materials, and even the most painstaking attempts by the most diligent design engineers will not prevent all failures. The designer is more likely to eliminate design hazards and material flaws if the following steps are added to the typical design sequence:

- Find and apply standards and regulations.
- Conduct formal design reviews.
- Carry out a formal hazard analysis.
- Carry out a formal failure analysis (when design complexity justifies this).
- Warn consumers and/or clients of hazards.
- Prepare and distribute instruction manuals.
- Use state-of-the-art design methods.
- Maintain complete design records.

Each of these points is explained at length in design textbooks[23] and is summarized in the file "Reducing Hazards: A Checklist for Designers and Manufacturers," included in Appendix G.

Manufacturer's Checklist

Manufacturers want people to buy and use their products. However, if products are unsafe (or even if consumers misuse or abuse good products), damage, injury, or death may occur.

Damage cases often end in court, where so-called "accident analysts" review the entire history of the product. Unless a safety or hazard analysis was carried out (and thoroughly documented) before the product was sold, the product will likely be judged unsafe and the manufacturer found liable. After the lawsuit, it is often evident that the problem could have been avoided if a fraction of the lawyer's fees had been spent on design safety. An old adage says that a dollar spent on design safety may save ten dollars on defence lawyers.

Manufacturers therefore have an incentive to help their engineers design safe products. The manufacturer's typical responsibilities are as follows:

- Establish safety as a company policy.
- Conduct adequate quality assurance and testing.
- Review warranties, disclaimers, and other published material for accuracy.
- Act promptly on consumer complaints.
- Inform designers of complaints and failures, so that the designer can find the causes, and incorporate changes in future designs. (This may seem obvious, but designers are not always informed properly.)
- Warn owners immediately of hazards.[24]

The Rivtow Marine case (below) explains the duty to warn users about hazards. In addition, each of the above points is explained briefly in the file "Reducing Hazards: A Checklist for Designers and Manufacturers," included in Appendix G.

CASE HISTORY 8.1

THE RIVTOW MARINE CASE: FAILURE TO WARN

This case history illustrates the legal necessity to warn of hazards. Rivtow Marine, a British Columbia logging company, chartered a logging barge from a British Columbia dealer named Walkem. The barge was fitted with a crane manufactured by Washington Iron Works, an American manufacturer. Washington had (earlier) constructed a similar crane, which had collapsed, killing the crane operator. When Rivtow Marine learned of that collapse, Rivtow stopped operating its own crane and inspected it. Serious cracks were found. These indicated that the Rivtow crane might soon collapse, so it was withdrawn from service. The barge and crane stood idle for some time, in the middle of a busy log-harvesting season, while repairs were made.

It was later learned that the dealer (Walkem) and the manufacturer (Washington Iron Works) had both been aware for some time of cracks on cranes of this type, yet neither had informed Rivtow Marine. So Rivtow sued both, alleging negligence for failing to provide a warning, and claiming damages for the cost of repairing the crane and for economic losses while the barge and crane were idle. The trial was appealed to the BC Appeal Court and then to the Supreme Court of Canada.

In a unanimous ruling, the Supreme Court awarded damages on the basis of negligent failure to warn. The defendants (Walkem and Washington) had "knowledge of the risk": they knew the business that Rivtow operated, they were aware that the crane was inadequate, they knew it was the busy season, and they knew what harm could arise from having the barge and crane out of operation. Furthermore, the loss was a "direct and foreseeable" consequence of the inadequacy of the crane, and not a remote or unforeseeable result, which might not have justified award of damages.

No personal injury occurred to Rivtow personnel; however, the failure to warn them caused economic loss. The Rivtow case is a key part of Canadian tort law, because, prior to this case, economic loss was generally considered to be too remote to justify an award. The Rivtow case has been cited and applied in many recent judgments.[25]

NATIONAL STANDARDS: THE STANDARDS COUNCIL OF CANADA (SCC)

Standards are useful publications, particularly for anyone involved in design, manufacturing, or professional practice. Curiously, we in Canada have come to expect products, services, machinery, and equipment to work properly and safely, so standards are most noticeable when they are absent. That is, when we are offered products or services that are unsafe, unreliable, inefficient, or are incompatible with other equipment, we recognize the importance of standards. Standards define dimensions, tolerances, strengths, voltages, computer protocols, and hundreds of other measurable factors for manufacturing high-quality items. In addition, many standards explain how to test, manage, or control quality. Standards are useful guides for the design, manufacturing, control, and operation of almost anything.

Fortunately, the Internet makes it easier to find standards than ever before, and you can usually download them immediately. A good starting point in a standards search is the Standards Council of Canada (SCC). The following information is taken from the SCC website, with permission:

The Standards Council of Canada: The Standards Council of Canada is a federal Crown corporation with the mandate to promote efficient and effective standardization. Located in Ottawa, the Standards Council has a 15-member governing Council and a staff of approximately 90 [people]. The organization reports to Parliament through the Minister of Industry, and oversees Canada's National Standards System. . . .

The National Standards System: The Standards Council has the mandate to coordinate and oversee the efforts of the National Standards System, which includes organizations and individuals involved in voluntary standards development, promotion and implementation in Canada. . . . Standards affect nearly every product or service we encounter in our daily lives. In Canada alone there are several thousand national standards, for everything from AC meters to zirconium.

Developing, maintaining and implementing that many standards is too big a job for any one organization. It requires a whole system—the National Standards System. For more than 25 years, the National Standards System has helped to ensure the safety and performance of products and services, helped to open the global marketplace to Canadians, and made Canada a leader in international standardization.

More than 15,000 Canadian members contribute to committees that develop national or international standards.

As well, more than 400 organizations have been accredited by the Standards Council. Some of these develop standards; others are conformity assessment bodies, which determine the compliance of products or services to a standard's requirements. . . .

Accredited standards development organizations may submit their standards for approval as National Standards of Canada. This designation indicates that a standard is the official Canadian standard on a particular subject. It also shows that the development process met certain specified criteria. National Standards of Canada may be developed in Canada or adopted, with or without changes, from international standards. . . .

International Standardization: Internationally, SCC manages Canada's participation in International Organization for Standardization (ISO) and the International Electrotechnical Commission (IEC), two of the world's most important voluntary standardization bodies. . . .[26]

The Standards Council of Canada performs many more tasks than this brief summary shows. For example, a recent SCC report summarizes Canadian laws (federal, provincial, and territorial) that apply to products used in the home by the typical consumer.[27] The report lists the applicable federal and provincial laws for packaging and labelling, textile labelling, hazardous products, energy efficiency, radio and telecommunications products, electrical safety, motor vehicles, components, and accessories to be sold through retail outlets. (It does not cover products such as food, drugs, medical products, and products intended for industrial or commercial applications.)

More information on their programs and services is available from their website at <www.scc.ca> (valid as of June 1, 2008).

INTERNATIONAL STANDARDS: THE INTERNATIONAL ORGANIZATION FOR STANDARDS (ISO)

International standards are particularly important for trade, since they help Canadian products to enter foreign markets. However, creating independent standards in every country would be a massive duplication of effort. When international standards are already established, they should usually be adopted.

The International Organization for Standards (ISO) was founded in 1947 with the mandate to "standardize the standards" among countries, and make them more available. By 1996, more than one hundred member countries were participating in its standards activities, including Canada, Britain, and the United States. In its present form, ISO is a network formed by the national standards institutes of 155 countries (with one member per country). The Standards Council of Canada (SCC) is Canada's representative. The ISO Central Secretariat, located in Geneva, Switzerland, coordinates the system. However, ISO is a non-governmental organization; in other words, each member represents his or her national standards organization, not the national government.

Whenever a new standard is proposed, ISO brings together a technical committee comprising experts from the various member countries. Typically, each nation that participates on the technical committee sets up an advisory group composed of experts from within its own borders. This group then generates a national consensus regarding the proposed standard.

The proposed standard must pass through three drafts. At each draft, members of the technical committee propose differing opinions and alternative wordings. The member countries then vote on the standard. If the final draft standard receives a two-thirds positive vote, it becomes an ISO standard and is translated into the three official ISO languages: English, French, and Russian. Each country can take a further step and adopt the ISO standard as a national standard, and publish the standard in the language of that country.

ISO has developed over 16,000 International Standards (as of 2006). These ISO standards may be searched through the Internet, and the title listings include a brief abstract (in many cases). (Note: Paper copies of ISO standards are usually more easily available through the Standards Council of Canada.) The ISO website is located at <www.iso.org> (valid as of May 31, 2008).

ISO 9000 and ISO 14000 Standards

ISO standards usually apply to specific products, but two standards are different: the ISO 9000 and 14000 series of "generic management system standards" apply to the management of an organization, rather than the products produced. The ISO 9000 and ISO 14000 families are among ISO's best known standards, ever. ISO 9001:2000 and ISO 14001 are implemented by almost 900,000 organizations in 161 countries.[28] These standards may be used by any organization, large or small, and are described briefly below.

ISO 9000—QUALITY MANAGEMENT AND QUALITY ASSURANCE STANDARDS

The ISO 9000 series of standards for managing a manufacturing corporation is intended to maximize the quality of manufactured products. The standard is very effective and has been widely adopted. Between 1987 (when the first version of ISO 9000 was released) and 1996, more than 100,000 corporations obtained ISO 9000 certification. The automotive industry now expects every supplier to be certified to ISO 9000 standards. ISO 9000 certification is the dominant quality management certification system in the world. It is estimated that an investment in ISO 9000 certification usually pays for itself within three years through increased productivity and reduced scrap.[29]

The ISO 9000 standard is very comprehensive. It requires a corporation to examine almost every aspect of its management, design, purchasing, inspection, testing, handling, storage, packaging, preservation, delivery, and documentation systems. Improving the quality of these systems enables effective evaluation of the manufacturing process and shows where quality improvements are required. An important part of ISO 9000 involves developing a "quality manual" to document the four key aspects of the certification process. This manual documents the following:

- Quality policies for every aspect of the corporation's operations.
- Quality assurance procedures, which involve 20 clauses in the ISO 9000 standard.
- Quality process procedures (or practices, or instructions), which include all of the company's production processes.
- Quality proof: a repository for all of the forms, records, and other documentation that give objective evidence—or proof—that the quality system is operating properly.

The ISO 9000 quality management system permits and encourages "certification" (internal and external audits). Certification is like an audit in a financial control system, where audits are established, accepted, and routine. Typically, to ensure impartiality, independent quality auditors, or "registrars," carry out the certification audit. Every aspect of a company's operations is examined in detail: fifteen-step processes, which may take more than a year to implement, are common. Re-certification audits should be carried out every third year.[30]

It is important to stress that certification is not necessary to use the ISO 9000 quality management system, but it provides an extra level of assurance. ISO 9000 is now the accepted world standard for quality management.

ISO 14000—ENVIRONMENTAL MANAGEMENT SYSTEMS

In view of the success of the ISO 9000 series of standards for quality management, many companies also adopted the ISO 14000 series of standards for environmental management. ISO 14000 was developed using the international consensus procedure (as for all ISO standards) and is very similar to ISO 9000 certification. The goal of the ISO 14000 standard is to minimize the company's harmful effects on the environment.

The ISO 14000 process requires the company to examine every function of its operations with the goal of identifying activities with a significant environmental impact; it then commits the company to preventing pollution in all of its forms. The standard does not set acceptable environmental levels—that is left to regulatory agencies. However, the standard does require that these environmental levels be determined and followed.

Monitoring and measurement are, of course, essential. Procedures for corrective action and emergency response are also required. Each of these activities may require setting performance criteria, defining responsibilities, assigning duties, providing training, and ensuring adequate communication. ISO 14000 does not require developing an environmental management manual. However, most companies would probably prepare it.

The ISO 14000 series was released in 1996, and many major companies immediately committed to implementing the standard.[31] Other standards in the ISO 14000 series concern environmental aspects of product standards, environmental auditing, environmental assessment of sites, environmental labels, environmental performance evaluation and life-cycle assessment, to mention only a few. The ISO 14064 standard for greenhouse gas accounting and verification was published in March 2006. ISO 14064 helps government and industry to manage programs for reducing greenhouse gas emissions and emissions trading.[32]

GOVERNMENT CODES AND STANDARDS

Designers use standards because it is good professional practice, but sometimes federal and provincial laws require the use of certain standards, and even some municipal bylaws may specify standards. Standards are cited in laws whenever the government wants to guarantee quality or uniformity. For example, laws typically require specific standards to be followed, to ensure that

- warning and danger signs have standard size, wording, and symbols,
- electric or electronic equipment is safe for public use,

- materials used in government contracts are of proper quality,
- public spaces are suitable for disabled access, and
- buildings, roads, water, sewage, and similar utilities are of adequate quality.

When the law specifies a standard, you must find and follow it. The most common government-specified standards are the building codes, discussed below. Note that "code" is virtually interchangeable with "standard." The difference is subtle: codes usually describe how something is constructed, and standards usually apply to the materials themselves. Therefore, codes often refer to standards, but both serve the same purpose: to improve quality, safety, and reliability.

BUILDING AND CONSTRUCTION CODES   Construction codes are often specified in laws. The National Research Council's Institute for Research in Construction (IRC) publishes these codes. For example, the National Building Code was developed in 1941 to consolidate the patchwork of provincial and municipal codes that existed across Canada in the 1930s. The code was so successful that it has been maintained ever since, and is revised every five years. IRC now publishes several other construction codes and user's guides (most of which are available on CD-ROM), including the following:

- National Building Code of Canada
- National Fire Code of Canada
- National Plumbing Code of Canada
- National Farm Building Code of Canada
- National Housing Code of Canada
- Model National Energy Code of Canada for Houses
- Model National Energy Code of Canada for Buildings

Building construction is under provincial control, of course, so the national codes are "model codes." However, all provincial governments have adopted these codes, or adopted them with modifications, or developed provincial codes based on the national codes. They are widely available from bookstores and the following websites: National Research Council <www.nationalcodes.ca>; Federal Publications Inc. <www.fedpubs.com>; Standards Council of Canada <www.scc.ca> (valid as of May 31, 2008).

OCCUPATIONAL HEALTH AND SAFETY (OHS) LAWS

Occupational Health and Safety laws protect workers from unsafe work conditions by requiring employers to follow safety regulations. In addition, Canada's Criminal Code now imposes criminal penalties when workplace safety violations result in injury or death. (The Criminal Code amendment was a consequence of the Westray mining disaster, discussed later in this chapter.)

These laws make the employer responsible for workplace safety, and require them to hire engineers or geoscientists to certify the safety of equipment or processes. To fulfill this important responsibility, engineers and geoscientists must be familiar with the relevant Occupational Health and Safety laws and regulations. These vary slightly across Canada:

> . . . Each of the ten provinces, three territories and the federal government has its own OSH [occupational safety and health] legislation. The federal government has responsibility for the health and safety of its own employees and federal corporations, plus workers in certain industries such as inter-provincial and international transportation (e.g., railways and air transport), shipping, telephone and cable systems, etc. Approximately 10% of the Canadian workforce falls into the federal jurisdiction. The remaining 90% of Canadian workers fall under the legislation of the province or territory where they work.[33]

Fortunately, a federal agency helps to make sense of this flood of information. The Canadian Centre for Occupational Health and Safety (CCOHS) provides direct Internet links to all provincial, territorial, and federal health and safety legislation through the CCOHS website, and provides an immense amount of safety advice, training, publications, and links to hazard databases. An auxiliary website, known simply as CanOSH, organizes laws by region; users can obtain the safety laws for any Canadian jurisdiction easily and directly.

To review the safety laws and regulations for your province, territory, or the federal government, contact either the CCOHS or CanOSH websites: Canadian Centre for Occupational Health and Safety (CCOHS), <www.ccohs.ca>; Canadian National Occupational Health and Safety (CanOSH) <www.canoshweb.org> (valid as of June 1, 2008).

Occupational Health and Safety Regulations

The Occupational Health and Safety (OHS) laws state that every Canadian is entitled to a safe and healthy work environment, and that the employer has a duty to provide it. The practical rules for workplace safety are usually found in the regulations made under the authority of the OHS law. The regulations are easily found through the CCOHS or CanOSH websites (above). Regulations are typically available free in a computer-searchable (pdf) format,[34] although some regulations are also available as an inexpensive pocket handbook.[35]

EMPLOYEES' RIGHTS
Federal, provincial, and territorial OHS laws typically give employees three basic rights:

- **The right to know.** Employees must be informed of workplace hazards, and must be properly trained to operate or manage dangerous machinery, equipment, processes, or substances.

- **The right to refuse dangerous work.** Employees have the right to refuse dangerous work without the risk of discipline or dismissal. Under specified circumstances, certain members of the health and safety committee can intervene to stop dangerous work.
- **The right to participate.** Employees have the right to participate in making the workplace safer through workplace health and safety committees.

EMPLOYERS' DUTIES

OHS laws place the responsibility for workplace safety firmly on the employer. The employer must ensure that the workplace is safe; must provide any needed protective devices, equipment, or materials; must ensure that they are used as prescribed by the regulations; and ensure that they are maintained in good condition. Specific guidance is given by comprehensive OHS regulations, which require many dangerous situations to be investigated and certified as safe by professional engineers or geoscientists. This means that professional engineers or geoscientists assume the responsibility for evaluating safety. Obviously, familiarity with OHS regulations is essential.

A typical list of topics in the OHS regulations is shown in Table 8.1, which gives only main headings. (The full regulations are typically 400 or 500 pages long.) As Table 8.1 shows, the OHS regulations establish the workplace health and safety committee and, under 40 or 50 sub-sections, give specific guidance from general safety procedures to those required in industries with specialized hazards. Each employer, employee and professional must be aware of the regulations that apply to their workplace.

CONTRACTS AND PRIME CONTRACTORS

Owners often hire contractors for specific projects. For example, an owner may hire a builder to construct a building. Contracts should specify clearly who is responsible for public safety and employee safety on the worksite. This person is called the *prime contractor* (or, in some provinces, the *constructor*). The prime contractor must comply with all the OHS regulations on the worksite as if the prime contractor were the employer. It is very important to ensure that the prime contractor is designated in the contract, because if none is designated (or if more than one is designated), then the responsibility for occupational health and safety falls back on the owner.[36]

ENFORCEMENT

The workplace health and safety committee is required to inspect the workplace regularly (typically monthly) to ensure safety and report potential hazards. In addition, government inspectors employed by the labour ministry visit workplaces to ensure that OHS regulations are being followed. Inspectors have extensive powers under the OHS law to enter and inspect workplaces, examine documents, test equipment, and so on. Where an inspector finds the OHS Act or an OHS regulation is being contravened, the inspector may order

TABLE 8.1 — Occupational Health and Safety Regulations

This table shows typical headings in OHS regulations.

Basic Requirements
- Definitions
- General Safety Precautions
- Joint Workplace Health and Safety Committee
- Hazard Assessment, Elimination and Control
- First Aid
- Emergency Preparedness and Response
- Workplace Hazardous Materials Information System (WHMIS)
- Specifications and Certifications

Personal Protection Requirements
- Personal Protective Equipment: Eye Protection; Flame Resistant Clothing; Foot Protection; Head Protection; Life Jackets and Personal Flotation Devices; Limb and Body Protection; Respiratory Protective Equipment.
- Toilets and Washing Facilities
- Noise Exposure
- Radiation Exposure
- Lifting and Handling Loads
- Violence
- Working Alone

Common Workplace Protection Requirements
- Fall Protection
- Ventilation Systems
- Entrances, Walkways, Stairways and Ladders
- Confined Spaces
- Chemical Hazards, Biological Hazards and Harmful Substances
- Tools, Equipment and Machinery
- Safeguards, Barriers and Shields
- Overhead Power Lines
- Locking Out Dangerous Equipment for Servicing
- Rigging (cables, wire rope, etc., for lifting)

Requirements for Industries with Specialized Hazards
- Forestry
- Oil and Gas Wells
- Residential Roofing
- Tree care operations
- Health Care and Industries with Biological Hazards
- Fire and Explosion Hazards (including welding vehicles and pipelines)
- Demolition
- Diving Operations
- Excavating and Tunnelling

Requirements for General Industrial Equipment
- Cranes, Hoists and Lifting Devices: Cantilever Hoists; Chimney Hoists; Hand-Operated Hoists; Material Hoists; Mobile Cranes and Boom Trucks; Overhead Cranes.
- Personnel Hoists: Roofer's Hoists; Tower and Building Shaft Hoists; Tower Cranes; Underground Shaft Hoists; Vehicle Hoists; Winching Operations.
- Powered Mobile Equipment: Forklift Trucks; All-Terrain Vehicles; Snow Vehicles; Pile Driving Equipment.
- Scaffolds and Temporary Work Platforms: Elevating Platforms; Aerial Devices.
- Explosives: Handling Explosives; Drilling; Loading; Firing; Destroying Explosives; Specific Blasting Activities.

Requirements for Mining
- Division 1—General: Fire Prevention and Emergency Response; Electrical Systems; Rubber-Tired, Self-Propelled Machines; Diesel Power; Conveyors.
- Division 2—Explosives: Transportation; Operational Procedures; Un-detonated or Abandoned Explosives; Blasting Machines and Circuits; Surface Mines; Underground Mines and Tunnels.
- Division 3—Underground Coal Mines: Mine Workers; Mine Equipment; Vehicles; Roof and Side Support; Ventilation System; Gas and Dust Control; Explosion Control.

the owner or prime contractor (or whoever is in charge of the workplace) to comply with the OHS Act or regulations within a specified time. Failure to comply with OHS laws or regulations may lead to prosecution in the courts and, upon conviction, fines and/or imprisonment.

ACCIDENT INVESTIGATION

Where an accident causing death or critical injury occurs on a worksite, the owner (or prime contractor) is required to render first aid, conserve the accident scene, and notify the labour ministry and the workplace health and safety committee. The labour ministry (or a related government agency) has authority under OHS laws to investigate workplace accidents, and employers must assist such investigations. Accident investigations may be regulated by the OHS Act, or under a separate law, such as the Workers' Compensation Act (discussed below).

Workers' Compensation Act

Each province and territory has a law, usually called the Workers' Compensation Act, that establishes a Workers' Compensation Board (WCB), although the name or law may vary in some jurisdictions. (For example, Ontario calls it the Workplace Safety and Insurance Board, and British Columbia combines both OHS and WCB legislation in a single Act). Workers' compensation is, in simple terms, a form of "no-fault" workers' insurance.

The Workers' Compensation Act typically requires employers to support the Workers' Compensation Board, and each employer is assessed (or taxed) as a proportion of their payroll. The funds so collected are typically called the "Accident Fund," and are used to pay wage, medical, disability, and related benefits to workers who are unable to work because of a workplace injury. That is, an employee injured in the workplace may collect compensation from the WCB without having to sue the employer. (In fact, the WCB pays even when the employer is insolvent.) In return for financing the WCB, employers are shielded from employee lawsuits.[37]

In addition, should a worker die because of a workplace injury, the worker's dependants may be entitled to receive benefits. In some provinces, the WCB administers both the Accident Fund and the Occupational Health and Safety laws.

Generally speaking, each Workers' Compensation Act authorizes the Workers' Compensation Board to require employers to report any accident that involves a serious injury or death, or a major structural failure, or the release of a hazardous substance. This Board has the authority to inspect workplaces, investigate incidents, require employers to improve worker safety, and impose penalties.

The CanOSH and CCOHS websites, cited above, are also good sources of information about Workers' Compensation Acts.

WORKPLACE HAZARDOUS MATERIALS INFORMATION SYSTEM (WHMIS)

Whenever hazardous materials are used in the workplace, they are under strict control by both federal and provincial (or territorial) OHS laws. Workers have a right to know if they are working with hazardous substances and to be assured that they are properly protected. In addition, the federal Hazardous Products Act and the Controlled Products Regulation (made under the Act) apply to material suppliers, importers, and distributors. These laws define which materials (or controlled products) are included in the Workplace Hazardous Materials Information System (WHMIS), and what information suppliers must provide to employers for controlled products used in the workplace. Suppliers who sell or import a controlled product for the workplace must provide a Material Safety Data Sheet (MSDS) for the controlled product. They must also ensure that either the controlled product or its container is labelled with all required information and hazard symbols.

WHMIS is an unprecedented example of cooperation among federal, provincial, and territorial governments, and many sources of information are available. Health Canada, a federal department, hosts a website devoted to WHMIS, with links to each provincial and territorial department that regulates hazardous materials. If you are involved with hazardous substances, check out these websites: Health Canada WHMIS site: <www.hc-sc.gc.ca/ewh-semt/occup-travail/whmis-simdut/index_e.html>; Government of Canada (federal Act and Regulations): Hazardous Products Act, ch. H-3 <http://laws.justice.gc.ca/en/h-3/text.html> (valid as of June 1, 2008).

CASE HISTORY 8.2

THE WESTRAY MINE DISASTER: FAILURE TO FOLLOW SAFETY STANDARDS

This case history shows how a disaster can result when managers sacrifice safety standards to maximize profit. Twenty-six men died when the Westray mine exploded—one of the worst industrial tragedies in Canadian history. As the Inquiry later noted: "Westray was an accident waiting to happen."

Introduction

Mines are dangerous, and coal mines are the most dangerous mines, because the rock is soft and because coal dust and methane explode if the mine is poorly ventilated. The first major Canadian mine disaster happened in 1873, and thousands of lives have been lost since then. In Springhill, Nova Scotia, 424 miners were killed in the mines between 1881 and 1969. Canada's worst coal mine disaster occurred in June 1914 in Hillcrest, Alberta, when an explosion killed 189 men.[38] As the years passed, people came to believe that modern ventilating, monitoring, and excavating methods made coal mines

Photo 8.2 — The Westray Mine. *The Westray coal mine explosion in 1992 killed 26 miners. In his report on the ill-fated mine, Justice Peter Richard blamed the coal company and the provincial government for the disaster, saying that the Westray mine operations were a "violation of the basic and fundamental tenets of safe mining practice."*

Source: CP/Andrew Vaughan.

safe. This belief was shattered on May 9, 1992, when an explosion killed 26 miners at the Westray mine in Plymouth, Nova Scotia. The Inquiry into this disaster revealed a "complex mosaic of actions, omissions, mistakes, incompetence, apathy, cynicism, stupidity, and neglect."[39]

Details of the Explosion

The Westray mine explosion occurred at 5:20 a.m. on a Saturday morning. The shaking of the earth was felt by most of the residents of Plymouth. Within hours, mine rescue experts had assembled from neighbouring towns. With oxygen tanks on their backs, they descended into the destroyed mine. It was soon clear that rescue efforts would be pointless—the explosion had killed everyone below ground. Nor could all the dead be retrieved. Ten dead miners are permanently entombed behind rock falls in the mine, much of which was flooded to prevent further explosions.

The Westray Inquiry

Within days of the tragedy, anecdotes about unsafe practices were widely reported. One miner described several infractions of the safety regulations: acetylene torches had been used in areas where methane levels could be dangerous; a supervisor had tampered with a methane level monitor to permit higher methane levels; and potentially explosive coal dust had accumulated so thickly that some machinery could not be operated. For fear of retaliation or intimidation, the miners rarely complained about the safety infractions, especially since management seemed to place production ahead of safety.[40]

In the midst of bitter accusations, the provincial government appointed Justice K. Peter Richard to carry out a far-reaching Inquiry into how and why the 26 miners died. Shortly afterward, the Royal Canadian Mounted Police opened a criminal investigation. In October 1992, the Nova Scotia Labour Department laid 52 non-criminal charges of unsafe practices against Curragh Resources, Inc., the company that owned the mine. These safety charges were later dropped to avoid jeopardizing the police investigation, which resulted in charges of manslaughter and criminal negligence being laid against Curragh Resources and two of its managers. These charges were later "stayed" (or effectively dropped). On appeal, the Supreme Court of Canada upheld an order for a new trial, but prosecutors decided that the evidence was insufficient to proceed, although the average person might think otherwise. No one was ever criminally prosecuted.

Throughout the court proceedings, the Westray Inquiry continued. Justice Richard's final report, *The Westray Inquiry: A Predictable Path to Disaster*, was published in December 1997. Justice Richard commented: "Westray is a stark example of an operation where production demands resulted in the violation of the basic and fundamental tenets of safe mining practice."[41] The following paragraphs constitute a synopsis of the key facts that the Inquiry brought to light. They are excerpted from the report's Executive Summary:

Prelude to the Tragedy

. . . In the rush to reach saleable coal, workers without adequate coal mining experience were promoted to newly-created supervisory positions. Westray did not train workers in safe work methods or in recognizing dangerous roof conditions—despite a major roof collapse in August. Basic safety measures were ignored or performed inadequately. Stonedusting, for example, a critical and standard practice that renders coal dust non-explosive, was carried out sporadically by volunteers on overtime following their 12-hour shifts. . . .

It is clear that the company was derelict in carrying out its obligations for training. . . . Quite simply, management did not instil a safety mentality in its workforce. Although it stressed safety in its employee handbook, the policy it laid out there was never promoted or enforced. Indeed, management ignored or encouraged a series of hazardous or illegal practices, including having the miners work 12-hour shifts, improperly storing fuel and refuelling vehicles underground, and using non-flameproof equipment underground in ways that violated conditions set by the Department of Labour—to mention only a few. Equipment fundamental to a safe mine operation—from the cap lamp to the environmental monitoring system—did not function properly.

It was equally clear that the Department of Labour was derelict in its duty to enforce the requirements of the two acts.

The Explosion: an Analysis of Underground Conditions

. . . [V]entilation is the most crucial aspect of mine safety in an underground coal mine. Methane fires and explosions cannot happen if the gas is kept from accumulating in flammable and explosive concentrations. . . . One of the principal functions of a ventilation system is to clear the methane at the working face of the mine and to exhaust it from the mine in non-explosive concentrations. It is clear that the Westray ventilation system was grossly inadequate for this task. It is also clear that the conditions in the mine were conducive to a coal-dust explosion. . . . The consensus of the experts suggests strongly that Westray was an accident waiting to happen. . . .

Responsibility

As the evidence emerged during this Inquiry, it became clear that many persons and entities had defaulted in their legislative, business, statutory, and management responsibilities. . . . [T]here is a clear "hierarchy" of responsibility for the environment that set the stage for 9 May 1992, and we ought not to lose sight of this hierarchy.

The fundamental and basic responsibility for the safe operation of an underground coal mine, and indeed of any industrial undertaking, rests clearly with management. The internal responsibility system merely articulates this responsibility and places it in context. Westray management, starting with the chief executive officer, was required by law, by good business practice, and by good conscience to design and operate the Westray mine safely. Westray management failed in this primary responsibility, and the significance of that failure cannot be mitigated or diluted simply because others were derelict in their responsibility.

The Department of Labour through its mine inspectorate must bear a correlative responsibility for its continued failure in its duty to ensure compliance with the *Coal Mines Regulation Act* and the *Occupational Health and Safety Act*. . . .

Compliance with the Coal Mines Regulation Act

Much has been said throughout this Inquiry about the inadequacy of the Coal Mines Regulation Act. As outdated and archaic as the present act is, it is painfully clear that this disaster would not have occurred if there had been compliance with the act. . . .

If the mine had been "thoroughly ventilated and furnished with an adequate supply of pure air to dilute and render harmless inflammable and noxious gases," then . . . the 9 May 1992 explosion could not have happened, and 26 miners would not have been killed. Compliance with these sections of the

Coal Mines Regulation Act was the clear duty of Westray management, from the chief executive officer to the first-line supervisor. To ensure that this duty was undertaken and fulfilled by management was the legislated duty of the inspectorate of the Department of Labour. Management failed, the inspectorate failed, and the mine blew up.[42]

The federal government heeded the lessons of the Westray disaster. As a result of the failure to convict anyone for the obvious negligence in this case, the Canadian Criminal Code was amended to hold corporations and individuals criminally responsible if they fail to provide a safe work environment. Section 217.1 was inserted into the Criminal Code:

> 2.17.1 Every one who undertakes, or has the authority, to direct how another person does work or performs a task is under a legal duty to take reasonable steps to prevent bodily harm to that person, or any other person, arising from that work or task.[43]

DISCUSSION TOPICS AND ASSIGNMENTS

1. This chapter suggests that the first step in a design project should include a search of the Internet for appropriate technical codes and standards. As an exercise, use the Internet to find at least one design standard for each of the following items: automobile tail-lights, elevators or escalators, buildings, pressure vessels, snowmobiles, children's toys, and the Canadian flag.

2. Using the Internet, obtain the OHS Act for your province or territory. (Try the CCOHS website above, or <www.canoshweb.org/en/legislation. html>. The Act may be included with the regulations or with WCB legislation, in some provinces.) Answer the following questions and where appropriate, quote the section in the Act:

   a. What is the precise name and website for the OHS Act?
   b. How does the Act define an "owner" and a "prime contractor" (also called a "principal contractor" or "constructor")?
   c. What is the stated purpose (or "object") of the Act?
   d. What is the maximum fine or penalty for contravening the Act?

Additional assignments can be found in Appendix E.

NOTES

[1]   B. M. Samuels and D.R. Sanders, *Practical Law of Architecture, Engineering, and Geoscience*, Canadian Edition, Pearson Prentice Hall, Upper Saddle River, NJ, 2007.
[2]   D. L. Marston, *Law for Professional Engineers: Canadian and International Perspectives*, 3rd ed., McGraw-Hill Ryerson, Whitby, ON, 1996.
[3]   A. M. Linden, *Canadian Tort Law*, Butterworths Canada, Markham, ON, 1993, p. 536.
[4]   P. H. Osborne, *The Law of Torts*, Irwin Law, Toronto, 2000, p. 8.
[5]   Ibid., p. 9.

[6]   Samuels and Sanders, *Practical Law of Architecture,* p. 130.
[7]   M. Kerr, J. Kurtz, and L.M. Olivo, *Canadian Tort Law in a Nutshell,* Carswell, Toronto, ON, 1997, p. 34.
[8]   J. D. Weir and S.A. Ellis, *Critical Concepts of Canadian Business,* Addison-Wesley, Don Mills, ON, 1997, p. 137.
[9]   Marston, *Law for Professional Engineers,* p. 46.
[10]   Regulation 941/90 under the *Professional Engineers Act,* RSO 1990, c. P.28, s. 72, available at <www.peo.on.ca> (June 1, 2008).
[11]   Weir and Ellis, *Critical Concepts of Canadian Business,* p. 139.
[12]   Linden, *Canadian Tort Law,* p. 215.
[13]   Weir and Ellis, *Critical Concepts of Canadian Business,* p. 161.
[14]   Ibid., p. 69.
[15]   Ibid., p. 73.
[16]   *Hazardous Products Act* (R.S., 1985, c. H-3) and *Motor Vehicle Safety Act* (1993, c. 16) Government of Canada, Department of Justice, available at <laws.justice.gc.ca> (May 19, 2008).
[17]   Marston, *Law for Professional Engineers,* p. 46.
[18]   Ibid., p. 51.
[19]   D. W. Noel and J.J. Philips, *Products Liability,* 2nd ed., West Publishing, Saint Paul, MN, 1981.
[20]   P.H. Osborne, *The Law of Torts,* 3rd ed., Irwin Law, Toronto, 2007, p. 139–140.
[21]   Samuels and Sanders, *Practical Law of Architecture,* p. 149.
[22]   Ibid., p. 150.
[23]   G. Voland, *Engineering by Design,* Addison Wesley, Don Mills, ON, 1999; J. Kolb and S.S. Ross, *Product Safety and Liability: A Desk Reference,* McGraw-Hill, New York, NY, 1980; Consumer Products Protection Commission (CPSC), *Handbook and Standard for Manufacturing Safer Consumer Products,* U.S. Government document, 1977; D.L. Goetsch and S.B. Davis, *Understanding and Implementing ISO 9000 and ISO Standards,* Prentice-Hall, Toronto, 1998; and G.C. Andrews, J.D. Aplevich, R.A. Fraser, and C. MacGregor, *Introduction to Professional Engineering in Canada,* 3rd ed., Pearson Education Canada, Inc., Toronto, ON, 2008.
[24]   C. O. Smith, "Products Liability: Severe Design Constraint," in *Structural Failure, Product Liability and Technical Insurance,* Proceedings, 2nd International Conference, July 1–3, 1986, Interscience Enterprises, Geneva, 1987, pp. 59–75.
[25]   Linden, *Canadian Tort Law,* p. 387; Marston, *Law for Professional Engineers,* p. 55.
[26]   Standards Council of Canada (SCC), 270 Albert Street, Suite 200, Ottawa ON K1P 6N7, SCC website at <www.scc.ca> (May 19, 2008). Excerpts reproduced with permission.
[27]   Standards Council of Canada (SCC), *Consumer Product Safety Legislation in Canada: An Introductory Guide,* Ottawa, ON, March 2003, available at <www.scc.ca> (May 19, 2008).
[28]   International Organization for Standardization (ISO) website at <www.iso.org/iso/iso_catalogue/management_standards/iso_9000_iso_14000.htm> (May 19, 2008).
[29]   Goetsch and Davis, *Understanding and Implementing ISO 9000,* p. 150.
[30]   Ibid., p. 151.
[31]   S. L. Jackson, *The ISO 14000 Implementation Guide,* John Wiley & Sons, Toronto, 1997, p. 1.
[32]   ISO website at <www.iso.org/iso/home.htm> (May 19, 2008).
[33]   "Jurisdictions," Canadian National Occupational Health and Safety website (CanOSH) at <www.canoshweb.org/en/map.html> (May 19, 2008).
[34]   *Occupational Health And Safety Code, 2006,* Alberta, available at <employment.alberta.ca> (June 1, 2008).

[35] *Pocket Ontario OH&S Act & Regulations 2008*, Ontario, available at <www.carswell.com> (June 1, 2008).

[36] Samuels and Sanders, "Contracts," *Practical Law*, p. 246.

[37] Samuels and Sanders, "Workers' Compensation," *Practical Law*, p. 246.

[38] H.A. Halliday and J. Joegg, "Mining Disasters," *The Canadian Encyclopedia*, available at <www.thecanadianencyclopedia.com> (June 1, 2008).

[39] Justice K. Peter Richard, "Executive Summary," *The Westray Story: A Predictable Path to Disaster*, Report of the Westray Mine Public Inquiry, published on the authority of the Lieutenant Governor in Council, Province of Nova Scotia (December 1, 1997). Copyright by the Province of Nova Scotia, 1997. The report is available on the Internet at <www.gov.ns.ca/lwd/pubs/westray> (June 1, 2008). Excerpts reproduced with permission from the Nova Scotia Department of Justice.

[40] M. MacIssac, "Miners Testify at Westray," *Maclean's,* January 29, 1996. See also "Westray Verdict," *The Canadian Encyclopedia*, available at <www.thecanadianencyclopedia.com> (June 1, 2008).

[41] Justice Richard, "Executive Summary."

[42] Ibid.

[43] Canadian Centre for Occupational Health and Safety (CCOHS), "Bill C-45—Overview," available at <www.ccohs.ca/oshanswers/legisl/billc45.html> (May 30, 2008).

Chapter 9

Computers, Software, and Intellectual Property

Computers are essential for design and analysis, and they are also indispensable for controlling manufacturing, exploration, refining, and process control. However, computers create unique liability issues. For example, if a key engineering or geoscience decision is based on faulty computer output, who is liable for the damage that may result? This chapter discusses professional liability for computer-generated errors, and suggests some simple procedures for validating computer software.

Computers also create new ethical problems, including vandalism, viruses, and software piracy. For example, computers are common tools for copyright infringement, perhaps because they make copying so rapid and convenient. This chapter therefore concludes with an overview of the laws for intellectual property, including copyright, patents, industrial designs, integrated circuits, and trademarks.

THE ROLE OF COMPUTERS IN ENGINEERING AND GEOSCIENCE

In the past 30 years, computers have drastically changed every phase of our lives, and professional practice is no exception. Recent university graduates may not realize how profound these changes have been.

A SHORT BUT AMAZING HISTORY Calculating devices such as the abacus, the slide-rule, and the adding machine have existed for centuries, but the first truly electronic computer was not built until 1945, at the end of the Second World War. The initial impetus came from the work of British mathematician Alan Turing, who developed primitive machines to assist in the decoding of German war messages at Britain's Bletchley Park intelligence centre during the war. This early computer development is not well known because of wartime secrecy, although the story of Turing's impressive achievements and tragic life is gripping and fascinating.[1]

The first computers were slow and expensive behemoths by today's standards; they filled rooms, yet they were capable of only primitive calculations. The invention of the transistor and large-scale circuit integration (LSI) permitted the miniaturization of electronic devices, and increased the reliability of components. A versatile personal computer (the Apple II) won over the market in 1977, and soon replaced the calculator and the slide-rule. Over the following two decades, the desktop workstation evolved, yielding immense, convenient computing power.

AN INCREDIBLE FUTURE   In the 21st century, professional engineers and geoscientists typically use desktop workstations connected to the Internet. This computing and communication power permits many of us to practise in areas that were at the cutting edge of research only a decade or so ago, such as earth modelling and visualization, computational fluid dynamics, and dynamic finite-element analysis, mechanism, and process simulation. (In fact, the common term "computer-aided design" is now out of date, since all design is now "computer-aided.") Many new fields of study have developed, such as digital control, mechatronics, and nanotechnology, all based entirely on digital devices.

The computer's incredible speed in analysis and visual design is creating a dynamic new age for engineering and geoscience. Tedious work is now done by hardware and software, freeing designers to be more creative. Today's workstations permit ideas to be visualized, simulations to be run, and alternatives to be analyzed in the earliest stages of any project. Calculations that were once laboriously prepared by slide-rule (and later, by calculator) are now displayed instantly, and drawings and maps that were once monotonously hand-drawn are now plotted in seconds.

However, computers are also creating new problems for professionals: problems such as copyright infringement, errors caused by flaws or "bugs" in computer programs, vandalism by hackers and crackers, computer-aided industrial espionage, and the growing problem of identity theft. These problems sabotage productivity, so engineers and geoscientists must be alert to them.

The danger of faulty computer software was emphatically illustrated three decades ago, when the Hartford Arena collapsed—an engineering disaster that was perhaps the world's first large-scale computer-aided failure. The arena's design was based on an erroneous stress analysis program, as explained in Case History 9.1, which follows.

CASE HISTORY 9.1

THE HARTFORD ARENA ROOF COLLAPSE: A COMPUTER-AIDED FAILURE

The Hartford Arena was a monumentally huge structure when it was completed in 1973. The arena housed a basketball court and seating for 5,000 spectators to watch the games. To minimize obstructions for spectators, only four columns supported the roof. Each column was near a corner of the building. The "space-frame" roof was a three-dimensional truss structure about 3 m (10 ft.) deep, and approximately 91 m by 110 m (300 ft. by 360 ft.) in plan size, suspended about 25 m (83 ft.) above the floor.

The Roof Collapse

At 4:15 a.m. on January 18, 1978, during a heavy snowfall, the huge roof suddenly and violently collapsed onto the central court, with the corners of the roof pointing up into the air. Fortunately, the collapse occurred in the middle of the night. Earlier in the evening, the arena had been packed with thousands of spectators, and all of them missed death or injury by a matter of a few hours.

The Cause of the Failure

During the investigation, the snow load at the time of the collapse was estimated to be less than half the rated load for the roof. Attention shifted to the design. The detail design of the structural steel had several gross errors, as described very well in the case study by Rachel Martin, which is available on the Internet.[2] However, the basic cause of the collapse was, as Henry Petroski stated, an "oversimplified computer analysis."[3] The Hartford Arena involved one of the earliest applications of computers to the analysis of complex space-frame structures, and the designers made a fateful error. Martin explains:

> The engineers for the Hartford Arena depended on computer analysis to assess the safety of their design. Computers, however, are only as good as their programmer and tend to offer engineers a false sense of security. The roof design was extremely susceptible to buckling which was a mode of failure not considered in that particular computer analysis and, therefore, left undiscovered.[4]

In other words, the stress analysis software overlooked the key idea that structural rods in compression buckle at a stress far lower than the yield strength of the steel, which is typically the limit for rods in tension. Any engineer could easily have discovered this error, at the earliest stages of the project, by comparing the stress calculated by the computer against the well-known Euler buckling equation—a simple calculation that can be performed in minutes.

Ethical Implications

Why the engineers neglected to perform such a simple, obvious check of their computer output is a mystery. Moreover, the design engineers had a very strong incentive to double-check their calculations. During the construction, the truss was assembled on the ground and hoisted into place. Large deflections were immediately apparent, and the engineers were informed. In fact, as Kaminetzky reports, the deformations were so much larger than expected that contractors could not insert the windows designed to fit below the girders. Even the ironworkers reported that the deformations were unreasonably large.[5] Nevertheless, the engineers ignored these warnings and did not double-check their work.

It should be clear that the actions of the design engineers were negligent or incompetent. The Hartford Arena engineers failed to validate the computer output adequately and subjugated their judgment to the computer. Computer

Photo 9.1 — Hartford Arena Roof Collapse. *The Hartford Arena was constructed in 1973, and housed a basketball court and seating for 5,000 spectators. On January 18, 1978, during a heavy snowfall, the huge roof suddenly collapsed, only hours after a well-attended game. The collapse was traced to an "oversimplified computer analysis." The arena is known as the first computer-aided failure.*

Source: © Bettmann/CORBIS.

program validation should be routine due diligence. The engineers compounded their negligence when they ignored the excessive deflection of the truss—a warning sign that something was wrong.

The details of the case were never revealed in court. After six years of legal preparation, an out-of-court settlement was reached, and a probing discussion of the causes was therefore precluded.

An engineer or geoscientist cannot guarantee that every project will succeed, just as a surgeon cannot save every patient, and a lawyer cannot win every lawsuit. However, what the engineer, geoscientist, surgeon, and lawyer must all guarantee is that they possess adequate knowledge, and that they will exercise reasonable skill, care, and expertise, appropriate to the profession, to carry out the client's wishes. In the case of computer-aided design, reasonable care requires validation of the computer software.

LIABILITY FOR SOFTWARE ERRORS

Software engineers aspire to high professional standards, but computer programs occasionally produce incorrect results, as the Hartford Arena collapse shows. The key question is: Who is liable if damage results from decisions based on faulty software?

Almost every commercial computer program includes a disclaimer stating clearly that the manufacturer and supplier are not liable for any damage arising from the program's use. Typically, the disclaimer specifically denies responsibility for direct or indirect damages, including loss of business profits, business interruption, personal injury, financial loss, and/or similar losses. In effect, this limits the manufacturer's liability to the price paid for the program.

This disclaimer shifts the responsibility to the user—a fact confirmed by the provincial Associations. For example, the PEG-NL (Newfoundland and Labrador) software guideline simply states: "Members are responsible for verifying that results obtained by using software are accurate and acceptable."[6]

The APEGGA (Alberta) guideline states the responsibility more thoroughly:

> Members are responsible for verifying that any results obtained from computer programs are reliable and valid. Professional members should: examine and understand the methodologies and input parameters, as well as the limitations of the results obtained; and verify, where appropriate, new software releases against a standard certified for general use.[7]

The PEO (Ontario) guideline defines the engineer's responsibility even more specifically. Under the heading "Use of Computer Software Tools by Professional Engineers," the guideline states:

> The engineer must have a suitable knowledge of the engineering principles involved in the work being conducted, and is responsible for the appropriate application of these principles. When using computer programs to assist in this work, engineers should be aware of the engineering principles and matters they include, and are responsible for the interpretation and correct application of the results provided by the programs.
>
> Engineers are responsible for verifying that results obtained by using software are accurate and acceptable. Given the increasing flexibility of computer software, the engineer should ensure that professional engineering verification of the software's performance exists. In the absence of such verification, the engineer should establish and conduct suitable tests to determine whether the software performs what it is required to do.[8]

Clearly, all of these guidelines hold the user responsible for verifying that the software is operating properly. This means that the user must test or verify the software before using the computer output in engineering design. Such tests, typically called *validation* tests, require independent calculations. Validation tests will vary, depending on the type of analysis and on whether the software was developed in-house (by the user) or was commercially purchased. (Typically, source code is not available for commercial software.) Let us consider these cases separately.

SOFTWARE DEVELOPMENT

Professional engineers and geoscientists often develop software—for themselves, or under contract for others. In fact, the need for skilled professionals in this critical field explains why software engineering is a licensed engineering discipline. When life, health, or public welfare is placed at risk, governments have a duty to regulate the discipline. Therefore, if you are developing software for internal company use, or under a software contract, or for sale to others, it is important that you follow accepted guidelines for accuracy, reliability, documentation, and testing. The first step is to specify the scope of the project; that is, to define precisely what is to be developed, and how it will be used.

Specifying the Scope of a Software Project

The PEO guideline is presently the most comprehensive provincial guideline for software development. It offers the following advice for specifying the scope of the project when negotiating contracts:

> An engineer embarking on the development of engineering software for a client runs the risk of liability if the software does not perform according to the client's requirements, or if its use causes harm to the client or the public. A well-drawn legal contract, which contemplates the development of engineering software for a client and its use by the client, can minimize the engineer's exposure to liability.

> It can also define the contractual rights and obligations between the parties to the contract. . . . [P]rovisions addressing at least the following concerns should be included in such a contract:

> - What is to be developed;
> - Deliverables;
> - Scope of use of deliverables;
> - Representations and warranties;
> - Ownership;
> - Limitation of liability;
> - Contract price, and
> - Maintenance and escrow.[9]

The contract terms for limiting liability are especially important. In the unlikely event that the contract should be breached, a clause limiting liability will be honoured, provided it is a thoughtful and reasonable estimate of the damages likely to result from the breach.[10] It is wise to consult a lawyer when you negotiate a contract with complex legal terms.

Software Testing

The PEO software guideline includes a lengthy discussion of several reviews and tests that should be followed during the software development. The following are suggested as a minimum:

- Software requirements review;
- Software design review;
- Code review;

- Unit testing;
- System integration testing, *and*
- Validation testing.[11]

The early reviews are important because they can save much development time. Also, the final test—that is, the validation—is especially important because it is the final verification step before the software is turned over to the user. The PEO software guideline defines *validation* as "testing the integrated system to ensure that it meets functional and conceptual design require-ments."[12] An old engineering adage puts it much more simply: "No impor-tant decision should ever be based on a single calculation." In other words, important calculations should always be independently double-checked. This adage dates back to slide-rule days, but applies equally to computer output. You must validate software before using it to make key decisions.

Software developers must ensure that their work follows a guideline such as the one published by PEO, or similar documentation for their province or specific discipline.

USING COMMERCIAL SOFTWARE

Many engineers and geoscientists use large commercial software packages for highly specialized analysis. Even commercial software may have flaws, but errors are more likely introduced by the user. For example, the user may

- use incorrect units for data input,
- apply the program to the wrong type of problem, one that is unsupported by the program theory (such as using a program intended for planar analysis in a 3-D application),
- set erroneous parameters (such as integration parameters) that result in incorrect computational accuracy,
- not understand the output display, and/or much more. In fact, users are notorious for misunderstanding software written by others.

To use commercial software properly, the first step is obvious: read the documentation. Introductory tutorials provided by the software developer are also very useful, and should be attended religiously. If questions arise, the developer's "help desk" should be consulted. Do not apply software if you have doubts about it, or unanswered questions.

In addition, you should always test new analysis software to validate it. Never assume that the commercial software "must be right." If a major tech-nical project fails because of software errors, the first question that a lawyer will ask you is: "What tests did you perform to ensure that the software was operating properly?"

If at all possible, validation should involve running at least the first three of the following tests, which are discussed roughly in order of increasing effort or complexity. In cases where failure could lead to injury, death, or serious financial or environmental disaster, all of the validation tests are essential. If the software fails any of the following tests, ask why, and don't go ahead until you have full confidence in the software!

- **Dummy runs.** Run a basic check on the program's computation, using nominal entries such as zeroes or ones, to get a known answer. A simple test is easy to imagine:
  - If zero loads are input to a stress program, the stresses calculated should also be zero. (Similar tests apply to electrical, thermal, pneumatic, and hydraulic programs, as well.)
  - If a file of identical numbers is input to an averaging program, the calculated mean, median, and mode must equal that number, and the standard deviation must be zero.
  - If a dynamic simulation is re-run with smaller integration parameters (a shorter time-step), the motion should be identical.
  - If an input file for a previous analysis program is fed to a new program (assuming that it is compatible), the programs should give the same output.

  These tests are a necessary (but not a sufficient) condition for validity. That is, any software that fails these simple tests is definitely unreliable, but passing the tests does not guarantee validity—more advanced analytical or theoretical tests are needed.

- **Approximate analytical checks.** Imagine a simplified configuration of your computer model that can be analyzed analytically. Apply analytical calculations to the simpler model, find an approximate answer, and compare it with the computer output. For example, a finite-element model of a complex structure can almost always be decomposed and approximated by simple beam and column equations. Take a most optimistic estimate and a least optimistic estimate and apply the analytical equations to each. The computer output should lie between these boundaries. This is a standard check. These tests are fairly quick; the results are approximate, but reassuring, although not totally conclusive.

- **Independent theoretical checks.** Make analytical computations using an independent theoretical basis. For example, dynamic simulations use numerical integration, but the integration can be checked by applying the laws of conservation of energy and momentum to the initial conditions and the final answers. Where such tests are possible, they are very convincing.

- **Advanced methods.** Clever and creative analysts can easily develop more advanced validation tests, unique to the discipline or specialty.

- **Complete duplication.** A full-scale duplication of the computation, using different software, hardware, and input files, is an expensive but convincing validation. Independent employees or consultants should conduct this test, if possible, to avoid systematic errors in the input data. This check is expensive, but it validates almost everything—input data, theory, and computation. If you have any doubts about critical analysis software, you should carry out this independent validation before making major expenditures. It is always cheaper to duplicate a computer calculation at the early stages of a project than to explain the omission to a board of inquiry after the project fails.

In summary, computer software is like any other tool: it must be used properly, and it must be calibrated (or validated). Validation tests are essential before output data is used for critical decisions.

COMPUTER SECURITY

Professional engineers and geoscientists have an enormous investment in computer hardware and software. In many companies, these are major assets, so professionals must be alert to any threats to this investment. The obvious risks are massive hardware or software failure, data loss, and unauthorized intrusion.

Computer Disaster and Recovery

The professional must provide routine maintenance for equipment, software, and data storage. In addition, it is wise to have a recovery plan for the possibility, however remote, of complete computer disaster, such as might occur in a fire or flood—the complete destruction of the computers and loss of the programs and data. Every professional practice should estimate the cost and impact of a computer disaster on the practice, and how long it would take to recover. For a small professional practice, the first line of protection is to have critical data and programs duplicated on backup disks, and stored in a safe, secure location. A plan for recovery, by buying or leasing alternative hardware and software, can easily be developed.

For large operations, backup procedures may be more complex. This topic is too specialized for inclusion here, but several books advertised on the Internet provide further advice on this subject. A simple search for "computer disaster recovery" will provide a wealth of information.

Internet Threats

The Internet is extremely useful to professional engineers and geoscientists for many purposes, such as finding codes and standards, health and safety regulations, and, as discussed later in this chapter, for searching the databases of patents, trademarks, and industrial designs. However, the Internet poses a security risk.

Thoughtless and selfish people, whether hackers, spammers, or vandals, pose increasing levels of threat by degrading e-mail service, by destroying or modifying data, or by creating denial-of-service (DoS) attacks. A typical DoS attack floods a computer with transmissions; this overwhelms the computer and effectively denies its services to legitimate users. To combat these threats, every professional office must have firewall and antivirus software.

- **Firewall.** This type of software guards a computer's "gates." That is, it guards your computer's connections to the Internet, and blocks or admits data transmissions according to the access rules you have set.
- **Antivirus software.** This software detects, identifies, and removes any viruses that have succeeded in breaching your firewall and entering your computer.

Firewalls and antivirus software protect your computer from a wide range of threats. A full glossary of these threats is available from developers such as Symantec, a provider of Internet security software.[13]

In summary, we must take the actions above to protect ourselves, but a larger question arises: How are we to balance the open freedom of the Internet with the potential for abuse, which is now growing out of control? A similar form of abuse occurred in the early days of radio broadcasting. National laws and international treaties now rigidly control radio frequencies, but these treaties did not exist in the 1920s. Early broadcasters simply selected their own frequencies and then increased their transmitting power until they drowned out the competition. This was unfair and unethical, and had to be remedied, and was. Our society (and software engineers in particular) must now develop rules to bring fairness to the Internet. The Internet is too important to our quality of life; we cannot allow vandals to destroy it.

PREVENTING SOFTWARE PIRACY

One of the most flagrant conflicts of interest today involves the copying of software, which is usually referred to as software *piracy*. Copying is so easy that, whether through ignorance or intention, the practice is widespread. When you purchase a computer program, you are not buying the right to duplicate that program, except for backup.

There are many good reasons why professional engineers and geoscientists should never use copied or pirated software.

- **Illegality.** The first and most obvious reason is that copying software is illegal. It violates the Copyright Act, which allows copying only for backup purposes. The Act forbids activities such as reverse engineering; here it differs from American law, which permits research on computer programs (including reverse engineering) under the "fair use" provisions. Proposals are being made to extend Canada's Copyright Act to include these permissions.[14] Generally speaking, however, software copying is illegal.
- **Unprofessional conduct.** Obviously, trying to run a professional practice with pirated software is very unprofessional. In fact, it could be interpreted as professional misconduct, and might result in discipline or loss of a professional licence.
- **Breach of contract.** The use of pirated software could result in a breach of any contract for which it is used. For example, a consulting contract could be breached if the client discovers that you are using pirated software.

- **No product support, documentation, updates, or patches.** Product support, documentation, updates, and patches are usually not available for pirated software.
- **Fines and embarrassment if caught.** In 1990 the software industry established the Canadian Alliance Against Software Theft (CAAST), which recently affiliated with the Business Software Alliance (BSA), an organization that runs educational, enforcement, and public policy campaigns to combat piracy in 80 countries around the world. BSA investigates allegations of software piracy and lists results of current infringement prosecutions. It estimates that Canada's economy lost more than $1.22 billion to software theft in 2008.[15]

Because of the extent of software piracy, this issue is addressed very pointedly in Ontario's *Guideline to Professional Practice:*

12. COPYRIGHT IN COMPUTER PROGRAMS

Under recent amendments to the Copyright Act, the uncertainty about copyright in computer programs has been eliminated by expanding the definition of "literary work" to include computer programs, which are broadly defined to include all computer programs, whether in source code or object code, regardless of how they are stored. Two exceptions under these amendments will allow certain uses to be made of computer software, which would otherwise be an infringement of copyright.

The first exception provides that it shall not be infringement for a person in lawful possession of a copy of a computer program to modify, adapt, or convert a reproduction of the copy into another program to suit that person's needs, provided:

- the modified program is essential for the compatibility of the computer program with a particular computer;
- the modified program is used only for the person's own needs
- not more than one modified copy is used by the person at any given time; and
- the modified copy is destroyed when the person ceases to be entitled to possession of the copy (i.e. upon expiry of a software licence).

The second exception provides that a person who is in lawful possession of a copy of a computer program, or of a modified reproduction of a computer program, may make a single backup copy of the program, provided the backup copy is destroyed when the person ceases to be the owner of the copy of the computer program.

The intention of these exceptions is to give the authorized software user a limited right to change the software, to ensure compatibility of the software with the authorized user's computer system, and to allow for the protection and security of the original program.[16]

The Alberta (APEGGA) guideline contains similar information:

2.1 Legal Considerations

Computer software is covered under the Canadian Copyright Act which provides for a financial penalty as well as a jail sentence for violation. The Copyright Act protects authors' legal rights and privileges to their creative works. It should be noted that a copyright in a work exists as soon as the work is created and there is no requirement to publish the work or to affix any special notice thereto. In addition to copyright considerations, usage of commercial software is also generally governed by contract law under the agreements of the software purchase contract and/or licence.

2.2 Ethical Considerations

The Code of Ethics establishes the duty of APEGGA members to enhance the dignity and status of the professions. APEGGA members shall conduct themselves with fairness and good faith toward other professional members and the public in the area of computer software usage to avoid conduct which would detract from the image of the professions.

In consideration of the Code of Ethics, APEGGA members must guard against any violations, real or apparent, of the Canadian Copyright Act and contract laws and the resulting legal and ethical consequences.

2.3 General Principles

All purchased/licensed computer software is subject to the full provisions of the agreements connected with the acquisition of the software and manuals associated therewith. All APEGGA members should be aware of the agreement provisions and abide by the terms of the agreements with particular regard to copying restrictions.

The use of copies of computer software or manuals that have been obtained in violation of copyright or trade secrets or in any other fraudulent manner is deemed unprofessional conduct on the part of an APEGGA member.

In addition to exposure to possible criminal prosecution, violation of copyrights or misappropriation of trade secrets associated with computer software by our members may result in disciplinary action by APEGGA.[17]

In view of the importance of copyright and its relevance to computer practice, the following paragraphs give an overview of Canada's intellectual property laws, which apply to copyright, patents, trademarks, industrial designs, and "integrated circuit topographies." Trade secrets are also discussed below, although they are not protected by these laws, and must be protected in other ways.

COPYRIGHT, PATENTS, TRADEMARKS, AND DESIGNS

Copyright, patents, trademarks, industrial designs, and integrated circuit topographies (more commonly known as integrated circuit designs) are known by the general term *intellectual property*. They are valuable and they may be bought, sold, or licensed, like any other property. However, intellectual property has one key difference: the ownership period is limited, and at the end of that period, the intellectual property becomes part of the public domain (except for trademarks, for which ownership may be extended).

The Canadian Intellectual Property Office (CIPO), an agency of Industry Canada, manages intellectual property in Canada. In particular, CIPO maintains a database, which can be searched through the Internet. In the United States, the Patent and Trademark Office (PTO) also maintains a database, searchable by Internet. These databases are open to anyone and are a useful source of valuable design information.

Importance of Intellectual Property

The basic principle for regulating intellectual property is to encourage creativity by protecting the rights of creative people, and to provide an orderly way to exchange creative ideas. Canadian and international laws therefore protect the rights of people who create intellectual property (for a specified period), so that good inventions and creative works will be disclosed to the public, and after the specified period, may be used by anyone. A basic knowledge of intellectual property law is essential, because

- professional engineers and geoscientists must be able to protect the intellectual property that they create (or that is created under their direction).
- everyone must respect the rights of others, so we must know these rights and avoid infringement.
- intellectual property represents a huge warehouse of ideas, technical knowledge, designs, and inventions that are available, free (or at very low cost), from the CIPO. Anyone can search the CIPO databases for existing patents, trademarks, and designs, and this information is a valuable aid in research, design, and marketing.

The basic legal concepts are discussed here and Table 9.1 summarizes the various forms of intellectual property and how long they may be protected. For more detailed information, or for answers to legal questions, consult the CIPO or Industry Canada, or see your lawyer.

General Rules for Intellectual Property

Five different types of intellectual property may be registered, but in general, only the owner of the intellectual property may register it. Usually, the person who created the work is the owner, unless the creator has sold, leased, or given the rights to another person. When the creator is an employee hired to create intellectual property, the employer is usually the owner. This fact should be defined in the employment contract. To avoid confusion, any debate over the ownership of the intellectual property should be settled before applying for registration.

Copyright

Copyright law enables owners to protect written works, such as literary, artistic, dramatic, and musical works, as well as aesthetic works, such as artwork and drawings (and many subcategories of all of these works). It is

TABLE 9.1 — Summary of Protection for Intellectual Property

Intellectual Property	What Is Protected	Duration of Protection
Copyright	Written literature and artistic, dramatic, and musical works, including computer programs, original drawings, paintings, sculpture, and similar works of art. Works reproduced mechanically or electronically (films, photos, recordings, communication signals, etc.) are given protection with a shorter duration. (Exceptions exist.) Copyright exists immediately, upon creation of the work. Registering is optional.	The life of the author/ creator, plus 50 years. Mechanically and electronically copied works are usually limited to 50 years.
Patents	New, useful, and innovative devices, machines, processes, or compositions of matter (or improvements to existing inventions). A patent protects the way something operates or is made. A patent must be granted to obtain protection.	20 years
Industrial Designs	The shape, configuration, pattern, or ornamentation applied to a finished article, made in quantities by hand, tool, or machine. An industrial design protects the appearance or ornamentation. (Some industrial designs may also qualify as trademarks.) Industrial designs must be registered.	10 years
Integrated Circuit Topographies	The patterns or configurations of components in integrated electronic circuits, including the three-dimensional geometry of the layers of semiconductors, metals, insulating layers, and other materials on a circuit board or sub-layer, which produce a known electronic function.	10 years
Trademarks	Logos, symbols, slogans, names, or designs (or any combination of these) used to identify a company's goods or services in the marketplace. Registration is needed for full protection.	15 years, renewable indefinitely for 15-year periods.
Trade Secrets	Manufacturing processes or material compositions may be kept secret, but secrecy contravenes the principle that rights are awarded for full disclosure, so no legal protection exists for trade secrets. If someone independently discovers the same secret, the person may patent it and prevent you from using it.	Uncertain.

important for professionals to note that technical reports, drawings, specifications, and computer programs are always protected by copyright.

- **How it is protected.** Under Canada's Copyright Act, only the owner of the work has the "right to copy" it (or to permit others to copy it). If someone infringes the copyright by unauthorized copying, the owner may use the law courts to enforce the owner's rights. Creative works are protected for the life of the author/creator plus 50 years beyond death. However, works that are reproduced mechanically or electronically, such as motion-picture films, photographs, recordings, communication signals, and so on, are given less protection—typically a maximum of 50 years from the date of creation. (Several exceptions to these general rules exist for specific types of creation.)

- **Obtaining your copyright.** Copyright differs from other forms of intellectual property because the creator (author, artist, performer, photographer, and so on) has copyright protection immediately and automatically upon creation of the work. Registering the copyright is optional; doing so simply gives more certainty in enforcing the rights. The work may be marked to identify the copyright owner (even if not registered) using a copyright symbol (a "C" inside a circle), the name of the owner, and the date of first publication (for example: © Jane Doe, 2009).

Illegal copying is, of course, the most common form of infringement. In past years, photocopy machines and video recorders were the means of illegal copying. However, the Internet is now the key tool for copyright infringement. It is remarkably easy to copy (and thus infringe the copyright) of written, visual, and audio material. Unscrupulous users can make digital copies of software, films, audio, and publications, very easily, on a flash drive, CD, or DVD. Such activities are illegal.

Technologies for protecting digital content are being developed, but are lagging far behind the technologies for stealing it. Software piracy and video and audio copying are costing large corporations many billions of dollars. For example, music companies in the United States have been unable to find any technical means to stop infringement and have begun suing to recover lost revenues. In 2003, people who downloaded music from the Internet were shocked to receive subpoenas to appear in court. Many computer users are astonished that such easy, simple copying could be illegal.

Copying is not acceptable in university, in employment, or in professional practice. A professional person must identify and cite the sources of any (and all) material taken from others. This law is particularly important when the material is included in a professional report or similar work. Failure to identify sources can have serious penalties.

Claiming credit for work that is not your own is called "plagiarism" and is usually both an infringement of copyright and a serious academic offence, especially in universities. Sources must always be cited. Copying a report and passing it off as one's own is flagrant plagiarism, and is subject to penalties. In universities, the academic penalty for plagiarism may include suspension or expulsion, depending on the severity of the offence. In the business world, the legal penalty for infringement depends on how much the copyright owner lost, financially, as a result of the infringement.

In summary, professional people must be alert to the consequences of illegal copying, software pirating, and plagiarism. Unauthorized use of the work of others, or claiming credit for the work of others, is not just unethical; these are copyright infringement and may lead to legal problems. For more information, the CIPO publishes *A Guide to Copyrights*.[18]

Patents

An invention must be new, useful, and innovative (or ingenious) to receive a patent. Improvements to inventions can also be patented; in fact, most patent applications are for improvements to existing devices. However, patenting an improvement does not give you the right to use the original invention. A patent protects the way that a device operates (unlike an industrial design, discussed below, which protects the appearance).

- **How it is protected.** Under Canada's Patent Act, the owner of a patented invention has the right to make, manufacture, use, or sell the invention for a period of 20 years from the date of the patent application. The patent owner can prevent others from making, using, or selling the invention, and may enforce this right in court.
- **Obtaining a patent.** Obtaining a patent is a fairly long process, but the key step is preparing the patent application. (An initial fee, and a small annual maintenance fee must be paid while the patent is in force; otherwise the patent enters the public domain.) The patent application has a standard format: petition, abstract, specification, claims, and drawings (where applicable). The petition is merely a formal letter, asking the Patent Office to grant the patent. The application itself becomes the patent after it has been examined and approved by the Patent Office. The four parts of the patent are described below.

    1. **Abstract.** The Abstract summarizes the patent in a concise form, suitable for publishing in the *Patent Office Record*, which is the official gazette of the Patent Office.
    2. **Specification.** The specification describes the invention, and usually answers four questions:
        - What problem does the invention solve?
        - What prior art exists, and why is it inadequate to solve the problem?
        - How does your invention work? (This is the "Disclosure.")
        - How is your invention new, useful, and ingenious, compared to prior art?
    3. **Claims.** The claims build a fence of words around your invention, by defining which features of the invention are to be protected as your property.
    4. **Drawings.** Where appropriate, drawings of the invention must be included.

The patented device, manufacturing method or material composition may be marked (if appropriate) with the word "Patented" and the patent number issued by the Patent Office.

At the start of a design project, the CIPO database should be searched, both as a source of useful ideas, and to avoid infringing existing rights. Once a new design or invention has been developed, the professional should consider registering it, to protect the inventor's (and the employer's) rights. For a nominal fee, patent agents and other private companies will assist in patent searches and applications. For more information on patent procedures, consult the CIPO publication *A Guide to Patents*.[19]

Industrial Designs

Industrial designs can be protected through a process similar to patent or copyright protection. In this case, however, only the aesthetic appearance is protected. Registration of an industrial design protects the shape, configuration, pattern, or ornament applied to a finished article, which is typically made in quantities by machine. For example, the pattern of decoration on the knives, forks, and spoons of a dinnerware set could be registered as an industrial design. An industrial design applies to the aesthetic (or artistic) appearance of manufactured articles, and differs from a copyright, which applies to original artistic works and written materials. For example, an original sculpture would be protected by copyright, but manufactured copies of the sculpture must be protected as an industrial design.

- **How it is protected.** Under Canada's Industrial Design Act, only the owner of a registered industrial design has the right to make, use, sell, rent (or offer to rent or sell) the design. The protection exists for a maximum of 10 years (from the date of registration). It is important to stress that the design must be registered to obtain this protection; designs that are not registered have no legal protection against imitation.
- **Obtaining industrial design protection.** Registering an industrial design is fairly simple. It requires an application form, and at least one drawing or photograph that illustrates the design, submitted with the appropriate registration fee. Marking the design gives you extra protection against copiers. An industrial design is denoted by a capital "D" inside a circle and the name of the design's owner on the article, its label, or its packaging.

If you develop original designs, you should protect your rights by registering them. After 10 years, design rights expire and the design enters the public domain. It is neither unethical nor illegal to use designs that have entered the public domain. In fact, designers are encouraged to use them, since this is often a fast and profitable way to develop attractive new products. For more information, CIPO publishes *A Guide to Industrial Designs*.[20]

Integrated Circuit Topographies

Integrated circuits are the basis for almost all modern communications, computers, and similar electronic equipment. Integrated circuit topographies are defined as the geometrical configurations of integrated electronic circuits, including the layers of semiconductors, metals, insulating layers, and other materials on a circuit board or sub-layer. The topography may indeed be three-dimensional. They are a special form of industrial design, they have their own registration process, and they can be extremely valuable. The law protects the geometry of the electronic circuit, but it does not prevent others from designing a different geometrical circuit that can perform the same electronic purpose. (In this case, patent protection may provide broader protection.)

- **How it is protected.** Under Canada's Integrated Circuit Topography Act, only the owner may make, use, or sell the registered topography, and the owner has the right to prevent others from making, using, selling, leasing, or importing the topography, or incorporating it in another integrated circuit. Registration provides legal protection for 10 years from the date of the application.
- **Obtaining design protection.** Integrated circuit topographies can be registered by submitting an application form, fees, a description of the function of the circuit, and a complete set of overlays, drawings, or photographs of the circuit. The registered circuit may be marked (when manufactured) with the alphanumeric title used to identify the topography on the application. For more information, CIPO publishes *A Guide to Integrated Circuit Topographies*.[21]

Trademarks

Trademarks are the commonly used logos, slogans, names, symbols, or designs that identify a company's goods or services in the marketplace.

- **How it is protected.** Under Canada's Trademark Act, trademark registration gives the right to use the trademark for 15 years. Unlike other forms of intellectual property, trademarks can be renewed indefinitely—as long as they still serve the purpose of identifying a company's goods or services in the marketplace. Trademark infringement is fairly rare in professional activities, although it occurs in retail sales. For example, high-priced consumer goods, such as prestige watches and fashion accessories bearing illegal trademarks, are occasionally found for sale in "discount" retail stores. Selling goods with a counterfeit trademark is illegal.
- **Obtaining trademark protection.** Before you apply to register a trademark for a service or product, you should first search the CIPO website to see whether the suggested trademark is already registered. This preliminary search may save time and effort. If your trademark does not conflict with any existing mark, an application form, a fee, and a drawing of the mark (if appropriate) are sufficient to start the examination and approval process. When a trademark is registered, the owner may warn copiers using the letter "R" in a circle (meaning "Registered"), or the letters "MD" (meaning "marque déposée"). Even if a trademark is not registered, owners may warn copiers with the letters TM (for "trademark") or SM (for "service mark") or MC (for "marque de commerce"). The Trademark Office recommends the use of such symbols, even though they are not essential to protect your rights.

A trademark cannot be identical (or deceptively similar) to existing trademarks or to "prohibited marks," which include symbols and logos of the Canadian government, the Royal Family, the armed forces, provinces, foreign countries, and many well-known international institutions. In addition, trademark applications must satisfy a series of simple rules to ensure that they do not create confusion over the goods or services being offered, and do not restrict the public's ability to use common language and geographical names. A summary of these rules and examples of satisfactory and ineligible trademarks are in the CIPO publication *A Guide to Trade-marks*.[22]

The following example illustrates the importance of searching the trademark database. The University of Waterloo recently introduced a Mechatronics Engineering program. The university contacted Engineers Canada (formerly the Canadian Council of Professional Engineers) to ensure that the program would satisfy the accreditation process and that the program had the legal right to use the term "engineering." (Engineers Canada is the registered trademark holder of the terms "engineer" and "engineering" as descriptors of services offered.) However, the term "mechatronics"—which indicates that the new program is a combination of mechanical, electronics, and robotics subjects—was not searched until after the program name had been widely advertised. University officials were surprised to learn that, about a decade earlier, a German company had registered the term in Canada as a trademark. (Negotiations over the use of the term were agreeably settled.)

Trade Secrets

Of course, you can protect intellectual property simply by keeping it secret. Trade secrets may be very effective for some inventions, such as manufacturing processes or material compositions (although they are irrelevant for trademarks, designs, or copyright). In general, secrecy is maintained by requiring employees to sign employment contracts with confidentiality clauses. Trade secrets have no legal status in patent law, so breaches of confidentiality must be enforced under contract law or tort law. For example, if an employee reveals a trade secret, the employer may be able to sue the employee for the loss under the terms of the employment contract. Trade secrets discourage disclosure, and are therefore contrary to the purpose of patent law, which gives protection in exchange for full disclosure.

CASE HISTORY 9.2

PATENT INFRINGEMENT

The following case history is adapted from an actual event in which the author played a role as an expert witness, but it has been drastically simplified to illustrate the key issues. Some facts, and all names, have been changed to avoid any embarrassment to those involved.

Background Information

An inventive construction worker designed a new winch device for raising materials on a construction site. The device, when installed on a building site, could deliver heavy materials (bricks, siding, windows and roofing), quickly, safely, and easily, to fairly high levels. The device was patented by the inventor, who sold the rights to a company owned by Engineers A and B. Engineers A and B reworked the design, making it stronger, more robust, and more attractive, but retaining the patented mechanism's function. The company thrived by manufacturing the winch device, named the "Winch-Atomic," and selling it to construction companies. Eventually, Engineer B wanted to move on to a new career, so Engineer A purchased B's shares in the company. The patent for the Winch-Atomic belonged to the company, so Engineer A gained control over these rights.

Patent Dispute

A few months after the sale of the company to Engineer A, Engineer B started a rival company, manufacturing a winch device called the "Liftodrome" that also lifted building materials. Upon inspection, it was clear that the Liftodrome was essentially identical to the Winch-Atomic, except that a key part of the winch was inverted. That is, key operating parts of the winch mechanism were similar, but installed upside-down.

Engineer A complained to Engineer B, stating that the Winch-Atomic patent had been infringed.

Legal Resolution of the Dispute

Engineers A and B could not resolve the dispute, so the matter was turned over to lawyers. Each side consulted expert witnesses, and a lawsuit was filed based on Engineer A's rights under the Winch-Atomic patent. The case was eventually referred to a judge. The key question was whether inverting the mechanism made it into a new invention.

It was argued that mechanisms are classified kinematically, depending on the number and types of members and joints, and turning a mechanism upside-down makes no difference to the classification of the mechanism. In fact, it is precisely to avoid such confusion that the topology (the type and degree of interconnection) is used to identify and classify mechanisms. Although the mechanism was inverted, it was still performing the same function. The judge ruled in Engineer A's favour, and Engineer B was required to pay substantial damages for the patent infringement.

Federal laws for intellectual property encourage creativity by protecting the rights of creative people, and provide an orderly way to exchange creative ideas. Professional engineers and geoscientists must be able to protect the intellectual property that they create, and must respect the rights of others. In addition, both the Canadian Intellectual Property Office (CIPO) and the

United States Patent and Trademark Office (PTO) maintain immense data-bases of valuable inventions and designs that are available, free, over the Internet.[23,24]

DISCUSSION TOPICS AND ASSIGNMENTS

1. You are a co-op student working in a communications company. You have access to electronic switching source code and you show it to a fellow student. Later, you suspect that your colleague may have logged on with your password and looked at the software. Although the material is read-only, you believe he may have copied the code onto a flash drive or CD. Although he is a great friend, you suspect that he may be trying to sell the code or use it for illicit purposes. What should you do?

2. Computer software is increasingly developed by a few eminent special-ists, and has features that no single practising professional may under-stand fully. Does this software increase the capability of the professional using it, or does it relegate the professional to the role of an input tech-nician, who knows how to operate the computer, but cannot comment on the theory in the program or the validity of its output? Is the future dim for the *average* technical professional, who will be demoted to oper-ating tools that are developed by specialized software experts? What should be done to ensure that this dismal prediction does not come true?

3. Using the Internet, search and find the Canadian intellectual property listed below. The Canadian patent database is available through the Canadian Intellectual Property Office (CIPO). Continue your Internet search on the U.S. Patent and Trademark Office. (The web addresses are in Notes 23 and 24, below.) Are the results identical?

   a. A patented apparatus for reducing noise in a jet engine, issued in the last 10 years.
   b. The owner of the trademark for the Blackberry wireless hand com-municator.
   c. The owner of the trademark: "Roll up the rim to win."
   d. The industrial design for the bumper or grill of any recent North American vehicle.

Additional assignments can be found in Appendix E.

NOTES

[1]   A. Hodges, *Alan Turing: The Enigma of Intelligence*, Burnett Books, London, 1983.
[2]   R. Martin, *Hartford Civic Center Arena Roof Collapse*, Dept. of Civil and Environmental Engineering (case study), University of Alabama at Birmingham, available at <http://matdl.org/failurecases/Building%20Cases/Hartford.htm> (June 15, 2009).
[3]   H. Petroski, *To Engineer Is Human: The Role of Failure in Successful Design*, St. Martin's Press, New York, NY, 1985, p. 199.
[4]   Martin, *Hartford Civic Center Arena Roof Collapse*. Excerpt used with permission.

[5]   D. Kaminetzky, *Design and Construction Failures: Lessons from Forensic Investigations*, McGraw-Hill, New York, NY, 1991, p. 224.

[6]   Association of Professional Engineers, Geologists and Geophysicists of Alberta (APEGGA), *Guideline for Relying on Work Prepared by Others*, 2003, p. 9, available at <www.apegga.org> (May 29, 2008).

[7]   Association of Professional Engineers and Geoscientists of Newfoundland (PEG-NL), *Guideline for the Use of Computer Software Tools by Professional Engineers and Geoscientists*, July 28, 1995 (1 page), available at <www.pegnl.ca/publications/index.html> (May 29, 2008).

[8]   Professional Engineers Ontario (PEO), *The Use of Computer Software Tools by Professional Engineers and the Development of Computer Software Affecting Public Safety and Welfare*, Toronto, 1993, p. 4, available at <www.peo.on.ca> (May 29, 2008). Excerpt reprinted with permission of PEO.

[9]   Professional Engineers Ontario (PEO), *The Use of Computer Software Tools*, p. 9. Excerpt reprinted with permission of PEO.

[10]   D.L. Marston, *Law for Professional Engineers: Canadian and International Perspectives*, 3rd ed., McGraw-Hill Ryerson, Whitby, ON, 1996, p. 153.

[11]   Professional Engineers Ontario (PEO), *The Use of Computer Software Tools*, p. 7. Excerpt reprinted with permission of PEO.

[12]   Ibid., p. 8.

[13]   Symantec, *Glossary*, available at <www.symantec.com/business/security_response/glossary.jsp> (May 29, 2008).

[14]   Industry Canada and Department of Canadian Heritage, *Copyright Reform Process* (website), Ottawa, available at <http://www.strategis.ic.gc.ca/epic/site/crp-prda.nsf/en/Home> (May 29, 2008).

[15]   The Canadian Alliance Against Software Theft (CAAST), now affiliated with the Business Software Alliance (BSA), available at <www.caast.org> or <www.bsa.org> (June 15, 2009).

[16]   Professional Engineers Ontario (PEO), *Guideline to Professional Practice*, Toronto, 1988, revised 1998, p. 21, available at <www.peo.on.ca> (May 29, 2008). Excerpt reprinted with permission of PEO. (PEO advises that this guideline is to be revised in the near future.)

[17]   Association of Professional Engineers, Geologists and Geophysicists of Alberta (APEGGA), *Guideline for Copying and Use of Computer Software*, Edmonton, AB, November 2005, V1.1, available at <www.apegga.org/Members/Publications/guidelines.html> (May 29, 2008). Excerpt reprinted with permission of APEGGA.

[18]   Canadian Intellectual Property Office (CIPO), *A Guide to Copyrights*, Industry Canada, Ottawa, available at <www.cipo.gc.ca> (May 29, 2008).

[19]   Canadian Intellectual Property Office (CIPO), *A Guide to Patents*, available at www.cipo.gc.ca (May 29, 2008).

[20]   Canadian Intellectual Property Office (CIPO), *A Guide to Industrial Designs*, available at <www.cipo.gc.ca> (May 29, 2008).

[21]   Canadian Intellectual Property Office (CIPO), *A Guide to Integrated Circuit Topographies*, available at <www.cipo.gc.ca> (May 29, 2008).

[22]   Canadian Intellectual Property Office (CIPO), *A Guide to Trade-marks*, available at <www.cipo.gc.ca> (May 29, 2008).

[23]   Canadian Intellectual Property Office (CIPO) website at <www.cipo.gc.ca> (May 29, 2008).

[24]   U.S. Patent and Trademark Office (PTO) website at <www.uspto.gov/patft/index.html> (May 29, 2008).

Chapter 10

Fairness and Equity in the Professional Workplace

Professions such as engineering and geoscience attract intelligent people with high personal standards, who expect to work in a professional environment. Everyone, in fact, has the right to work in a fair and equitable workplace. Furthermore, a positive work environment usually improves employee performance, so fairness and equity pay off by creating a more successful enterprise.

This chapter discusses the importance of fairness and equity in the professional workplace, starting with the basis for these rights, and investigates how well the workplace meets the standards that should be expected, and special problems unique to the Canadian workplace. The chapter closes with three case studies illustrating the harmful effects of discrimination in the professional workplace.

ACHIEVING FAIRNESS AND EQUITY

To achieve fairness and equity in the workplace, the first steps are to recognize the special problems that women and members of minority groups may face within the organization and to ensure that effective policies are in place for hiring and promoting employees. Communicating these policies, their meaning, and their purpose to all staff is the key to success. If some issues arouse anxiety or anger in a particular group, it is often because of misconceptions, and these can be defused and eliminated through open dialogue and information sessions.

Unfair and unethical behaviour, such as discrimination or harassment based on race, national or ethnic origin, colour, religion, sex, age, mental or physical disability, or sexual orientation has no place in any profession—or anywhere in a civilized society. In Canada, this type of behaviour is illegal under the Criminal Code and human rights legislation. Some professional

This chapter was contributed by Dr. Monique Frize, P.Eng., O.C., Former NSERC/Nortel Chair for Women in Science and Engineering (Ontario), Professor in the Faculty of Engineering at Carleton University and University of Ottawa.

engineering and geoscience Acts have included clauses to prevent discrimination within our professions. For example, the *Guideline on Human Rights in Professional Practice* published by Professional Engineers Ontario (PEO) specifically states that discrimination and harassment are professional misconduct, and subject to discipline.[1] Similarly, the Alberta (APEGGA) Code of Ethics forbids discrimination, and APEGGA also publishes a *Guideline for Human Rights Issues in Professional Practice*.[2] Both of these comprehensive guidelines are freely available on the Associations' respective websites.

Nevertheless, unfair practices persist in the profession because they are subtle or systemic, or perhaps because we are unaware of them and of the destructive impact that they can have on individuals. These practices and their underlying causes are discussed in this chapter, so that readers can recognize and eliminate them.

DEFINITION OF DISCRIMINATION

Dictionaries define *discrimination* as "the action of discerning, distinguishing things or people from others, and making a difference." In recent years the term has also come to be associated with segregation, which is defined as "the act of distinguishing one group from others, to its detriment." It is the harmful aspect of discrimination that this chapter addresses.

THE CANADIAN CHARTER OF RIGHTS AND FREEDOMS

The Canadian Charter of Rights and Freedoms sets out some basic principles that should guide the daily life of every citizen.

- Clause 7 states: "Everyone has the right to life, liberty and security of the person and the right not to be deprived thereof in accordance with the principles of fundamental justice."[3]
- Clause 15 (1) defines equality rights: "Every individual is equal before and under the law and has the right to the equal protection and equal benefit of the law without discrimination and, in particular, without discrimination based on race, national or ethnic origin, colour, religion, sex, age or mental or physical disability."[4]
- Clause 15 (2) also addresses the right to have programs of affirmative action in cases where improvement and more balance in the participation of under-represented groups are needed. It reads: "Subsection (1) does not preclude any law or program or activity that has as its object the amelioration of conditions of disadvantaged individuals or groups."[5] For example, the federal government created a Chair (or professorship) for women in engineering in 1989, and added five Chairs for women in science and engineering in 1997.[6] The Chairs have the mandate to study and develop strategies to increase the participation of women at all levels in engineering and science disciplines where the enrollment of women is low.

The Charter is important to professionals: all contracts, collective agreements, work protocols, and handbooks must be consistent with provincial human rights legislation and with the Canadian Charter of Rights and Freedoms. Discrimination is against the law. Contracts, including collective agreements, can be rescinded, and statutes and regulations can be nullified if found to be discriminatory. The main difference between the Canadian Charter of Rights and Freedoms and other federal and provincial human rights legislation is that the Charter applies to all levels of government, including agencies directly controlled by governments. In contrast, provincial human rights legislation applies to matters under provincial jurisdiction. Within their own jurisdictions, both the Charter and the provincial human rights Acts are *primacy* legislations. This means that the human rights Acts supersede all other laws of that jurisdiction, unless expressly declared otherwise by an Act of Legislature or Parliament.[7,8]

ENROLLMENT PATTERNS IN UNIVERSITY PROGRAMS

Women are a majority in the Canadian population, but their under-representation in professional engineering and geoscience raises obvious questions. Figure 10.1 shows the enrollment patterns for both men and women in engineering undergraduate programs between 1991 and 2005. The average proportion of women in engineering undergraduate programs in 1985 was 11 percent; by 1991 it was 15 percent, increasing to 19 percent in 1995 and 21 percent in 2001. However, as Engineers Canada notes:

> The proportion of female undergraduate engineering students declined consistently from 2002 to 2005. This represents a troubling turnaround in a trend that saw female engineering enrolment grow for a full ten years prior to 2001. Similar trends have been detected in the U.S., where engineering enrolment of minority groups, including women, has been trending downward. One explanatory hypothesis is that occupational opportunities for women have grown rapidly, while males have not expanded their participation in non-traditional occupations. As a result, female university students are represented in increasing numbers in many other science disciplines that require fewer course prerequisites and offer equally challenging careers. . . . Female undergraduate enrolment in engineering peaked at 20.7 percent in 2001 and has declined to 17.5 percent in 2005. It is not simply that female enrolment has grown more slowly than male enrolment. The absolute number of women studying engineering at the undergraduate level declined by 7.0 percent over the past five years, while male enrolment increased by 14.6 percent.[9]

In postgraduate studies, total female enrollment was fairly stable, as the decrease in female master's students was matched by an increase in female doctoral candidates. In full-time engineering masters' programs, female enrollment was only 10 percent in 1989; by 1995, the figure was 20 percent; by 2001 it was 24 percent; but in 2005, it declined to 23 percent. For doctoral programs, the female enrollment was 6 percent in 1989; 13 percent in 1995; 17 percent in 2001, and 18.7 percent in 2005. Women comprised 2.2 percent of engineering faculty members in 1991 and 7.6 percent in 2001.[10]

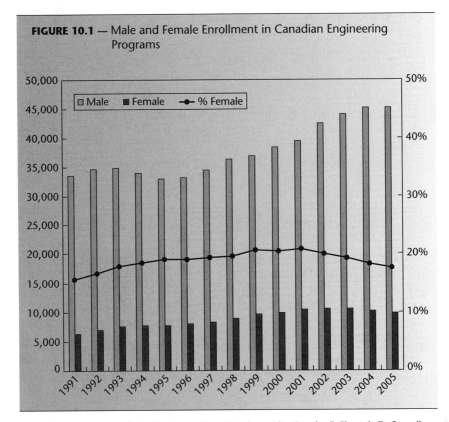

FIGURE 10.1 — Male and Female Enrollment in Canadian Engineering Programs

Source: Engineers Canada, "Undergraduate Enrolment by Gender," Chart 1–7, *Canadian Engineers for Tomorrow: Engineering Enrolment and Degrees Awarded, 2001 to 2005,* Ottawa, 2006, p. 4, available at <www.engineerscanada.ca/e/files/report_enrolment_eng .pdf>. Chart 1–7 reprinted with permission of Engineers Canada.

Universities vary in their ability to attract women to technical programs. Enrollment of women also varies by discipline, with higher enrollments in chemical, environmental, biomedical or biological, and industrial engineering and lower enrollments in electrical, computer, and mechanical engineering. In the past decade, other professions such as medicine, dentistry, veterinary medicine, and law have reached female enrollments of 50 percent; many have even surpassed this level. The low percentages in engineering suggest that the profession is not yet fully open to women.

Moreover, as Figure 10.1 shows, since 2003 there has actually been a decrease in the enrollment of women in engineering undergraduate programs in Canada. This new trend must be reversed if the profession is to achieve a reasonable representation by gender. At the same time as this decline, the enrollment of international students (mostly male) in Canadian engineering study programs increased from 6.0 percent of the student body in 2001 to 9.5 percent in 2005.[11]

SOCIALIZATION IN EARLY CHILDHOOD

A major reason for women's low participation in engineering is the perpetuation of gender stereotypes—that is, the myth that certain jobs and careers are appropriate for women and others, for men. Traditional choices for women are teaching, nursing, hairdressing, and secretarial jobs; men become engineers, pilots, firefighters, and dominate the trades. The media have a large influence in perpetuating these stereotypes; however, some teachers, parents, and guidance counsellors also condone them. A British Columbia provincial report explains how girls and boys begin to form gender-role stereotypes at a very early age and examines the effects of socialization (i.e., how children are brought up) and self-esteem on the education and training choices of girls and women.

> Significant gender inequalities persist in Canadian society and are reflected in and reinforced through the formal and informal processes of socialization. Gender socialization begins at birth, intensifies throughout childhood and adolescence, and continues as part of lifelong learning. From the moment we are born we begin the process of learning how to be women and men. We learn what attitudes, values, and behaviours are acceptable for each gender in our society. We learn what is expected of us, what roles we can play, how to exercise self-control, how to live in a community. Social scientists call this learning process socialization. When this process is applied to how women and men are expected to behave, this is called gender socialization.[12]

In *Failing at Fairness*,[13] the Sadkers report how teachers tend to pay more attention to boys. Boys are reprimanded more, but they are also given more encouragement to answer questions; they are challenged more than girls, and they are praised more for their answers. In contrast, girls tend to be praised for neatness and good behaviour. It is common to hear young women describe how they were discouraged from studying mathematics and science.[14] Some guidance counsellors make a special effort to encourage girls and boys to consider non-traditional career choices. However, others— women and men—discourage young people from doing so. In high school, few young women study physics, which is a prerequisite for engineering studies at university.

Lupart and Cannon found that by grade 7, students have decided on careers that either include information technology or exclude it. Boys selected careers in information technology (IT) as their first choice; girls selected IT as their sixth and last choice.[15] In her study of enriched mini-course attendance in engineering and science by gender, J.M.J. McDill found that "age 13 appeared critical for girls with respect to choices made for the future . . . and little change is expected after that age. Teenage girls appear to have formed strong opinions by this point, not only with respect to general subject area, but also for specific topics within those areas."[16] These two studies show that girls make early career choices; clearly, for the greatest impact, intervention is needed when girls are between the ages of 12 and 15.

In mathematics and science courses, girls now show equally strong academic performance to that of boys, yet most of them avoid choosing careers that involve these topics. For boys, the problem is different. Many boys believe that good marks are not as important as excelling in sports, and their parents often agree with this belief. This attitude obviously harms their chances of getting the marks they need for university entrance and of winning scholarships.

The media, parents, and schools must stop perpetuating stereotyped expectations for both sexes in order to ensure that all students reach their full potential. The result of gender socialization in schools leads to systemic discrimination in study choices. Sadly, as a result of poor counselling, some young people end up studying subjects that seem appropriate but are not what they would have chosen—based on their true interests and skills—had they been given more information and encouragement. Clearly, there is a need to eliminate this sort of misdirection.

RETENTION ISSUES

Retaining women and minority students in university engineering or geoscience programs is now less difficult than it was in the past, but retention rates vary greatly among universities. An important factor, especially for women, is the masculine culture of engineering and geoscience. Robinson and McIlwee[17] describe the culture of engineering and geoscience as one that emphasizes the importance of technology over personal relationships, formal abstract knowledge over inexact humanistic knowledge, and male attributes over female ones. Hall and Sandler[18] describe the chilly climate for women faculty and students. These are difficult aspects to change, since women and minorities who want to be accepted will generally help maintain the status quo. The culture tends to be preserved by faculty members, department heads, and deans who may not understand the issues, or who may not want to make changes for the sake of a small group of students. One incentive to do so is that making the environment friendlier for women and minorities also makes it friendlier for everyone.

Some improvements have been observed in recent years. For example, there is definitely more sensitivity regarding the use of inclusive language in classrooms; fewer sexist and racist remarks are heard. Obviously, the learning experience of women and members of minority groups varies from class to class, and from university to university. Some inappropriate acts and behaviour still occur. For example, a professor may ignore a female or minority student in a class, or may not intervene when the behaviour of some students is demeaning to others. Whether it is inadvertent or not, this attitude on the part of the professor adds to the problem instead of solving it. Professors and instructors must do more than simply show sensitivity to social diversity. As role models, they must participate actively in advancing the concepts of fairness, equity, and respect.

Universities have made special efforts to ensure fairness and equity, but if real change is to occur, those policies must be explained and enforced.

Individual professors and deans can be highly effective in bringing about improvements. The first-year experience is the most critical, and this is when special attention must be paid to retaining students. Some students drop out because they do not receive adequate encouragement; their career expectations are thereby severely reduced. This is also a loss for the university. Some universities have increased levels of retention and morale by instituting peer mentoring programs. Scholarships based on entrance qualifications, but with some directed to high-achieving women, are effective in attracting more women to engineering and geoscience. When an equitable balance is achieved, these programs may be eliminated.

WHAT UNIVERSITIES CAN DO

In addition to staging effective recruitment programs, such as summer camps and various outreach activities, faculties of engineering can increase the enrollment of women beyond current levels by designing curricular content carefully, and by developing teaching styles that are more appealing to women. This means including societal relevance in the content of courses and using less-traditional teaching styles, such as self-learning.

Attracting more women into graduate programs and hiring more women faculty members would have a positive impact. Universities need to create policies that allow young faculty members—female and male—to balance family and career. The criteria used for assessing faculty performance, based on decades of tradition, should be re-examined to see if they are still relevant. This applies not only to how merit is defined and measured, but also to how awards, appointments, and promotions are given.[19] For example, examining the quality of publications (such as the number of citations) rather than their quantity, and looking at the potential of candidates instead of what they have accomplished by the time of the interview, would be fairer to candidates who have taken a career break to have children.

Outdated stereotypes and biases can affect the success rates in competitions for scholarships, fellowships, awards, grants, jobs, promotions, and the allocation of research chairs. These biases can be reduced through training and sensitization programs, by ensuring fair gender representation on committees making decisions, and by making a pro-active effort to find qualified women for the positions or awards.

WHAT IS SEXUAL HARASSMENT?

Sandler and Shoop define sexual harassment as follows:

> Unwelcome sexual advances, requests for sexual favours, and other verbal or physical conduct of a sexual nature constitute sexual harassment when any one of the following is true:
>
> • Submission to such conduct is made either explicitly or implicitly a term or condition of a person's employment or academic advancement.

Photo 10.1—Hawker Hurricane Fighter Aircraft. *During the Second World War, the Hawker Hurricane was a combat fighter manufactured by Canadian Car & Foundry Co. in Thunder Bay (then Fort William). The aircraft was designed in Britain but produced in Canada under the direction of Elizabeth (Elsie) MacGill, the University of Toronto's first woman electrical engineering graduate. She received a master's degree in aeronautical engineering in the United States in 1929 and became chief aeronautical engineer for Canadian Car in 1938. She designed the Maple Leaf Trainer, a small training plane, before assuming responsibility for producing the Hawker Hurricane. Total Hurricane production was 2,000 aircraft—a significant contribution to the Canadian war effort.*

Source: Photograph courtesy of the Canada Aviation Museum, Ottawa.

- Submission to or rejection of such conduct by an individual is used as a basis for employment decisions or academic decisions affecting the person.
- Such conduct has the purpose or effect of unreasonably interfering with a person's work or academic performance or creating an intimidating, hostile, or offensive working, learning, or social environment.[20]

Briefly, the three key characteristics of sexual harassment are therefore that the behaviour is unwanted or unwelcome; the behaviour is sexual or related to the sex or gender of the person; and the behaviour occurs in context of a relationship where one person has more *formal* power than the other, or more *informal* power. Examples of a formal power relationship include: a work supervisor and an employee; a faculty member and a student; a physician and a patient. An informal power relationship exists where one peer or colleague, while nominally equal in the hierarchy, nevertheless exerts an influence over another.[21]

An earlier study reported that, although sexual harassment in the engineering workplace is not exclusively a problem for women, it might be more difficult for them, because only a few women may work at a site or in a firm.[22] They may feel isolated and perhaps find little support or sympathy for what may be perceived as a "boys will be boys" attitude. Companies vary greatly in their policies and procedures for dealing with harassment. Education is at the heart of the solution. To eliminate harassment in the workplace, companies need a fair investigation procedure that ensures that accusations are adequately verified, but does not victimize the complainant.[23]

When sexual harassment does occur, providing moral support for complainants can help reduce the stress they feel. Generally, informal investigations can accomplish more than formal ones, which are more confrontational. That said, in some situations a formal complaint is the only possible approach. Most organizations should provide both mechanisms when developing policies and procedures.

THE BENEFITS OF DIVERSITY

Women and minorities can bring new ideas to the professions. In a 1996 Canadian study, Ann van Beers[24] interviewed 20 female and 20 male engineers in the Vancouver area. She found that some of the women, when they became well established in their careers, were able to develop their own style and approach to problem solving. For example, they used a more contextual approach, had good communication and people skills, and liked the writing part of the work. In her study of science and engineering students at universities in British Columbia, Vickers and colleagues[25] found that a substantially larger proportion of females, compared to males, stated that making a contribution to society was an important criterion when choosing a career.

Human experience can contribute to good design. For example, a woman engineer in New Brunswick installed the first baby diaper change tables in ferry terminals, in both men's and women's washrooms. Her experience as a mother, combined with her civil engineering skills, added useful features to the previous models of terminal design. This feature is now found in airports and train stations everywhere.

Women's affinity for a consultative style of working is very much in tune with today's management philosophy. Their verbal and interpersonal skills, when combined with a solid technical education, become real assets, especially for smaller firms whose engineers must interact with suppliers, clients, and regulatory agencies. Similarly, individuals from other under-represented groups who have been raised in different cultures may bring different and innovative solutions to engineering problems. Everyone benefits when diverse, gender-balanced teams design products or solve environmental challenges.

ELIMINATING DISCRIMINATION

As mentioned earlier in this chapter, discrimination is unlawful. The Canadian Charter of Rights and Freedoms, provincial Human Rights Codes, and several provincial licensing Acts or regulations prohibit it. Employers, managers, and employees should be aware of these laws and ensure that they are followed. If further assistance is needed, many guides are easily found on the Internet. In particular, the PEO (Ontario) and the APEGGA (Alberta) guidelines (both cited earlier) discuss discrimination from the point of view of the professional, and advise employers and managers on ways to avoid it.

Discrimination against Women

A recent survey carried out for the Women in Engineering Advisory Committee of PEO (WEAC/PEO), which polled both female and male engineers across Canada, found that

> [w]orkplace challenges continue to exist for female engineers. Women feel they face at least some attitudinal barriers from their superiors, and a substantial proportion of men share that view. In particular, women are concerned about opportunities to network and to gain entry to executive levels in their organizations.[26]

The study concluded that "the workplace is changing in positive ways for women, but old, lingering beliefs held by even a few can act as barriers to full participation."[27]

In a large American study,[28] the authors interviewed recipients of prestigious fellowships in science and engineering to assess the career success of men relative to women. The study found that, on average, men were offered appointments at the Associate Professor level, while equally qualified women were offered appointments at the Assistant Professor level, from the highest rated colleges in the U.S. This was true for all fields except biology, where women were in larger numbers than in other fields. Other studies have also clearly demonstrated that bias exists in evaluating candidates for postdoctoral awards, or in selecting candidates for jobs. In the first study, Wenneras and Wold[29] demonstrated that men's performance had been overestimated, while women's had been underestimated; the women needed far more publications than the men to obtain the same postdoctoral fellowship. In the second study, men and women applicants had equal qualifications for 20 summer engineering jobs. However, male selectors chose 16 men and 4 women as the best candidates, whereas female selectors chose 11 women and 9 men as the best candidates for the 20 positions.[30]

Discrimination against Minority Groups and Disabled People

Discrimination takes many forms, and may be direct or indirect. Direct discrimination is usually easily identified, but indirect discrimination is subtler, and even the person who is discriminating unlawfully may not be aware that such conduct is illegal. Indirect discrimination occurs when an employer enforces a rule that has a greater effect on one group than another, for no good reason. For example, it is not discriminatory to expect an employee to be adequately experienced; however, a hiring rule that requires every applicant to have 10 years of experience in Canada may discriminate against recent immigrants. Similarly, requiring an applicant to produce a valid driver's licence may discriminate against disabled people (if driving is not a requirement of the job). Similarly, any restrictions on an applicant's height, weight, or strength would be discriminatory (provided that these criteria are not essential to perform the job).[31]

Employers and managers should get advice from the human rights department if there is any question that a specific job requirement might unfairly affect some groups more severely than others.

Advice to employees is clear: If you should undergo (or witness) direct or indirect discrimination, report the incident, following your company's internal policy. If your company does not have a policy for human rights complaints, your provincial Human Rights Commission will receive it. In most provinces, the licensing Association will also investigate allegations of harassment or discrimination involving a professional engineer or geoscientist.

FAIR PRACTICES IN THE WORKPLACE

Realistic objectives for hiring people from under-represented groups should be based on achieving better than, or at least the level of, the availability of each under-represented group in the pool of candidates. Creating a committee to identify diversity issues and to design strategies to increase the presence of personnel from under-represented groups can be highly effective.[32] Some good hiring practices are suggested below.

- Jobs should be advertised widely and externally besides being posted internally.
- A strong effort must be made to encourage qualified members of under-represented groups to apply. This means finding and contacting them, as they may be few in number and may not consult the mainstream advertising channels.
- Unbiased interview techniques should be used. People involved in hiring must be trained to recognize inappropriate and illegal interview questions and the importance of treating all applicants with fairness and respect. One way to test the appropriateness of a question is to ask whether every applicant—male, female, or a member of a minority group—should be asked precisely the same question. If the answer is no, the question should not be asked.

Discriminatory practices in hiring are often evidenced by a predominantly female (or visible minority) staff, supervised by a predominantly male (or white) senior management. For example, in the 1940s, bank tellers were almost exclusively female, while branch managers and senior executives were exclusively male. Since then, women with management potential have been identified and assisted, through pro-active training, to qualify for promotions. This process provides female role models for younger women and helps to integrate women's values into the corporate culture. In turn, the women who are promoted must be committed both to the organization and to the goal of employment equity, if real change is to occur.

Unexpected new benefits and insights result from hiring women and members of minority groups, especially when they are encouraged to introduce diversity and innovation, and when they feel their attributes and values are respected.

Employment Equity

Instituting an employment equity program to achieve a fairer societal representation is different from applying affirmative action or quotas. A Saskatchewan government brochure defines employment equity as follows:

> Employment equity is a comprehensive pro-active strategy designed to ensure that all members of society have a fair and equal access to employment opportunities. It is a process for removing barriers that have denied certain groups equal job opportunities. . . . Employment equity programs encourage employers to hire, train, and promote members of these groups.[33]

Employment equity is often misunderstood as being identical to a quota system. The latter refers to hiring from one group until the desired number is reached, whether the candidates are fully qualified or not. This approach can lead to frustration and anger among groups that feel disadvantaged. Employment equity is much fairer, and leads to a positive outcome. A survey of engineering graduates from 1989 to 1992 found that most female graduates had been hired by organizations with employment equity policies.[34,35] These same organizations also hired male engineers in proportion to the existing pool of male graduates. This showed that employment equity policies removed obstacles for under-represented groups, without adding barriers for the majority group (in this case, male engineers).

Fairness in Employee Performance Assessment and Promotions

DiTomaso and Farris[36] conducted a study of employee performance of American scientists and engineers in high-tech companies. The study included Caucasian men, foreign-born men, U.S.-born men from minority groups, and Caucasian women. The study showed significant differences in the ways that managers rated the performance of men and women, and in the expectations that men and women had of their managers. The study showed that, to avoid unfairness in performance ratings, managers should put more effort into developing objective and measurable criteria for assessments. They must also focus more attention on the type of feedback they provide, communicate the rules clearly, and test whether these rules have been understood.

Fairness in Awards and Tributes

In the past, it was rare to see a woman being nominated for an award or a prize, or for a woman to be invited as a keynote speaker, or for a woman to serve on an expert panel in a technical society or professional association. In the first decade of the 21st century, there are still professional events where only men appear in visible roles. For the sake of fairness, organizations should monitor the proportion of women and other under-represented groups on executive committees, in governing councils, and in those receiving awards or invited as speakers. If the quality of the candidates is really the criterion for rewards, women and minority professionals should be making the list more often.

INTEGRATING IMMIGRANTS INTO THE PROFESSIONS

Professionals should do more than merely obey laws against discrimination— we should give the less advantaged a helping hand. That is the Canadian way. Engineers Canada has taken a positive step to assist internationally educated graduates (IEGs) with a national project to remove barriers to the engineering profession in Canada.[37]

The Importance of Integration

Integration of immigrants is essential to Canada's future economic success. As mentioned briefly in an earlier chapter, the "baby boomers" (the generation born after the end of the Second World War in 1945) are beginning to retire, and the retirement rate will begin to peak in 2010, creating vacancies and opportunities for promotion. However, retirements will cause shortages of trained professionals, because Canadian birth rates have dropped in recent decades, and university enrollments are not adequate to fill the vacancies. This prediction was studied by Engineers Canada, which observes:

> Virtually every report produces a similar forecast: regardless of current labour force con-
> ditions, the engineering profession is facing a crisis of numbers. . . . Labour force ana-
> lysts have identified the solution for engineering and many other professions facing the
> same demographic imperative. If our population and education system cannot produce
> sufficient engineers to meet the needs of the economy, then immigration and licensing
> of engineers from other countries must be encouraged.[38]

Engineers Canada concluded that we must improve the way that Canada accepts and validates IEGs.

The Integration Project

In 2003, Engineers Canada started a national project called From Consideration to Integration (FC2I), to explore and overcome obstacles faced by IEGs entering the Canadian work force. The first phase of the three-phase project involved collecting information and consulting with government agencies, provincial and territorial licensing bodies, immigration agencies, universities, and employers. New immigrants must, of course, learn English and adapt to Canadian culture, but the most serious obstacles to success for IEGs are our procedures for validating foreign credentials and experience, and obtaining a first Canadian job. The first job is critical, because it is the key to getting work experience recognizable to Canadian employers.

The second-phase FC2I report made 17 recommendations to improve the integration of IEGs, without reducing education, experience, or language standards. These 17 recommendations are directed toward federal and provincial governments, licensing bodies, universities, and immigrant settlement groups, and call for streamlining of licensing requirements, adoption of best methods, cross-cultural adaptation, assisting the transition, and removing unnecessary obstacles.[39]

The final phase of the FC2I project, which is now underway, is to implement these recommendations. The average professional engineer or geoscientist should support these initiatives to integrate IEGs into the professional workplace. On a personal level, mentoring new immigrants is a vitally useful and important activity. Immigrant settlement organizations across Canada offer language training, skills upgrading, job support, and many other necessary tools for IEGs, but closer connections with the licensing Associations are important to disseminate information, to mount professional development courses, to help IEGs satisfy licensing requirements, and to help them get the key first Canadian job.

What We Can Do to Help Immigrants

Cross-cultural adaptation is mainly the burden of the immigrant, but employers and managers must encourage the adaptation. The average professional engineer or geoscientist can also assist by responding positively to IEGs in the workplace—a very small effort. Fortunately, guidance is available for everyone. For example, an excellent book, *Managing Cultural Diversity in Technical Professions,* by Lionel LaRoche, gives much useful advice for human resources managers on cross-cultural training.[40]

LaRoche illustrates how minor cultural differences can lead to serious misunderstandings. For example, in an appendix containing North American idioms, LaRoche lists many common sports sayings (from football and baseball, although surprisingly, he overlooked hockey and golf) that are common in daily conversation, and points out that these sayings may be meaningless to people who have never played these sports. (The term "home run" may be as meaningless to an immigrant as "Hit them for six!" is to the average Canadian.)

The concept of teamwork and methods of communication are subtly different in other cultures. For example, humour is common in the Canadian workplace, and generally makes work more enjoyable; however, an innocent joke, if misinterpreted, may cause serious offence to immigrants who come from cultures where work is not a place for humour. Although LaRoche's book is aimed mainly toward employers and managers, it is fascinating reading for any professional and, in particular, would help immigrants to decode the Canadian workplace and adapt to the new culture.

CONCLUDING COMMENTS

Employers and managers must recognize the benefits and new perspectives that women and minority groups bring to the organization, and must not underestimate the performance of these employees. People from diverse backgrounds and perspectives enrich and improve the organization's performance. Every organization must set its own goals, of course, but all should aim to create an environment where the work is challenging, inclusive, comfortable, and productive, for all employees.

Women are a majority in the Canadian population and their underrepresentation in professional engineering and geoscience still poses a challenge. Some relevant workplace issues and their solutions are discussed in greater detail in the report *Women in Engineering: More Than Just Numbers.*[41]

The rights of minorities and the disabled are clearly specified in the human rights legislation discussed earlier in this chapter. The special problem of integrating new immigrants into the professional workplace is an important challenge, and the pro-active initiative by Engineers Canada, discussed above, is to be lauded; it is worthy of our full support.

As our professions continue to integrate diversity, many of the obstacles and challenges mentioned in this chapter will eventually disappear. The following case studies (based on real situations) illustrate the challenges that exist. The names of the individuals and companies have been changed to protect the privacy of those concerned.

CASE STUDY 10.1

DISCRIMINATION ON THE BASIS OF GENDER

This case study is based on the "Case of the Mismanaged Ms.," in the *Harvard Business Review.*[42]

STATEMENT OF THE PROBLEM

You are an engineer and the chief executive officer of a profitable company called the Exeter Corporation. You are contacted by Susan Smith, a highly valued sales manager at Exeter, who has been "passed over" for promotion to director of product development. The promotion was given to Sam Brown, on the recommendation of the vice president of marketing, Peter Young. Smith considers this decision to be discrimination, and is threatening to sue Exeter for unfair practices. She asks you to respond to her concerns within 24 hours.

If you do not, you will probably lose a valuable employee, and her lawyer will be exploring the possibility of a settlement through the Human Rights Commission or the courts.

You arrange a meeting with Young and the human resources director. At the meeting, you ask why Brown was selected over Smith. You are told that the difference between the two candidates was marginal. Young's explanation for his recommendation includes both objective and subjective criteria. His first comment is that Brown's experience, seniority, and familiarity with the industrial sector weighed slightly in his favour. Young adds that, through Brown's greater participation in company social events and in the squash ladder, he was better known to all the vice presidents, who said that Brown "looked like a winner." They could not say the same about Smith because she was less well known.

When prodded by the human resources director, Young suggested an additional list of problems and shortcomings he attributed to Smith:

- "Mark Tannen, vice-president of manufacturing, is thought to be having an affair with Smith, and he is pushing her for promotion."
- "If Smith was promoted, Exeter might be liable to discrimination charges placed by Brown because of Mark Tannen's push for promotion for his honey."
- "The director of product development is a man's position. Human resources—soft, person-to-person stuff—is for women. Factories are for men."
- "Exeter clients prefer to deal with men. They know how to relate to their wives, mothers, and girlfriends, but not to women product development managers."
- "Women are undependable. They get married, get pregnant, want time off, and are less committed to the job."

Young provides no evidence to support these assertions. However, it is clear that they have influenced his decision to appoint Brown. He believes his decision made good business sense.

After the meeting, you reassess the situation. According to the objective data presented, Brown and Smith were both qualified for the position. Smith has shown excellent achievement as a product line manager. The same could be said about Brown. Understandably, choosing between the two would be difficult. Ignoring Young's subjective evaluation of Smith's "shortcomings," the choice was either promoting a woman (Smith) to a higher management level or promoting a man (Brown), who has marginally more experience.

You review your company's existing employment equity policies and current equity situation. Although one-quarter of the employees at Exeter are women, there are no women at the executive level and none on the board of directors. Recently, you and the human resources director issued a policy stating that the company would make great efforts to ensure equity and fairness in the manner in which employees are recruited, trained, and promoted.

Therefore, although you have no hard evidence, you worry that gender inequity may permeate the organization. Also, if Smith pursues her lawsuit, you wonder whether it may encourage other women to come forward and state similar experiences. You realize that if Peter Young's comments on Smith's "shortcomings" were repeated in the courts, Exeter would surely be found guilty of discrimination. Thus, the firm would experience both a financial loss and the loss of an excellent employee.

QUESTIONS

1. What criteria should have been used to select the new director of product development? Are these the same as Young's criteria?

2. Who should have been appointed to the job, based on your criteria: Sam Brown or Susan Smith? Explain your answer.

3. Since you are the CEO, what should you do in the next 24 hours regarding the potential lawsuit threatened by Smith?

4. As CEO, what long-term issues do you face if you want to ensure employment equity at Exeter, and what steps should you take to put this equity process into place?

5. Does your Association's Code of Ethics address this type of issue? Does it make a difference whether Peter Young, Sam Brown, and Susan Smith are also professional engineers or geoscientists? If your answer is yes, quote the appropriate sections of the code. If not, should the code provide guidance for dealing with this case, and what should it say?

CASE STUDY 10.2

SEXUAL HARASSMENT

STATEMENT OF THE PROBLEM

Michelle Kirkland has been a mechanical engineer in a consulting firm for four years. Recently, she wrote to a senior female engineer to discuss a serious work-related problem and to ask for advice on how to solve it. Below are extracts from her letter:

> In my academic years I never had any problems being a woman in a male-dominated environment, and therefore very naively entered the work force with a very positive and healthy attitude toward men in engineering. Today, unfortunately, that is no longer the case. After four years of verbal abuse and three incidents of sexual harassment from my immediate supervisor, I have become so cynical about men that I no longer enjoy my work. Most men quite naturally treat women without respect and as second-class citizens without even being aware of it.
>
> The worst part of the situation is that I feel I cannot talk to anyone about this. In our corporation, female managers are practically unheard of and men seem

to stick together like glue. Their attitude is that everything seems to be my fault: "Women are more sensitive" and "Women are less reliable" are the most recent comments that I bluntly received from my manager.

I was considering leaving the profession at one point, but meeting other women engineers motivated me to fight back harder and try again. Should I transfer to another department? Should I leave the company? (But are there any better ones out there?) Should I leave the profession and let my daughter solve the problems? I really do not know what to do. Sticking it out means additional stress in an already stressful job, headaches, and more anger. On the other hand, leaving means letting "them" win.

The senior engineer has sufficient personal knowledge of Kirkland and the atmosphere in Kirkland's workplace to believe that these allegations are true.

QUESTIONS

1. What would you suggest Kirkland do?

2. How does this work atmosphere of verbal abuse and harassment within the company affect the company's effectiveness and profitability? Is this a "professional" environment? Explain and justify your answer.

3. Was the manager's behaviour in violation of your province's Code of Ethics? If so, quote the appropriate sections. If not, what new clauses would you add to the code to deal with specific issues of harassment?

CASE STUDY 10.3

DISCRIMINATION ON THE BASIS OF RACE

STATEMENT OF THE PROBLEM

Assume that you are a professional engineer or geoscientist working for the Canadian branch of a large company with a head office located in a foreign country. A huge international technical Congress is to be held in your city, including a convention, technical seminars, and a trade show. Your company wants to present its products and services at the trade show, and you are in charge of designing, constructing, and staffing several product displays. The plan includes a reception area and a booth where a receptionist will greet people and direct them to the displays.

On the first day of the trade show, you have everything ready—displays, booths, and literature—and you have arranged with a temporary employment agency to hire a receptionist. The receptionist arrives early, and she appears to be suitable for the job. However, before you can explain her duties, your boss calls you aside, observes that the receptionist is Black, and states, "Company image is critical to the chief executives." He explains that Black people are very rare in the country where the head office is located, and when the chief executives arrive, they definitely will not expect a Black receptionist to welcome them and other visitors to the company's displays. Your boss instructs you to "Express our regrets to the receptionist, pay her for the day, and call the employment agency for another person to fill the job."

QUESTIONS

1. What action should you take?
2. What laws or Codes of Ethics may be violated if you follow your boss's directions? Does a foreign-based company have the right in Canada to specify the colour of people in key jobs? Does anyone?
3. If you follow your boss's directions, what action would (and should) the "receptionist" take for being offensively denied an opportunity to work?

Compare your answers with a very similar case, that of *Payne v. OHRC et al.*[43]

DISCUSSION TOPICS AND ASSIGNMENTS

1. You are the senior engineer responsible for employing and orienting new engineers and geoscientists in a large consulting firm. What policies would you expect the firm to have in place for interviewing, hiring, and promoting employees and for resolving internal disputes in order to ensure fairness and equity?
2. Assume that you are the senior partner in an incorporated mid-size professional consulting firm, and are in the process of hiring a new professional member. It is likely that the chosen person will stay with the firm for a long and profitable career. The top candidate for hiring has excellent university grades, three years of relevant experience since graduation, and has applied for a licence, although it has not yet been awarded. Before you make your final decision, you discuss the candidate's qualifications with other members of your firm. In each of the following cases, discuss the human rights issues, and whether the facts should affect the person's hiring:

   a. The candidate took a routine medical test, and was ruled as medically fit, but the physician reported that the candidate has a minor spinal deformity, and also was once treated for drug dependency.
   b. The candidate is a member of a visible minority, and other members are concerned about whether the candidate will "fit in" with the rest of the consultants.
   c. The candidate is a member of a religious group that celebrates different religious holidays than the other members of the firm, who are concerned about sharing the workload.
   d. The candidate was charged with welfare fraud as a student, but the charges were dropped.
   e. The candidate was convicted of welfare fraud, is presently awaiting trial for drug trafficking, and award of a professional licence is therefore in doubt.

3. You have heard a rumour that a supervisor is harassing a young person in your company. You are the senior person responsible for the department in which these two people work. What will you do about this situation? What measures should be in place to eliminate such a situation in the workplace? Should your actions be the same or different (and in what way) for the following four cases:

   - A female employee being sexually harassed by her male immediate supervisor.
   - A male employee being sexually harassed by his female immediate supervisor.
   - A visible minority employee being verbally harassed by a White supervisor.
   - A White employee being verbally harassed by a visible minority supervisor.

4. Sexual harassment, as defined in this chapter, includes "unwelcome sexual advances, requests for sexual favours, and other verbal or physical conduct of a sexual nature." In other words, it means that someone is annoying you by unwelcome actions of a sexual or gender-related nature. Would the following be considered sexual harassment, and if so, what action would be appropriate, under the circumstances?

   a. An older, senior female boss makes grossly offensive jokes and remarks about men—their activities and anatomy—during work hours. You must work with her on a fairly close basis, over a period of several days, during budget preparations, which occur quarterly. You find her conduct is unprofessional and working with her is unpleasant.

   b. An employee in the machine shop has sexually explicit posters on the full length of a private locker door in a change room. The locker is closed throughout the day, but is open at starting time and quitting time, when employees must change into protective clothing to work in an electronics room. You do not use the locker room, and no one has complained, but you find the posters to be pornographic and unprofessional.

5. Prepare a display or talk for a local high school to show the engineering or geoscience activities at your company (or opportunities for these studies at your university). Ensure that your message conveys the idea that the career paths and opportunities are open to all, regardless of race, sex, religion, or disability, and so on. Consider using recent videos about careers in engineering and geoscience and other means of attracting young people to these fields. You especially wish to portray how engineers and geoscientists apply their knowledge for the benefit of humankind, to solve problems and to design the world in which we live and work. What will you say, and how will you say it?

Additional assignments can be found in Appendix E.

NOTES

[1] Professional Engineers Ontario (PEO), *Guideline on Human Rights in Professional Practice*, June 2000, available at <www.peo.on.ca> (May 29, 2008).

[2] The Association of Professional Engineers, Geologists and Geophysicists of Alberta (APEGGA), *Guideline for Human Rights Issues in Professional Practice*, V1.1, October 2005, available at <www.apegga.org/Members/Publications/guidelines.html> (May 29, 2008).

[3] Government of Canada, *The Canadian Charter of Rights and Freedoms*, Schedule B to the Canada Act 1982, clause 7, available at <http://www.laws.justice.gc.ca/en/charter> (May 29, 2008).

[4] Ibid., clause 15(1).

[5] Ibid., clause 15(2).

[6] M. Frize, C. Deschênes, E. Cannon, M. Williams, and M. Klawe, "A Unique Project to Increase the Participation of Women in Science and Engineering (CWSE/Canada)," presented at the Engineering Foundation conference on Women in Engineering, Mont Tremblant, Quebec, July 14–18, 1998.

[7] Government of Ontario, *Human Rights Code*, R.S.O. 1990, Chapter H.19, available at <www.e-laws.gov.on.ca/html/statutes/english/elaws_statutes_90h19_e.htm> (May 29, 2008).

[8] APEGGA, *Guideline for Human Rights Issues in Professional Practice*, p. 8.

[9] Engineers Canada (formerly Canadian Council of Professional Engineers— CCPE), *Canadian Engineers for Tomorrow: Trends in Engineering Enrolment and Degrees Awarded 2001 to 2005*, p. 4, available at <www.engineerscanada.ca/e/pu_enrolment.cfm> (June 15, 2009). Excerpt reproduced with permission of Engineers Canada.

[10] Engineers Canada, *Canadian Engineers for Tomorrow*.

[11] Ibid., p. 5.

[12] R. Coulter, *Gender Socialization: New Ways, New World*, British Columbia Ministry of Equality, Victoria, BC, 1993.

[13] M. Sadker and D. Sadker, *Failing at Fairness: How America's Schools Cheat Girls*, Charles Scribner's Sons, New York, 1994.

[14] W.H. Peltz, "Can Girls + Science − Stereotypes − Success? Subtle Sexism in Science Studies," *The Science Teacher*, December 1990, pp. 44–49.

[15] J.L. Lupart and M.E. Cannon, "Gender Differences in Junior High School Students Towards Future Plans and Career Choices," *Proceedings of the 8th CCWEST*, St. John's, 2000.

[16] J.M.J. McDill, "Participation Trends in and Lessons Learned From Outreach," *WEPAN 2003 Conference Proceedings*.

[17] J.G. Robinson and J.S. McIlwee, "Men, Women and the Culture of Engineering," *Sociological Quarterly*, vol. 32, March 1991, pp. 403–421.

[18] R. Hall and B. Sandler, *The Classroom Climate: Projecting the Status and Education of Women*, Association of American Colleges, Washington, DC, 1982.

[19] P. Caplan, *Lifting a Ton of Feathers: A Woman's Guide to Surviving in the Academic World*, University of Toronto Press, Toronto, 1992; and Natural Sciences and Engineering Research Council (NSERC), *Towards a New Culture: Report of the Task Force on How to Increase the Participation of Women in Science and Engineering Research*, NSERC, February 1996.

[20] B.R. Sandler and R.J. Shoop (eds.), *Sexual Harassment on Campus: A Guide for Administrators, Faculty, and Students*, Allyn & Bacon, Boston, 1997, p. 4.

[21] Sandler and Shoop, *Sexual Harrassment on Campus*.

[22] C.M. Caruana and C.F. Mascone, "Women Chemical Engineers Face Substantial Sexual Harassment: A Special Report," *Chemical Engineers Progress*, January 1992, pp. 12–22.

[23] M. Frize, "Eradicating Sexual Harassment in Higher Education and Non-Traditional Workplaces: A Model," *Proceedings*, Canadian Association against Sexual Harassment in Higher Education Conference, Saskatoon, November 1995, pp. 43–47.

[24] A. van Beers, *Gender and Engineering: Alternative Styles of Engineering*, master's thesis, Department of Sociology and Anthropology, University of British Columbia, Vancouver, July 1996.

[25] M. Vickers, H.L. Ching, and C.B. Dean, "Do Science Promotion Programs Make a Difference?" *Papers and Initiatives* (ed. M. Frize), More than Just Numbers Conference, University of New Brunswick, Fredericton, May 1995.

[26] Professional Engineers Ontario (PEO), *National Report of Workplace Conditions for Engineers*, Women in Engineering Advisory Committee, Toronto, ON, available at <www.peo.on.ca/publications/reports.html#Report> (May 29, 2008).

[27] Ibid.

[28] G. Sonnert and G. Holton, "The Career Patterns of Men and Women Scientists," *American Scientist*, January–February 1996.

[29] C. Wenneras and A. Wold, "Nepotism and Sexism," *Nature*, May 1997.

[30] M. Foschi, L. Lai, and K. Sigerson, "Gender and Double Standards in the Assessment of Job Applicants," *Social Psychology Quarterly*, vol. 17, April 1994, pp. 326–339.

[31] Professional Engineers Ontario (PEO), *Guideline on Human Rights in Professional Practice*, p. 10.

[32] N. DiTomaso and G.F. Farris, "Diversity in the High-Tech Workplace," *Spectrum*, June 1992, pp. 21–32.

[33] Government of Saskatchewan, *Employment Equity (Women in the Workplace)*, Brochure, Women's Secretariat, Regina.

[34] M. Frize, "Remarks on What is Happening to UNB's Engineering Graduates?" *APENB Bulletin*, June 1993, pp. 9–10.

[35] M. Frize, "The Engineering Profession: Is It Friendlier for Women?" *IEEE Engineering in Medicine and Biology Magazine*, vol. 13, no. 1, March 1994, pp. 15–17.

[36] DiTomaso and Farris, "Diversity in the High-Tech Workplace."

[37] The textbook author, Dr. G. Andrews, contributed this section on integrating immigrants.

[38] Engineers Canada, *Canadian Engineers for Tomorrow*, p. 11.

[39] Engineers Canada, *From Consideration to Integration (FC2I)*, Phase II Report, July 2004, p. 11, available at <http://fc2i.engineerscanada.ca/e/about_overview.cfm> (June 15, 2009).

[40] L. LaRoche, *Managing Cultural Diversity in Technical Professions*, Butterworth-Heinemann (Elsevier), Burlington, MA, 2003.

[41] Canadian Committee on Women in Engineering, *Women in Engineering: More Than Just Numbers*, Report of the Canadian Committee on Women in Engineering, University of New Brunswick, Faculty of Engineering, Fredericton NB, 1992.

[42] This case is based on S. Seymour, "Case of the Mismanaged Ms.," *Harvard Business Review*, November–December 1987, pp. 77–87. See also A. Mikalachki, D.R. Mikalachki, and R.J. Burke, *Teaching Notes to Accompany Gender Issues in Management: Contemporary Cases*, McGraw-Hill Ryerson, Toronto, 1992, pp. 5–8.

[43] *Payne v. OHRC et al.*, Human Rights Tribunal of Ontario, 2002, available at <www.ohrc.on.ca/en/resources/annualreports/AnnualReport23/pdf> (June 15, 2009).

Chapter 11
Principles of Ethics and Justice

Engineers and geoscientists are well-trained to solve technical problems; however, many technical problems have ethical consequences. To ensure that your solutions are both technically correct and ethically right, you need a basic knowledge of the principles of ethics and justice.

In this chapter, we examine four ethical theories that have been used for centuries as fundamental guides for solving ethical problems. These ethical theories must be applied carefully, because solutions that are ethically right in theory may sometimes be unfair in practice. To ensure fairness, we must also apply basic principles of justice.

This chapter also discusses professional Codes of Ethics, which are practical guides to professional conduct. The chapter concludes by describing a general strategy for solving ethical problems. Readers will find this problem-solving strategy informative and reassuring. The ethics and justice concepts in this chapter are applied to practical case studies later in this textbook.

ETHICS AND PROBLEM SOLVING

Engineers and geoscientists are trained to be problem solvers. When we are faced with a technical problem, we solve it using well-known theorems and laws from mathematics and science. It is reassuring to know that basic theories exist to solve ethical problems, also. These ethical theories—developed over the centuries—form the basis of our laws, regulations, and Codes of Ethics. Let us begin this overview by defining our terms.[1]

Ethics is one of the four branches of philosophy (according to one system in common use). Each branch investigates different fundamental questions. The four branches are:

Ethics: The study of right and wrong, good and evil, obligations and rights, justice, and social and political ideals.

Logic: The study of the rules of reasoning. For example, under what conditions can an argument be proved true?

Epistemology: The study of knowledge itself. What is knowledge? Can we know anything? What can we know? What are the sources of knowledge?

Metaphysics: The study of very basic ideas such as existence, appearance, reality, and determinism. Metaphysics asks questions about the most abstract and basic categories of thought: thing, person, property, relation, event, space, time, action, possibility/actuality, and appearance versus reality.[2]

Ethics and logic have many practical applications in our lives. Ethics helps us to distinguish right from wrong, an ability that signalled the dawn of civilization. Ethical writings can be traced back over 3,000 years. In fact, many ethical concepts that we commonly apply today are older than our basic mathematical and scientific disciplines (such as calculus, statics, dynamics, stress analysis, and so forth), which originated in the 17th century. Logic is also important to engineers and geoscientists, because it is the basis of mathematical derivation. By contrast, epistemology and metaphysics are highly theoretical, and have very few practical applications.

FOUR ETHICAL THEORIES

Many prominent philosophers have devoted their lives to developing ethical theories, and a thorough discussion of their thought would fill a thousand textbooks. We can hardly hope to condense this treasury of philosophical thought into a single chapter. However, let us discuss a few basic ethical theories that apply directly to common ethics problems, and which are already embedded in our customs, laws, and Codes of Ethics.

Many ethical theories have evolved over the centuries, but the four theories selected below are likely the best known. They differ significantly, and none of them is universally superior to the others; even so, it is startling to see how well they agree when applied to certain ethical problems. Each theory is identified by the name of its proponent, but many earlier philosophers contributed to formulating the theories, and many modern philosophers have suggested improvements.

- Mill's utilitarianism
- Kant's formalism, or duty ethics
- Locke's rights ethics
- Aristotle's virtue ethics

Mill's Utilitarianism

John Stuart Mill (1806–1873) was the major proponent of utilitarianism, which states that, in any ethical problem, the best solution produces the maximum benefit for the greatest number of people.[3] This theory is probably the most common justification for ethical decisions in engineering, geoscience, and, indeed, in modern society. Democratic government itself can be justified on utilitarian grounds, since it permits the maximum good (control over government) for the maximum number of people (the majority of voters).

The difficulty of applying the utilitarian principle lies in quantitatively calculating the "maximum benefit." Mill proposed that three key factors should be considered in determining the maximum benefit: the number of people affected, the intensity of the benefit involved, and its duration (or, conversely, the severity and duration of the pain to be avoided). For example, consider automobile seat-belt legislation: All drivers and passengers endure some brief inconvenience when they buckle up, whereas only a few people obtain the benefit (the avoidance of injury or death, when they are involved in accidents). However, we would agree that the intensity and duration of the distress (injury or death), avoided by relatively few people, is so severe that it outweighs the brief inconvenience (buckling-up) endured by everyone else.

In evaluating benefits, it is important that we apply certain criteria:

- The benefit to oneself must not be given any greater value or importance than the same benefit to any other individual.
- No preference should be given to personal friends or favoured groups. All benefits should be awarded without regard to race, creed, colour, language, sex, and so on.
- Benefits must be distributed equally. That is, when selecting a course of action, an equal distribution of benefits is preferable to an unequal distribution.

In summary, utilitarianism states that the best course of action in an ethical problem is the solution that produces the maximum benefit for the greatest number of people, with the benefit most equally divided among those people.

Utilitarianism is very valuable in making ethical decisions. The utilitarian theory is easily understood; it is consistent with the concept of democracy; and in many cases, it is easy to apply. For example, income tax is easily justified by utilitarian theory. A modest hardship (paying tax) is imposed equally on all residents (as a percentage of income). This yields an immense benefit to society, because the tax dollars support hospitals, schools, and essential infrastructure. If we were to eliminate income tax, individuals would have to provide their own private health care, schooling, and roads—a virtually impossible task. We may sometimes disagree with the details, such as tax rates and exemptions, but income tax yields the maximum benefit to the greatest number of people, with the hardship fairly equally distributed.

Kant's Formalism, or Duty Ethics

The theory of duty ethics, or "formalism," is based on the work[4] of Immanuel Kant (1724–1804), who proposed that every individual has a fundamental duty to act in a correct ethical manner. This theory evolved from Kant's belief or observation that each person's conscience imposes an absolute "categorical imperative" (or unconditional command) on that person to follow those courses of action that would be acceptable as universal principles for everyone to follow. For example, everyone has a duty not to tell lies, because if we tolerated lying, then no promises could be trusted, and our society would be unstable. This makes sense to most people; almost everyone has this innate sense of duty, and believes that rules of conduct should be rules that everyone should follow.

Kant believed that the most basic good was "good will," or actively seeking to follow the categorical imperative of one's conscience. This is in marked contrast to Mill, who believed that universal happiness was the ultimate good. In Kant's philosophy, happiness is the result of good will: the desire and intention to do one's duty.

Kant emphasized that it was the intention to do one's duty that was significant, not the actual results or consequences. One should always do one's duty, even if the short-term consequences are unpleasant, since this strengthens one's will. For example, even "white" lies should not be tolerated, since they weaken the resolve to follow one's conscience. The formalist theory contends that in solving an ethical dilemma, one has a duty to follow rules that are generated from the conscience (the categorical imperative), and that if a person strives to develop a good will, happiness will result. Many of the rules that support this universal concept are well known—"Be honest," "Be fair," "Do not hurt others," "Keep your promises," "Obey the law," and so on—and not surprisingly, our happiness would certainly increase if everyone followed them.

Kant also stated that a consequence of following the categorical imperative would be an increased respect for humanity. Life should always be treated as an end or goal, and never as a means of achieving some other goal. Kant's formalism would condemn water or air pollution as unethical, and would include any activity that endangered life, regardless of the purpose of the activity. In Kantian formalism, everyone (and each engineer or geoscientist, in particular) has an individual duty to prevent harm to human life and to consider the welfare of society to be paramount. (As you will see later in this chapter, this axiom from Kant is the first rule in almost every Code of Ethics.)

In sum, Kant's formalism emphasizes the importance of following universal rules, the importance of humanity, and the significance of the intention of an act or rule rather than the actual outcome, in a specific case. The only problem with applying formalism relates to its inflexibility—duties based on the categorical imperative never have exceptions. Fortunately, we can obtain further guidance by considering the other ethical theories.

Locke's Rights Ethics

The rights-based ethical theory comes mainly from the thought and writings of John Locke (1632–1704).[5] Rights-based theory states that every individual has rights, simply by virtue of his or her existence. The right to life and the right to the maximum possible individual liberty and human dignity are fundamental; all other rights flow out of them. Each individual's rights are basic; other people have a duty not to infringe on those rights. This contrasts with Kant's duty-based ethical theory, which contends that duty is fundamental; in the rights-based theory, duties are a consequence of personal rights.

Locke's writings had a powerful impact on British political thought in the 1690s; they also motivated the French and the American revolutions. Basic human rights are embedded in the U.S. Constitution and in the Canadian Charter of Rights and Freedoms. The Charter recognizes that everyone has the following rights:

- Fundamental freedom of conscience, religion, thought, belief, opinion, expression, peaceful assembly, and association (clause 2).
- Democratic rights to vote in an election (or to stand for election) of the House of Commons or of a legislative assembly (clause 3).
- Mobility rights to enter, remain in, and leave Canada (clause 6).
- Legal rights to life, liberty, and security of the person and the right not to be deprived of these rights except in accordance with principles of fundamental justice (clause 7).
- Equality rights before and under the law and the right to equal benefit and protection of the law (clause 15).[6]

We must recognize that everyone has these basic rights and that they should not be infringed upon. However, the Charter does not contain every right that should exist—just the fundamental rights that have been hammered out in Parliament and in the courts of law over the past two centuries. Other rights have evolved from Locke's theory. Some of these rights are enacted in other legislation. Some exist in grey areas of uncertainty, and some "rights" are claimed as a cloak for selfishness. Examples of these three types follow:

- Locke's theory suggests that everyone has the right to a working environment that is free from sexual harassment or racial discrimination. This right would appear to be common courtesy. Few would challenge it, and it is generally included in provincial labour laws. The right to a smoke-free environment is not in the Charter, but many municipal and provincial laws now guarantee it.
- However, many people claim rights that are not in the Charter or in other legislation. These rights fall into grey areas and must be examined individually. For example, it would seem to be common courtesy to extend the right to private e-mail, the right not to be subjected to loud noise, the right not to be insulted or bullied on the Internet, and so on. These rights are based on human dignity and individual liberty, and they should be respected unless it can be proved that denying such rights satisfies a greater good.
- Rights-based ethical theory does have limits. As an example, consider income tax again. Even today, some people challenge the concept of income tax, claiming that it infringes on the individual's right to retain his or her property. Others insist that they have a right to smoke in public buildings, even though such behaviour has been ruled illegal in most Canadian cities.

Clearly, rights-based arguments cover a spectrum. Some rights are indisputable and are embedded in law; other rights fall into a grey area, suggesting that they should be respected wherever possible, but are not absolute.

However, some people claim rights that are really selfishness in disguise. In summary, rights-based ethics has an important place in resolving ethical dilemmas, but the theory is not sufficient to deal with every situation.

Aristotle's Virtue Ethics

Aristotle (384–322 B.C.) was one of many early Greek philosophers whose thoughts are still relevant over two millennia later. Aristotle observed that the quality or goodness of an act, object, or person depended on the function or goal concerned. For example, a "good" chair is comfortable, and a "good" knife cuts well.[7]

Similarly, happiness or goodness will result for humans once they allow their specifically human qualities to function fully. Aristotle observed that humans have the power of thought—the one sense that animals do not have. Therefore, he postulated that humans would achieve true happiness by developing qualities of character using thought, reason, deduction, and logic. He called these qualities of character "virtues," and he visualized every virtue as a compromise between two extremes or vices.

His guide to achieving virtue was to select the "golden mean" between the extremes of excess and deficiency. For example, modesty is the golden mean between the excess of vanity and the deficiency of humility; courage is the golden mean between foolhardiness and cowardice; and generosity is the golden mean between wastefulness and stinginess.

Aristotle's virtue-based philosophy is admirable, and most people have an innate ambition to lead a virtuous, balanced life. Although the concept of virtue is subjective, open to interpretation, and not a universal rule, it still has some obvious applications. In particular, Aristotle's concept of the golden mean is extremely useful in solving ethical problems by considering the extremes and seeking the compromise—the golden mean, or the "happy medium"—between the extremes. This approach is often useful in ethical problems.

AGREEMENT AND CONTRADICTION IN ETHICAL THEORIES

The four theories described above have survived the test of centuries, and all of them are useful in finding fair solutions to ethical problems. Table 11.1 offers a brief summary of the theories. Each theory has a wide range of applications, but none is superior in every situation. Philosophers have long been seeking the one universal principle on which all ethical thought is founded, but a single unifying concept has yet to emerge.

In many applications, all four theories agree completely. Sometimes, however, they contradict, and each theory yields its own unique answer to the same problem. We call this contradictory type of ethical problem a "dilemma." A dilemma is commonly defined as an ethical problem that requires a person to choose between two opposing courses of action. (Note: We often use the term "dilemma" in this textbook for problems with more than two possible outcomes.)

TABLE 11.1 — Summary of Four Ethical Theories

	Statement	Conflict
Mill's Utilitarianism	An action is ethically correct if it produces the greatest benefit for the greatest number of people. The duration, intensity, and equality of distribution of the benefits should be considered.	A conflict of interest may arise when evaluating the benefits, or when distributing them equally. Benefits must not favour special groups or personal gain.
Kant's Duty-Based Ethics	Each person has a duty to follow those courses of action that would be acceptable as universal principles for everyone to follow. Human life should be respected, and people should not be used as a means to achieve some other goal.	Conflicts arise when following a universal principle may cause harm. For example, telling a "white" lie is not acceptable, even if telling the truth causes harm.
Locke's Rights-Based Ethics	All individuals are free and equal, and each has a right to life, health, liberty, possessions, and the products of his or her labour.	It is occasionally difficult to determine when one person's rights infringe on another person's rights. Also, people occasionally claim self-serving "rights."
Aristotle's Virtue-Based Ethics	Happiness is achieved by developing virtues, or qualities of character, through deduction and reason. An act is good if it is in accordance with reason. This usually means a course of action that is the golden mean between extremes of excess and deficiency.	The definition of virtue is occasionally vague and difficult to apply in specific cases. However, the concept of seeking a golden mean between two extremes is often useful in ethics.

As an example of agreement between the theories, consider the Golden Rule: "Do unto others as you would have others do unto you." This is a clear statement of Kant's formalism: it imposes a duty on the individual to respect human life as a goal rather than as a means to achieve some other goal. On the other hand, it could be considered a utilitarian principle, since it brings the maximum good to the maximum number of people. Any inconvenience to the individual is balanced by the equal or greater benefit to the people with whom that person comes into contact. The proponents of rights-based ethics would agree with the Golden Rule, but would claim that the duty of the individual to act fairly comes from the rights of others to be treated fairly. Finally, Aristotle would recognize "fairness" as a virtue. The four ethical theories are therefore consistent in identifying the Golden Rule as a good maxim for guiding human behaviour, as we would expect.

Similarly, the basic ethical precepts of most religions are supported by all four ethical theories. Consider the Ten Commandments from the Book of Exodus, which are the ethical basis of Judeo-Christian religions. Each of the commandments clearly imposes a duty on the individual and at the same time grants rights to others, requires virtuous behaviour, and creates a stable environment that yields the maximum benefit for all. An investigation of the basic precepts of all the great religions would show similar agreement.

Ethical theories agree remarkably well in many cases, but they may contradict in other situations. For example, consider the following hypothetical case.

An Ethical Dilemma

BACKGROUND INFORMATION
Professional engineers Smith and Legault are both fairly senior employees, with over 10 years of experience. They are part of a 10-person team assigned to develop and test a massive software control system for an electrical power generating plant, which is under construction. They are good friends, and occasionally party together after work. Smith drinks heavily and often takes illicit hard drugs. Legault suspects that Smith has an addiction or dependency on alcohol. At times, Smith has wide mood and attitude swings. The project manager has cautioned Smith for absenteeism, but no disciplinary action was taken. Legault occasionally conceals minor errors and "covers" for Smith's absences.

As a friend, Legault is concerned that Smith's erratic behaviour will eventually result in discipline of some sort. Equally important, Legault is worried that Smith's alcohol and drug abuse is affecting Smith's work, and that the software may be faulty. Legault has repeatedly tried to convince Smith to seek treatment, but Smith denies that any problem exists. Legault hesitates to take any further action because of their close personal friendship.

QUESTION
Today, the control software failed a preliminary test. Legault has checked the data dump, and it appears that Smith's coding is the likely cause of the failure. The entire team is dismayed. Legault faces a dilemma: should Legault continue to protect Smith as a friend, or should Legault report Smith's drug use and suspected alcohol dependency?

AUTHOR'S ANALYSIS
In a real situation, you would have much more information, but a few facts are obvious: Every project must meet tight deadlines, so faulty software could cause an expensive delay. The software will be fully tested for safety before release, but sloppy coding might cause inefficient operation. Let us apply the ethical theories to the dilemma.

- **Duty theory.** As a friend, Legault has a duty to help Smith overcome the dependency, and must not act on unproven allegations. The problem statement implies that these duties have been met: the abuse is well known and Smith has refused assistance. Legault also has a duty to colleagues, whose jobs may be jeopardized if the project fails. Legault also has a duty to the general public to ensure that the software is developed professionally, runs efficiently, and does not contain hidden bugs. In fact (as discussed later) every Code of Ethics states that the public interest should come first. The duty-based theory overwhelmingly indicates that Legault must insist that Smith seek treatment, even if it means reporting the problem to management.

- **Rights theory.** Conversely, the rights-based theory would say that Smith's health is a private matter. Smith has a right to personal privacy, and Legault has no right to investigate Smith's health or to discuss it with anyone.

Obviously, the duty-based and rights-based theories yield simple, clear rules, but those rules contradict each other. We must examine the other theories to hear what they say. The utilitarian and virtue-based theories require a subjective judgment, so more information is usually needed before we can apply them. In this case, the degree of danger to others, the seriousness of the abuse or dependency, and Smith's willingness to seek treatment are relevant factors.

- **Utilitarianism.** The utilitarian theory balances the risk of harm to the project and to the public (if Legault does not intervene), against the risk of harm to Smith's career (if Legault exposes the addiction). The estimated intensity of such harm is a factor. If the software fails the final validation test, the project will be delayed, the employer will suffer a loss, the whole team may suffer, and Smith's health problems may become known anyway. Legault's failure to act may simply have delayed the inevitable, and made the outcome worse for everyone. The utilitarian theory—even based on such meagre information—would favour intervention, because the greatest good, for the greatest number, would outweigh Smith's potential loss.
- **Virtue.** The virtue-based theory would recognize drug and alcohol dependency as extreme and undesirable. The golden mean between abstinence and addiction is moderate use. The virtue-based theory would condemn Smith's abuse, and therefore imply that some action should be taken to alleviate it.

SUGGESTED DECISION

Even with the limited information provided, three of the four theories clearly recommend intervention. However, while this may be the end of the ethical discussion, it is not the end of the problem. Knowing the right course of action, finding the courage to implement it, and doing so fairly are separate matters.

Ideally, the process must be fair, and must preserve Smith's dignity and self-respect. Legault might still convince Smith to take sick leave and enter a recovery program, thus salvaging Smith's career and finances. Since a large corporation typically has an Employee Assistance Program (EAP) to help employees with serious personal problems, contacting the EAP would be a good start. Other help may be available. As a last resort, Smith should be reported to the department manager. The role of Legault, as a friend, is not to conceal the problem, nor is it to be a snitch; rather, it is to apply the decision fairly, with a minimum of personal chaos.

In summary, examining a dilemma using the four ethical theories usually gives the right solution. When theories contradict, then the theory that is *most* appropriate must be followed. This requires a value judgment, and is therefore subjective. The good news is that the person making the subjective judgment will have a clear conscience if the decision is made in an orderly process, is consistent with a recognized ethical theory, and is made fairly. Fairness is explored in the next section.

PRINCIPLES OF JUSTICE

If you are faced with an ethical dilemma, and your decision agrees with one of the ethical theories above, it is probably right, but is it fair? Curiously, an ethical decision may be unfair, even if it agrees with ethical theories. In a word, you must also seek justice, which is a key subdivision of ethics. A legal dictionary gives the following definition of justice: "A state of affairs in which conduct or action is both fair and right, given the circumstances."[8]

In other words, your decision must satisfy both the test of "rightness," by agreeing with the ethical theories, and also the test of "fairness." Fairness is harder to identify, although its opposite, "unfairness," is usually obvious. Let us examine the concept of justice, and consider how unfairness might arise.

Justice may be categorized in several ways. The following overview considers four basic categories of justice, depending on the application.[9]

1. Procedural Justice—Fairness in the Decision-Making Process

According to a well-known saying: "Justice must be done, but it must also be *seen* to be done." That is, the decision-making process itself must be fair, and must treat those involved with dignity and respect. For example, consider a supervisor who fires an employee for incompetence, based on rumours from co-workers, without discussing the reasons with the employee. Even if the employee deserved to be fired, everyone would see this as an unfair process. In other words, even if the decision was right, the process was unfair.

Fortunately, fair procedures have evolved over the centuries based on two principles that are usually called "natural justice." The two principles are: the right to be heard, and the right to be judged by an impartial person. These principles are called "natural," because they are obviously fundamental and self-evident. Like axioms in mathematics, they do not need any further proof.

- **The right to be heard.** This principle requires that a person must be informed when the person's rights or property are in jeopardy, and must be permitted to defend himself or herself. In law, this principle is best known as *habeas corpus*, which was a British Act passed in 1679. Habeas corpus entitles people imprisoned without charge to challenge their detention—a vital personal right. It is still a fundamental part of our law, and is cited in the Canadian Charter of Rights and Freedoms.[10]

- **The right to be judged by an impartial person.** This principle requires ethical decisions to be based solely on the merits of the case. Obviously, a judge or decision maker must be unbiased, and must have no personal interest or involvement in the case. Otherwise, the judge would have a conflict of interest, and the outcome would not be seen to be fair. When a judge (or decision maker) has a conflict of interest, the judge must declare the conflict, step aside, and turn the role over to an impartial judge.

Over the years, many more specific rights have been established, based on these two basic axioms of natural justice. For example, a person charged with a criminal offence has the right to require the prosecutor to disclose all facts and documents upon which the charge is based. Natural justice requires this full disclosure to allow the person to present a proper defence.

Similarly, in professional discipline hearings (discussed in Chapter 4), a person who investigates an allegation of misconduct is prohibited from sitting on the Discipline Committee that decides the outcome. It might be argued that a person who is already familiar with the case would make a better judge (or a faster decision). However, we exclude anyone with prior knowledge of the case to avoid any perception of bias that would violate the principles of natural justice.

GUIDANCE FOR THE PROFESSIONAL

Professional engineers and geoscientists may need to resolve internal disputes in their companies, or may be asked to serve on Complaint or Discipline Committees within the profession. In doing so, the professional must always follow the principles of natural justice. That is, you must always

- **Get all of the information**. Allow each person to state their point of view, and to challenge statements made by others. The goal is to ensure that you obtain all of the information needed to make the right decision. When possible, it is fairer to ask those concerned to discuss the conflicts directly with one another, rather than interviewing the people separately. Direct, face-to-face exchanges help to reduce misunderstandings and inaccuracies. Make certain that all documents are disclosed to everyone involved.
- **Act impartially and consistently**. Do not let any personal benefit, conflict, or bias affect the outcome. Avoid any emotional involvement or personality differences (wherever possible). Maintain a positive attitude, and never pre-judge the outcome. In other words, you must act in good faith. When a conflict of interest is unavoidable, you must disclose the conflict. If the conflict of interest is serious, you must step aside, and let someone else (usually a superior) resolve the problem.

2. Corrective Justice—Fairness in Rectifying Wrongs

When someone harms a person or damages a person's property, ethical theories agree that the person has the right to rectification, replacement, or repair. (Alternatively, the person who caused the harm or damage has a duty to rectify it.) The rectification must be fair; this is corrective (or retributive) justice. Canada's criminal law, tort law, and professional disciplinary powers are practical applications of corrective justice.

Fairness is difficult to achieve in serious criminal cases involving murder, rape, or bodily harm, where no reparation relieves the harm done. In these cases, the offender is punished, and the most common punishment is imprisonment. What length of sentence is fair for a given crime? Is the purpose of the imprisonment to remove the perpetrator from society, to punish the perpetrator, to rehabilitate the perpetrator, or to deter others from similar behaviour? These questions of fairness arise often in criminal justice, but are beyond the scope of this textbook. However, tort law and professional discipline are certainly relevant, as we see below.

- **Tort law.** The word *tort* means *injury* or *damage*, and tort law requires engineers and geoscientists to be responsible for their actions and decisions. This aspect of corrective justice is discussed at length in Chapter 8 of this textbook.
- **Professional discipline.** The Associations that license professional engineers and geoscientists protect the public by deterring unlicensed persons from practising, and by disciplining licensed professionals. Under the Act, each Association must respond to complaints and must discipline professionals who are found guilty of misconduct, which usually includes incompetence, negligence, and breaches of the Code of Ethics. This aspect of corrective justice is discussed at length in Chapter 4.

3. Distributive Justice—Fairness in Social Benefits and Burdens

Distributive justice addresses the question: How should the benefits and burdens of our society be distributed? The ethical theories give slightly different rules: Rights, duties, and virtue theories would require benefits and burdens to be shared equally. Utilitarianism requires the greatest good for the greatest number, equally distributed. However, any theory can be applied unfairly. For example, in the days of slavery, utilitarianism was occasionally misused to suggest that imposing severe hardship on a minority (slaves) was balanced by a great benefit to many others (slave-owners and their families). Obviously, such unfair distortion illustrates the dangers in careless application of the ethical theories.

Moreover, even if we accept that benefits and burdens should be shared equally, what is the fairest interpretation of equality? For example:

- Canada's income tax system is "graduated"; the tax rate increases with income. That is, the wealthy pay a higher rate than the poor. Some advocates recommend a "flat" income tax that requires everyone to pay the same percentage of his or her income. The flat tax is more "equal," so which is fairer?

- Ethics dictates that everyone should be equal before the law, but in practice, wealthy people can afford to use the courts, poor people can get legal aid from the government, but middle-class people hesitate to use the courts because they must spend their life savings before they get legal aid. Is this equality? Is it fair?

Distributive justice is a major concern for professional engineers and geoscientists. We are presently faced with the environmental problems of pollution, greenhouse gas emission, global warming, "peak" oil, and sustainable development. Who gets the benefit and who carries the burden in these matters? For example:

- Is it fair for North Americans to use fossil fuels indiscriminately, when the resulting global warming, over the next few decades, will cause sea levels to rise and flood low-lying countries such as Holland and Bangladesh, thus creating a generation of "environmental" refugees?
- Is it fair for the current generation to consume non-renewable resources inefficiently and thoughtlessly, thus depriving future generations who may need these resources desperately for simple survival?

These environmental problems are modern forms of the "Tragedy of the Commons," and are discussed in Part 4 (Chapters 15 to 17) of this textbook.

4. Political Justice—Fairness in Political Rights and Duties

Political justice is concerned with fairness in political rights and duties, and asks questions such as: How should we divide powers between federal, provincial, and municipal governments? How do we ensure that everyone has an equal vote in elections? Ethical theories generally support democracy as the best form of government, but which is most democratic: our traditional majority rule, or proportional representation (which has been proposed and defeated in two provincial referenda), or multi-step elections, in which the less popular candidates are eliminated in "run-off" elections, thus guaranteeing a winner with majority democratic support?

Political justice is one of the most important areas of justice, and it affects all citizens. Although many Canadians may find constitutional matters boring, we may need political intervention to solve environmental problems. For example, if carbon emission should be rationed or curtailed to prevent global warming, political justice must prevail—everyone must be treated fairly.

Some Final Comments on Justice

We seek justice in every ethical decision. Classifying justice into its different types is helpful to gain insight. Applying an ethical theory indiscriminately may result in decisions that are "right," but unfair and unjust. To solve ethical

Photos 11.1a and 11.1b — The CN Tower, Toronto. *(Left) The CN Tower, built in 1976, was the word's tallest building and freestanding structure for over 32 years, and an impressive icon on the Toronto skyline. The tower rises to a height of 553 metres (1,815 feet) and is the centre of telecommunications for Toronto. Antennae on the tower broadcast signals for many Canadian radio, television, and communication companies. The CN Tower is a tribute to Canadian ingenuity. Its construction required innovative techniques. The concrete core and three curved supporting arms were formed using a novel "slip-form" supported by hydraulic jacks, which moved upwards, gradually decreasing in size, to create the tower's elegantly tapered shape. In 1995, the American Society of Civil Engineers (ASCE) declared the CN Tower one of the Seven Wonders of the Modern World.*
(Right) The design of the CN Tower required extensive analysis and testing. The photo shows a 1:500 aero-static scale model of the CN Tower being tested in the boundary layer wind tunnel at the University of Western Ontario.

Sources: (Left) Corel "Toronto" CD #462019; (Right) Courtesy of The Boundary Layer Wind Tunnel Laboratory, University of Western Ontario. © Canada Lands Company (CLC) Limited. Reprinted with permission.

problems, we need the ethical theories (they are our guides to the right course of action), but the process and the decision must meet the test of fairness. Ethical principles and fairness lead to justice, and help us to avoid ethical pitfalls.

CODES OF ETHICS AS GUIDES TO CONDUCT

To put ethics into practice, most people want clear rules, not philosophical theories. Over the centuries, many laws and regulations have evolved based on ethics and justice but giving clearer guidance. For example, criminal and civil laws are effective guides to personal conduct. These laws are surprisingly consistent around the world, in spite of the different political systems, cultural influences, and religious beliefs.

Professional Associations are empowered to write and enforce Codes of Ethics that guide the professional conduct of engineers and geoscientists. These codes are also surprisingly consistent. Infringements can lead to penalties enforced by the provincial justice system or by the Association (as explained in Chapter 4). The Codes of Ethics for the provincial and territorial Associations are found in Appendix B.

General Principles

Codes of Ethics usually include statements of general principles, followed by instructions for specific conduct that emphasize the duties to society, to employers, to clients, to colleagues, to subordinates, to the profession, and to oneself. Although the various codes express these duties differently, their intent and the results are very similar. The following paragraphs summarize what the Codes of Ethics have in common.

DUTY TO SOCIETY A professional engineer or geoscientist must consider his or her duty to the public—or to society in general—as the most important duty. In other words, professionals have a duty to protect the safety, health, and welfare of society whenever society is affected by their work. This goal is achieved through professional self-regulation. That is, the government delegates its authority to the Associations, which define standards of admission, discipline licensed members, and regulate the profession. This arrangement benefits society, because the Associations ensure that professionals are competent, reliable, up-to-date, and ethical.

DUTY TO EMPLOYERS A professional engineer or geoscientist must act fairly and loyally to the employer, and must keep the employer's business confidential. Furthermore, a professional is obliged to disclose any conflict of interest.

DUTY TO CLIENTS A professional engineer or geoscientist in private practice has the same obligations to clients as an employee has to the employer.

DUTY TO COLLEAGUES A professional engineer or geoscientist must act with courtesy and good will toward colleagues. This simple statement of the Golden Rule is supported by all four ethical theories. Professionals should not permit personal conflicts to interfere with professional relationships. Most Codes of Ethics state specifically that fellow professionals must be informed whenever their work is reviewed.

DUTY TO EMPLOYEES AND SUBORDINATES A professional engineer or geoscientist must recognize the rights of others, especially if they are employees or subordinates.

DUTY TO THE PROFESSION   A professional engineer or geoscientist must maintain the dignity and prestige of the profession, and must avoid scandalous, dishonourable, or disgraceful conduct.

DUTY TO ONESELF   Finally, a professional engineer or geoscientist must ensure that the duties to others are balanced by the individual's own rights. A professional person must insist on adequate payment, a satisfactory work environment, and the rights awarded to everyone through the Charter of Rights and Freedoms. The professional also has a duty to strive for excellence and to maintain competence in the rapidly changing technical world.

A Comparison of Codes of Ethics

The provincial and territorial Codes of Ethics are compared and discussed below. The seven general duties are in each code, but may be expressed differently. Some Codes of Ethics may impose additional duties. To read the Code of Ethics for your Association, visit your Association's website, listed in Appendix A. The codes are also reproduced in Appendix B.

ALBERTA   Alberta's Code of Ethics was revised in 2000 and is now summarized in a brief preamble and five simple rules of conduct. The five rules instruct professional engineers, geologists and geophysicists as follows:

- In their areas of practice, to hold paramount the health, safety and welfare of the public and to have regard for the environment.
- To undertake only work that they are competent to perform by virtue of their training and experience.
- To conduct themselves with integrity, honesty, fairness, and objectivity in their professional activities.
- To comply with applicable statutes, regulations, and bylaws in their professional practices.
- To uphold and enhance the honour, dignity, and reputation of their professions and, thus, the ability of the professions to serve the public interest.

Alberta's code focuses on rules of personal conduct, rather than on duties to others. This brevity makes the code easier to read and to understand. Although the code does not explicitly include all of the duties described above, the APEGGA *Guide for Ethical Practice* expands on the code and, in 23 sub-clauses, covers all of the seven duties common to the other codes.[11]

BRITISH COLUMBIA   British Columbia's Code of Ethics was revised in 1994. It is clear and succinct, with 10 clauses (similar to those of several other provincial codes) that specify the duties that professional engineers and geoscientists owe to society, to clients or employers, to colleagues, to the profession, and to themselves. In contrast to some codes, clause 1 specifically includes protection of the environment and the need for workplace health and safety. Clause 9 instructs professionals to "report to their association or other appropriate agencies any hazardous, illegal or unethical professional decisions or practices by members, licensees or others."

MANITOBA Manitoba's Code of Ethics was revised in 2000. It specifically states that breaches of the code may be considered unskilled practice or professional misconduct and thus subject to disciplinary action under the Act. This code is arranged as five basic "Canons of Conduct" to guide professional behaviour. The five canons comprise 38 very specific clauses that offer useful advice on various aspects of professional practice. Some of the clauses in Manitoba's code do not appear in other codes, but are good advice for professional practice anywhere. Examples:

- **Canon 1** contains 3 clauses that specifically require practitioners to obey the laws of the land. It also notes some specific areas where compliance is necessary.
- **Canon 2** contains 9 clauses to help practitioners protect the physical, economic, and environmental well-being of the public—which is defined as the "prime responsibility" of practitioners—and emphasizes the importance of quality in professional work.
- **Canon 3** contains 13 clauses explaining how to apply skill and knowledge to satisfy the needs of a client or employer in a professional manner. Several clauses describe specific conflicts of interest, and explain that they are not acceptable.
- **Canon 4** contains 5 clauses describing how to uphold the honour, integrity, and dignity of the profession. Among other things, these clauses relate to the exchange of information, proper advertising, and the reporting of people who are violating the Act.
- **Canon 5** contains 8 clauses that give very meaningful advice to practitioners to be fair to colleagues and to support their professional development.

NEW BRUNSWICK New Brunswick's Code of Ethics is found in Section 2 of the 2005 APEGNB By-laws. The code has five parts.

- **Foreword.** This concise introduction summarizes the duties of engineers and geoscientists.
- **Professional Life.** These 9 clauses set some guidelines for professional practice and personal conduct. This part (like a similar one in British Columbia's code) specifically includes protection of the environment and the need for workplace health and safety. A clause unique to this code advises professionals to "observe the rules of professional conduct which apply in the country in which they practise and, if there are no such rules, observe those established by this Code of Ethics."
- **Relations with the Public.** These 7 clauses focus on dealings with the general public. One of these clauses is unique in that it forbids discrimination.

- **Relations with Clients and Employers.** These 17 clauses define many key aspects of professional practice—accepting tasks, ensuring safety, avoiding conflict of interest, considering environmental effects, receiving payment, and so on.
- **Relations with Engineers and Geoscientists.** These 14 clauses concern relations with colleagues. Some of these clauses give very specific guidance; for example, they uphold the principle of adequate compensation, prohibit members from using free engineering designs from suppliers in return for specifying their products, and instruct members not to associate with any enterprise that does not conform to ethical practices.

NEWFOUNDLAND AND LABRADOR   This province's Code of Ethics comprises an introduction and three sections that specify the duties of the professional engineer or geoscientist to the public, to the client or employer, and to the profession. It is clear, concise (22 clauses), and easily understood.

NORTHWEST TERRITORIES   NWT's Code of Ethics has a preamble and 12 Rules of Conduct that encompass all of the seven duties alluded to in the previous section. The preamble to this code entreats members to serve in public affairs when their professional knowledge may benefit the public and to demonstrate understanding for members-in-training under their supervision.

The first rule specifically requires engineers, geologists, and geophysicists to have proper regard for the physical environment. The code includes the duty to advise the provincial registrar of any practices by other members of the Association that are contrary to the Code of Ethics.

NOVA SCOTIA—ENGINEERING   The Nova Scotia engineering code is called the Canons of Ethics and is an appendix to the By-Laws. The code is comprehensive (28 clauses) and contains clauses and duties typical of the other Acts. This code is unique, however, in that it instructs engineers to refrain from conduct contrary to the public good, even if directed by the employer or client to act in such a manner, and similarly instructs employers not to direct employees to perform acts that are unprofessional or contrary to the public good.

NOVA SCOTIA—GEOSCIENCE   The Nova Scotia geoscience Code of Ethics has an introduction and 12 clauses that cover all of the duties, expressed in familiar terms. Several less common terms include duties to report any illegal or unethical geoscience decisions, and to assist geoscience societies, schools, and the scientific geoscience press.

NUNAVUT   The federal government created the territory of Nunavut in 1999, and the Association of Professionals, Geologists and Geophysicists of the Northwest Territories (NAPEGG) was granted authority to regulate the professions of engineering, geology, and geophysics in Nunavut. Until Nunavut enacts its own Code of Ethics, readers should consult the Code of Ethics for the Northwest Territories.

ONTARIO—ENGINEERING   Ontario's Code of Ethics for engineers contains all of the duties listed above, including a few unique clauses, such as requiring the engineer to display his or her licence at the place of business, and requiring moonlighting engineers to inform their clients that they are employed and to state any limitations on service that may result from this status. Ontario's Professional Engineers Act is unique, since infractions of the code are not enforceable under the Act unless the infractions are considered "disgraceful, dishonourable or unprofessional" conduct, which qualifies them as professional misconduct. A separate regulation defines professional misconduct, and includes harassment, negligence, and incompetence, and several forms of conflict of interest. The guide to conduct is therefore in two parts (the Code of Ethics and the definition of professional misconduct). One professional misconduct clause that seems to be unique to Ontario is "permitting, counselling or assisting a person who is not a practitioner to engage in the practice of professional engineering."

(*Note:* As of June 2009, Professional Engineers Ontario is currently surveying all professional engineers to determine whether to change the regulations to make the Code of Ethics fully enforceable under the Act. Check the PEO website for recent information on this decision.)

ONTARIO—GEOSCIENCE   The Ontario Code of Ethics for geoscientists is very comprehensive. It contains an introduction, eleven sections, and many subsections. The eleven sections, which encompass all of the duties typically found in Codes of Ethics, are titled as follows:

- Service and Human Welfare
- Public Understanding
- Business Ethics
- Duty to Others and the Environment
- Competence and Knowledge
- Signing and Sealing of Documents
- Faithful Agent or Trustee
- Conflict of Interest
- Overruling of Judgment
- Professional Advertising
- Breach of Code

A breach of Ontario's geoscience Code of Ethics is specifically defined as an act of professional misconduct. (This contrasts with the Ontario Professional Engineers Act, which defines professional misconduct separately.)

PRINCE EDWARD ISLAND   The engineering Code of Ethics for PEI is clear and comprehensive, consisting of a foreword and 26 clauses, under the headings of Professional Life, Relations with the Public, Relations with Clients and Employers, and Relations with Engineers. All of the duties mentioned above are included. Since geoscience is unregulated in PEI, that province has no code for geoscientists.

QUEBEC—ENGINEERING   Quebec's Code of Ethics for engineers is the longest of any province (over 70 clauses) and has a unique arrangement. The basic duties noted earlier are all represented in the code. Besides these, the code has clauses that define additional duties for engineers and that also provide useful professional advice. The Quebec code is far too complex to summarize here, but is included in Appendix B. Engineers from other provinces may well find its detailed clauses of interest and value.

QUEBEC—GEOSCIENCE   The Quebec Geologists Act came into force in August 2001. The Order of Geologists of Quebec prepared a draft Code of Ethics, which is presently being studied for adoption by the Quebec government. The draft code is very long compared to other codes (except the OIQ code), and includes all of the duties suggested above, to the public, the client, the employer, and the profession. One section concerns rules for advertising.

SASKATCHEWAN   Saskatchewan's Code of Ethics is one of the shortest (2 main clauses and 9 sub-clauses). Nevertheless, it includes all of the typical duties specified earlier, including the duty to report any possible illegal practices, professional incompetence, or professional misconduct to the Association, and the duty to be aware of the societal and environmental consequences of projects.

YUKON   Yukon's Code of Ethics for Engineers has 29 clauses, which cover the seven basic duties noted in the earlier section. Since geoscience is unregulated in Yukon, no Code of Ethics exists for geoscientists.

This overview shows that the codes for each province and territory are similar, but not identical. The codes are useful guides to personal conduct. Adherence to a province's Code of Ethics is not voluntary, nor is it a lofty ideal that would be "nice, but not essential" to achieve. In every jurisdiction (except Ontario), the Code of Ethics is enforceable under the Act, and clear violations of the code may result in disciplinary action in the form of a reprimand, suspension, or expulsion from the profession, as described in Chapter 4. In Ontario, Regulation 941 under the Professional Engineers Act defines professional misconduct separately from the Code of Ethics.

TECHNICAL SOCIETIES AND CODES OF ETHICS

Almost every technical society publishes a Code of Ethics. Several of these codes are in Appendix C, and the code for the National Society of Professional Engineers (NSPE) is in Appendix D. The NSPE code is very similar to the provincial codes, and has been endorsed by many engineering societies. However, these codes are voluntary guides to conduct and should not be confused with the Association Codes of Ethics (discussed above), which are enforceable under provincial or territorial Acts. In technical societies, an infraction of the Code of Ethics may lead to reprimand or expulsion from the society, but such penalties are extremely rare. In Canada, the licensing Associations are far more effective in enforcing their Codes of Ethics, and in regulating professional conduct.

A STRATEGY FOR SOLVING COMPLEX ETHICAL PROBLEMS

The Codes of Ethics are useful guides for professional practice. Most ethical decisions are simple, and we solve them by using common sense. However, some ethical problems can be more challenging, perhaps because logic suggests one decision, but your intuition (or your emotions) tells you another. A strategy for solving these ethical dilemmas is useful. Even if you never use the formal strategy for ethical problems, it is reassuring to know it exists.

Engineers and geoscientists have an advantage in resolving ethical dilemmas, since problem-solving and decision-making techniques are a routine part of their professional education. The design process provides the framework for an ethical problem-solving strategy.

The Design Process

New designs do not spring fully developed into the mind of the designer. Creation requires a series of steps, including inspiration and deduction, in the right order and at the right time. The process usually begins with a vaguely perceived need, and ends by implementing a plan that satisfies the need. The steps in the design process are typically as follows:

1. Recognize that a problem or need exists, and gather information about it.
2. Define clearly the problem to be solved, and any constraints that limit the solution.
3. Generate or propose alternative solutions or methods to achieve the goal (*synthesis*).
4. Evaluate the benefits and costs of alternative solutions (*analysis*).
5. Choose the optimum design (or re-define the problem and repeat the process).
6. Implement the best solution.

When faced with a problem or need, many people want to "jump in" and solve it immediately. However, the early steps are critical; action without thought is usually foolish.

Applying the Design Process to Ethical Problems

The design process is actually a methodical problem-solving technique. It can be modified into a strategy for solving ethical problems, as follows:

1. RECOGNIZE THE EXISTENCE OF A PROBLEM AND GATHER INFORMATION

Ethical problems may be difficult to recognize, especially in the early stages. Regrettably, some professionals are simply unaware that they are responsible for the safety of their work and for avoiding adverse impacts on society and the environment until they are charged with infractions of the Code of Ethics.

When a problem is recognized, it is usually vaguely defined. We must investigate it thoroughly, and solve it impartially. (Note that these are the key aspects of natural justice.) We typically ask the questions that all news reporters are taught to ask: Who, What, Where, When, Why, How? For example:

- *Who* is involved? (Identify the individuals involved, and their motivations.)
- *What* type of harm or damage has occurred (or may potentially occur)?
- *Where, when, why,* and *how* has harm occurred (or may potentially occur)?

For example, a manufacturing process, once thought to be harmless, may be suspected as a cause of cancer or toxicity in the workplace. This is an ethical (and perhaps a legal) problem, because workers cannot be placed at risk in this way. However, more information is needed. A supervisor must meet with the people in the workplace, and document any incidents that might support this allegation. Information about the manufacturing process and the toxic agent must be gathered and verified to confirm this suspicion. If the information shows a clear and present danger, then action must be taken to protect those at risk (as discussed in Chapter 8). However, in most cases, no action should be taken until all of the relevant information has been collected and examined, because the proper course of action may be quite different from what was initially expected.

2. DEFINE THE ETHICAL PROBLEM

The next step is to define the ethical root of the problem. You must ask yourself: What exactly is wrong? What is unfair about the situation? Do any actions contravene the law, the Association's Code of Ethics, or any other laws or regulations? Often what first appeared to be an ethical problem was really a rumour, a misunderstanding, or a disagreement over technical facts. When the facts are fully known, action may be unnecessary or obvious, and you can skip directly to the implementation step. However, in some cases, the ethical problem is indeed an ethical dilemma with no clearly correct answer. Your next job is to generate possible solutions.

3. GENERATE ALTERNATIVE SOLUTIONS (SYNTHESIS)

This phase of problem solving requires creative thought and is usually difficult. Many alternatives may exist, and they must all be considered if we are to find the optimum solution. The creative methods used for solving design problems (such as brainstorming) apply equally to ethical problems. Avoid an ethical dilemma with two equally undesirable courses of action. Creative thinking will almost always yield a better alternative.

4. EVALUATE ALTERNATIVE SOLUTIONS (ANALYSIS)

The courses of action are then analyzed and compared. As a first test, the solution (or decision) must satisfy the law and the Code of Ethics. If the laws and the code do not apply to the case, then we would look to the ethical theories. Finally, in our evaluation, we would ask whether the solution has any unfair side effects. The analysis can be summarized in the following questions:

- **Legality.** Does the solution satisfy the law and the Association's Code of Ethics?

- **Utilitarian ethics.** What benefits accrue as a result of this solution, and for whom? What hardships are involved? Are the benefits and hardships equally distributed?
- **Duty-based ethics.** Can this solution be applied to everyone equally? Can the solution be published and withstand the scrutiny of your colleagues and the public?
- **Rights-based ethics.** Does this solution respect the rights of all participants (or "stakeholders")? Does anyone suffer harm?
- **Virtue-based ethics.** Does the solution develop or support virtues? Is the solution a golden mean between unacceptable extremes? In particular, does the solution maintain the ideals of the profession?

5. DECISION MAKING AND OPTIMIZATION

SELECT THE OPTIMUM   During the analysis, one course of action usually satisfies the above questions better than any other; this course of action is the best or *optimum* solution.

RECONSIDER ANY DEADLOCKS   In some cases, the alternatives are equally balanced, and no course of action is clearly superior. When this happens, you must trace back through the process from the beginning, and answer the following questions:

- Has all the necessary information been gathered?
- Was the problem defined clearly?
- Have I sought advice from the people concerned?
- Has an alternative or compromise solution been overlooked?
- Have all the consequences of each alternative been fully evaluated?
- Is a personal benefit or conflict of interest affecting my judgment?

RESOLVE DEADLOCKS   If the above questions can be answered satisfactorily and there is still no optimum course of action, you should probably select the course of action that does not yield a benefit to you, as the decision maker. In a closely balanced decision, you must avoid any personal benefit, because it will appear to be bias.

If no personal benefit is involved, then you must select and follow the ethical theory that you consider most appropriate. This involves a personal value judgment, but you will have a clear conscience, since you made the decision carefully, logically, and ethically.

FAIRNESS CHECK   Finally, in our evaluation, we would double-check that the solution was fair. That is, does it impose any unfair conditions or unexpected side effects on those concerned? If the solution passes the fairness test, then you can defend your decision fully.

6. IMPLEMENT THE SOLUTION

Implementing the decision is the final step. Although the appropriate action will vary from case to case, it is usually advisable to act quickly and decisively when ethical decisions are required, especially if health, safety, or someone's reputation is at stake.

A BRIEF DISCUSSION OF THIS STRATEGY

This strategy may seem too formal, but try it a few times. After applying it rigorously to a few practical problems, readers find that they recognize ethical patterns, and can select appropriate solutions quickly and intuitively. The formal strategy is needed only for difficult cases.

- **Utilitarian bias.** This strategy is sometimes criticized for being weighted too heavily toward utilitarianism. It aims for solutions that satisfy all of the four ethical theories (or at least most of them); in that sense, this search for relative balance is itself a form of utilitarianism. However, this should be an advantage, because engineers and geoscientists are accustomed to utilitarianism; after all, it is at the root of most applied science.

- **Ethics, morality, and religion**. Readers will note that the term "morality" is rarely used in this textbook, even though *morality* and *ethics* mean virtually the same thing. The term "ethics" is therefore used for consistency. Also, religion is an important aspect of ethics and morality. Many of the ethical theories have their roots in religious teaching, and when applied to ethical problems, religious precepts usually agree with the ethical theories. For many Canadians, religion is the most important guide. However, in keeping with the Canadian Charter of Rights and Freedoms, professional decisions must be independent of any specific religion.

- **Other methods of applied ethics.** Many other formal methods have been proposed for ethical problem solving, but none is universally recognized. Methods for "applied ethics," originally called "moral casuistry," go back over three centuries. Casuistry began in the 17th century and was a "sincere effort to apply rigorous standards of critical argument to the questions of moral conduct."[12] This older method is very different from the problem-solving technique discussed here. In fact, the term "casuistry" is now a very disparaging label for arguments that are poorly formed or hypocritical.

The strategy proposed in this chapter is merely one more tool—a tool, however, that engineers and geoscientists may find useful for solving ethical problems.

CONCLUDING COMMENTS ON ETHICS AND JUSTICE

Many people have an intuitive sense of ethics; they apply their Code of Ethics easily, and they have the innate creative ability to generate ethical and fair solutions. For the rest of us, the methodical strategy explained in this chapter may help to develop the ability. The strategy may be applied to personal ethical problems (also called "micro-ethical" problems) or to larger societal, environmental, or humanitarian ethical problems (also called "macro-ethical" problems).[13]

In ethics, as in any discipline, practice makes perfect. That is, testing your problem-solving ability will improve it. The next three chapters illustrate applications of ethics and justice using case studies from professional employment, management, or private practice.

DISCUSSION TOPICS AND ASSIGNMENTS

1. Codes of Ethics typically forbid statements that maliciously injure the reputation or business of another professional. Some people might consider a prohibition against malicious statements to be an infringement on freedom of speech. For example, if a colleague makes a serious error in judgment, or even a serious calculation error, pointing it out in good faith would be beneficial. Write a brief discussion to define the boundary between statements that are justified criticism and those that are clearly malicious. In your answer, define the terms "malicious," and "good faith."

2. Some Codes of Ethics include duties that do not appear in other codes. Examine the Code of Ethics for your Association. Does it require professionals to

   - advertise in a dignified, professional manner?
   - report infractions of the Code of Ethics to the registrar of the Association?
   - set fees based on the difficulty of the work and the degree of responsibility?
   - refuse to pay commissions or reduce fees to obtain professional work?
   - obey the law of the land?

   For each of the above clauses: Is this duty specified in your Code of Ethics? If not, do you agree that the clause is an appropriate guide for professional conduct? Explain your reasoning briefly.

3. Critics of Mill's utilitarianism say that this ethical theory is *relativistic*, since it depends on subjective comparisons and could yield different results under slightly different conditions. Even worse, utilitarianism can encourage narrowly practical and/or selfish attitudes because acts are seen as ethical if they increase pleasure or reduce pain. Conversely, Kant's duty-based ethics is based on *absolute* rules that encourage honesty, unselfishness, ethical character, and self-sacrifice, and Kant expects everyone to follow the same rules, thus creating an ethical consistency. Debate the benefits and disadvantages of these two theories of ethics. Compare and contrast the relative nature of utilitarianism with the absolute nature of duty-based ethics by providing at least one example where each would differ and where each would agree.

Additional assignments are located in Appendix E, and case studies illustrating ethics and justice in employment, management, and private practice are found in the next three chapters.

NOTES

[1]   This chapter is based on hundreds of sources. Publications are cited only for the main work supporting each of the four ethical theories. Copies (or reproductions) of these works are readily available from various publishers, and several are available in full on the Internet.

[2]   J.T. Stevenson, "Philosophy," *The Canadian Encyclopedia*, Copyright, Historical Foundation of Canada, Toronto, available at <www.canadianencyclopedia.ca> (May 24, 2008).

[3]   J.S. Mill, *Utilitarianism,* published by Parker, Son and Bourn, London, 1863, reproduced in *Utilitarianism, On Liberty* and *Considerations on Representative Government* (ed. H.B. Acton), published by J.M. Dent & Sons Ltd., London, 1988, also available in full at <http://books.google.ca> (May 28, 2008).

[4]   I. Kant, *The Metaphysic of Ethics*, 1797, translated from German by J.W. Semple, published by Thomas Clark, Edinburgh, 1836, available in full at <http://books.google.ca> (May 28, 2008).

[5]   J. Locke, *Two Treatises of Government,* 1690, reproduced by C. Baldwin, printer, London, 1824, available in full at <http://books.google.ca> (May 28, 2008).

[6]   Government of Canada, *The Canadian Charter of Rights and Freedoms*, Schedule B to the Canada Act 1982, available at <http://www.laws.justice.gc.ca/en/charter> (May 28, 2008).

[7]   Aristotle, *Ethica Nichomachea*, c.322, ethics lectures transcribed by Aristotle's son, Nichomachus. The work, or reviews of it, are widely available; for example, see *Masterpieces of World Philosophy* (ed. F.N. Magill), HarperCollins Publishers, 1991.

[8]   L. Duhaime, "Justice," *Legal Dictionary*, available at <www.duhaime.org/LegalDictionary.aspx> (May 28, 2008).

[9]   L. Solum, "Legal Theory Lexicon 018: Justice," posted on *Legal Theory Lexicon*, January 11, 2004, available at <lsolum.typepad.com/legal_theory_lexicon/ 2004/01/legal_theory_le_2.html> (June 15, 2009).

[10] Government of Canada, *The Canadian Charter of Rights and Freedoms*, Clause 10.

[11] Association of Professional Engineers, Geologists and Geophysicists of Alberta (APEGGA), *Guide for Ethical Practice,* available at <www.apegga.com> (May 28, 2008).

[12] J. Haldane, "Applied Ethics," in *The Blackwell Companion to Philosophy* (ed. N. Bunnin and E.P. Tsui-James), Blackwell Publishing, Oxford, 2003, pp. 494–498.

[13] J.R. Herkert, "Microethics, Macroethics, and Professional Engineering Societies," *Emerging Technologies and Ethical Issues in Engineering: Papers from a Workshop,* National Academy of Engineering, October 14–15, 2003, pp. 107–114, available at <www.nap.edu/catalog/11083.html> (May 28, 2008).

Chapter 12
Ethics in Professional Employment

Most engineers and geoscientists are professional employees, working in teams on challenging projects. An overwhelming majority of them are satisfied with their working conditions, according to a comprehensive survey by Engineers Canada (summarized in Chapter 6).[1] However, working conditions are not under the control of the licensing Associations; individuals must negotiate, and must insist on, professional status and working conditions.

In this chapter, we discuss practice and ethics issues that arise in professional employment. These issues include technical authority, workplace disputes, and conflicts of interest. We also examine the professional response to unethical managers, the limits of an employer's authority, and whistle-blowing. Five case studies illustrate these workplace issues and ethical dilemmas, and the chapter closes with the *Challenger* space shuttle disaster: a tragedy that could have been prevented, but the technical advice of an engineer was over-ruled.

TECHNICAL AND MANAGEMENT AUTHORITY

Accepting an employment offer creates a contract. The employer acquires the authority to set legitimate goals, and instruct the employee; the employee agrees to use his or her professional ability to achieve the employer's goals, and in return, obtains appropriate compensation and working conditions. However, sometimes the employer and employee disagree over professional or ethical issues.

The need for employer authority is obvious; lack of direction could lead to chaos and bankruptcy. The employer must have *management* authority to direct the company's resources, whereas the professional must have *technical* authority to apply engineering or geoscientific knowledge and skills. In a well-run organization, the distinction between management and technical authority is well understood; professionals show mutual respect, and cooperate to achieve the employer's goals.

However, what would you do if your employer over-ruled you, or asked you to perform activities that are illegal or unethical? These problems are not common, but do occur. Let us consider a range of cases.

Over-Ruling Technical Recommendations

Occasionally an employer or client over-rules the technical recommendation of a professional engineer or geoscientist. For example:

- The operations manager in an oil exploration company insists on purchasing expensive equipment for oil and gas exploration, even though the geoscientist knows the equipment is outdated and unnecessary, and has advised the manager against buying it. The manager believes that the equipment might be useful in future, and should be bought now, because of exchange rates. If the manager over-rules the geoscientist, and purchases the expensive equipment, it will sit idle. Senior management and company shareholders will want to know why. They will presume that the geoscientist requested the purchase, and the geoscientist's perceived poor judgment might affect his or her career prospects. The purchase is not unethical, but it is a technical error. What should the geoscientist do?

Many ethical guides specifically mention this type of case, because it is a fairly common problem.[2] The proper action when an employer or client over-rules the advice of a professional is to

- explain your advice to the employer or client in writing, and also explain clearly the consequences of ignoring your advice.
- get a response in writing, to ensure that the advice and the consequences have been understood. This is particularly important when the employer or client is not a technical professional.
- the employer or client now has full responsibility for the decision and any consequences. You have satisfied your responsibilities, unless the decision involves illegal or unethical activities. (In that case, the professional still has an obligation to public welfare, as discussed below.)

Illegal Activities

On rare occasions, a professional engineer or geoscientist may be asked to engage in an activity that is clearly contrary to the law. The law may be a criminal law, a civil or business law, or a regulation made under the authority of an Act (such as an environmental regulation). Often, this occurs because pressure on management to generate profits is converted into pressure on the professional to "cut corners," For example:

- A geoscientist is asked to approve the improper disposal of industrial wastewater that is known to contain toxic or carcinogenic chemicals. This disposal is contrary to provincial environmental regulations.
- An engineer and his employer meet with a municipal government committee to discuss tenders for construction of a new town hall. After the meeting, the clerk returns the engineer's preliminary plans. However, the clerk inadvertently includes the preliminary plans of a rival consultant. In

the few minutes available, the employer asks the engineer to photocopy the rival's plans. Copying the plans is contrary to the Copyright Act, which forbids such unauthorized copying (although reading the plans is not an offence).

In situations like this, the necessary action is clear: the engineer must advise the employer that the action is illegal and must not break the law. Employers do not have the authority to direct an employee to break the law.

Activities Contrary to the Code of Ethics

An engineer may be asked to perform an action that, while not clearly illegal, is a breach of the Code of Ethics of the Association. For example:

- Using the ISO standard for gear design, a machine design engineer calculates that a gear train has a factor of safety of 1.82 against fatigue failure. The employer, who is not a technical person, asks the engineer to "round off" the factor, so that the gear train meets the client's specifications, which require a factor of safety of 2.0. This is technically wrong, because rounding to two figures gives 1.8 (not 2.0). Moreover, the fatigue life is inadequate, although this fact may not become evident for many years. The designer must refuse, because the employer's request contravenes the Code of Ethics, which requires competence and integrity (or honesty, depending on the code). The request might also be attempted fraud, which is illegal. In design, integrity is important; we can forgive errors, because we all make them, but we cannot forgive intentional fraud.
- A professional geoscientist works for a company that was recently taken over by a larger corporation. The companies' two exploration and analysis units are being combined, and some employees will be surplus to needs. The unit manager asks the geoscientist to review and comment secretly on the competence of a colleague, and provides copies of the colleague's reports and expense accounts. The manager's final comment is "One of you is on the way out." This request is clearly contrary to the Code of Ethics. Every code requires professionals to treat colleagues with courtesy, fairness, and in good faith (or similar wording, depending on the code). Most Codes of Ethics specifically forbid secret reviews of colleagues. Secret reviews are contrary to the concept of natural justice.

In these cases, the professional should decline to act on the employer's request. An employer cannot direct a professional engineer to violate the Code of Ethics. The employer may be unaware that the Code of Ethics has legal significance under the professional engineering and/or geoscience Act. It may be sufficient simply to inform the employer of this fact. An employer, manager, or client who is also a professional engineer or geoscientist is equally bound to follow the Code of Ethics (and many other professions have similar Codes of Ethics).

Activities Contrary to the Conscience of the Professional

A professional may be asked to perform an activity that, while not illegal, and not a violation of the Code of Ethics, nevertheless contravenes his or her conscience. Conscientious objection has a long and honourable history in Canada, and objections to slavery, nuclear weapons, and unjust wars are well known. Conscientious objection is not common in the professions, because professionals simply avoid working in unethical (or marginally ethical) industries. Examples of such industries are:

- Industries that produce non-essential and potentially harmful or addictive products, such as distilleries, breweries, wineries, and tobacco processors;
- Industries that ostensibly entertain, but may cause addictive behaviour and serious financial loss, such as casinos, slot machines, lotteries, and similar gambling;
- Industries that manufacture landmines, weapons, ammunition or explosives; and
- Industries that pollute or create dangerous by-products.

Each individual must consult his or her conscience to decide whether an ethically marginal industry benefits society enough to justify working for it. Government prohibitions of alcohol and gambling in the 1900s were unsuccessful. Therefore, Canada has chosen (along with most Western countries) to balance the ethical equation by controlling these industries more stringently, and by alleviating the misery of pollution or addiction by regulations, welfare, and social programs. Nevertheless, many Canadians face a dilemma when asked to work in such industries. For example:

- A software engineer is hired to develop banking software but, during a slow period, the employer asks the engineer to develop computer graphics for slot machines in a gambling casino. Slot machines are legal (in casinos and licensed restaurants, depending on the province), but many people consider them to be deceptive and unethical. Slot machines do not reveal the true odds of winning. Casinos themselves are designed with a deliberately distracting atmosphere. The maze-like, windowless, noisy surroundings deter gamblers from realizing the time of day, the weather, or even the way out. In addition, a significant proportion of gamblers become addicted to betting, and lose everything: their family life is affected; their standard of living drops; and some require counselling or social assistance. The software engineer does not want to put his or her creative effort into promoting gambling, so what should he or she do?

In this case, laws and Codes of Ethics do not give much guidance. The ethical strategy in the previous chapter might yield a compromise solution— simply to move the software engineer to another project, unrelated to gambling. However, if no alternative can be found, the engineer may face a serious ethical dilemma: ignore his or her conscience, or find other employment.

Ethical problems sometimes have unfair personal consequences. Obeying your conscience and refusing to follow the employer's directive may result in disciplinary action or dismissal. The possibility of dismissal, the consequences of unemployment, and the remedies for wrongful dismissal (discussed in the next chapter) should always be considered.

PROFESSIONAL EMPLOYEE GUIDELINES

Working conditions are not under the control of the licensing Associations, so Codes of Ethics do not cover typical employment problems, such as salaries, benefits, hours of work, hiring or terminating professional employees, and so on. Some of these issues are explained in law textbooks[3] and are regulated by provincial labour laws, but labour laws usually set minimum values that rarely apply to professional employees.

The Canadian Society of Professional Engineers (CSPE) was established to provide professional engineering employment advice. CSPE is intended to be a national federation of advocacy groups, as explained in Chapter 6, but the Ontario Society of Professional Engineers (OSPE), established in April 2000, is the only provincial branch under the CSPE umbrella (as of 2008). An American advocacy group, the National Society of Professional Engineers (NSPE), has been established for several decades, so NSPE information may be of use to Canadian engineers and geoscientists.

In 1973, NSPE published a set of guidelines for professional employees and employers. The crisis that initiated the guidelines was the U.S. government's cutbacks on aerospace expenditures in the late 1960s, including cancellation of the proposed supersonic transport aircraft (SST). Many engineers and scientists were suddenly and unexpectedly unemployed, and suffered severe financial hardship in the following years. In 2006, NSPE re-published the guidelines in a fourth revision, and it is still a useful guide to the professional workplace—for both employees and employers.[4]

The purpose of the NSPE guidelines is to establish a professional workplace based on "ethical practices, co-operation, mutual respect, and fair treatment." Additional objectives are to safeguard the public, to encourage professionalism and professional growth, and to combat discrimination based on age, race, religion, political affiliation, gender, or sexual preference. The guidelines contain more than 60 detailed clauses, which are divided into four sections: Recruitment, Employment, Professional Development, and Termination. Although the NSPE guidelines do not have any legal authority in either the United States or Canada, they give valuable advice for all professional employees. The guidelines are available online from NSPE, and are also reproduced in Appendix D of this textbook.

PROFESSIONAL EMPLOYEES AND LABOUR UNIONS

Professional engineers and geoscientists have a right to negotiate pay scales, hours of work, and other conditions of employment. Ideally, the professional employee should have a personal contract specifying salary, hours of work, overtime requirements, benefits (sick leave, vacations, pensions, professional insurance, and so on), regular review of performance and working conditions, promotion or raises based on merit, terms for permanent employment after a probationary period (or alternatively, terms for contract renewal), and so forth.

However, sometimes an employer may refuse to negotiate these basic conditions. The engineer is then faced with a difficult choice: accept unfair conditions, resign, or take part in "collective action" (group negotiation or unionization) against the employer. Professional employees have a duty to the employer, but also have a duty to themselves, and to the profession as a whole, and must not accept unprofessional working conditions or inadequate pay.

Labour legislation, in both Canada and the United States, established long ago that professional employees have the right to take collective action, and even to form or join unions. Every province has a Labour Board and a Labour Relations Act that can provide guidance and assist employees who are contemplating collective action. Of course, professionals who are also company managers are not permitted to join unions, and it would be illogical for them to do so, but other professional employees have no such prohibition.

However, forming a union involves confrontation, generates bureaucracy, and takes time and effort. Therefore, unionization should be a last resort. Professional employees should first try to negotiate personal (or group) employment contracts, because contracts are simpler and, with effective advice and negotiation, can be made as legally binding as union contracts.

Professional employees who are unable to negotiate personal or group employment contracts may be forced to unionize. In this case, they should form a collective group composed entirely of professionals, if possible. It is usually unwise for professional employees to join an existing staff or labour union, because they usually will be a minority in the union, and perhaps become obliged to support labour action that is not in their interests.

When professional employees resort to collective action, they do so reluctantly. Such action is not unethical. The need to unionize usually indicates that the employer failed to set fair policies and negotiating procedures.

UNETHICAL MANAGERS AND WHISTLE-BLOWING

A professional employee who finds evidence of illegal or criminal activities in the workplace, such as fraud, theft, misrepresentation, or destructive environmental practices, has a duty to remedy the situation. The proper action depends on the case; usually the professional would report the facts to his or her employer (or supervisor) for action. Quick reporting is important, because delay might be interpreted as condoning the illegal activities. Depending on the organization, the employer (or supervisor) may need to obtain authority from a senior manager or owner, who is ultimately responsible for the organization.

Further action is urgent when the illegal activity is a hazard to the public. For example, if an audit of toxic materials reveals that negligent storage practices are permitting poisonous liquids to seep into the environment, the professional must ensure that the employer remedies the hazard and notifies regulatory agencies and others who may be affected.

However, what do you do if the employer ignores a hazardous problem? This creates a dilemma: your duty to the employer conflicts with your duty to the public welfare. The public welfare must take precedence, according to the Code of Ethics, so the professional employee is usually faced with three possible courses of action: correct the problem, blow the whistle, or resign in protest.

- **Correct the problem.** The employee should first try to correct the problem and change company policy. This is usually the most effective course of action, especially if the illegal actions are minor and/or the employer is open to improvement and change. If the professional's immediate supervisor disputes the situation, the professional has a duty under the Code of Ethics to inform the employer of the potential consequences when a professional opinion is disregarded or over-ruled. If the supervisor still refuses to act, the professional would likely "go over the supervisor's head" to senior management. Further decisions would depend on the facts of the case.
- **Blow the whistle.** The engineer could alert external regulatory agencies that the company is acting dishonestly. This is called whistle-blowing, and it is an unpleasant and unfriendly act. Companies usually find reasons to dismiss whistle-blowers. However, in rare situations, where a professional employee has informed employers and senior management of a clear and serious hazard to the public, but they have refused or neglected to correct it, then whistle-blowing may be the only option. Whistle-blowing is not recommended until all other routes have been tried. (Whistle-blowing and its problems are discussed in Chapter 15.)
- **Resign in protest.** The employee could resign in protest. This course of action may be necessary in serious cases, where remaining with the company might imply collusion in the illegal activities. In such cases, a professional employee should always consult a lawyer before resigning. There may be grounds for considering a forced resignation as wrongful dismissal. (Wrongful dismissal is discussed in Chapter 13.)

CONFLICT OF INTEREST—AN OVERVIEW

Professional people, whether employees, managers or consultants, must aim for a high level of personal conduct. Conflicts of interest must be avoided, because they are unethical, they are almost always unprofessional, and they occasionally may be criminal.

General Definition of Conflict of Interest

In general, a conflict of interest occurs whenever a professional receives any benefit or has any relationship that interferes with the duty owed to the client or employer. In simpler terms, a conflict of interest occurs whenever an employee secretly receives a benefit or payment from more than one person for the same activities. Secret payments are also called "secret commissions" in the Criminal Code.

Every Code of Ethics requires disclosure of any conflict of interest. Secret payments are unethical, contravene the Code of Ethics, and are illegal in many cases. However, conflicts are not always about money. A conflict of interest could arise over a benefit, such as hockey tickets, free trips, or sexual favours. The benefit might be indirect, such as secretly assisting a close friend or relative, or something intangible, such as an opportunity to meet a celebrity.

Common Conflicts of Interest

Conflicts of interest may occur in several ways. In their book *The Responsible Public Servant*, Kernaghan and Langford define seven common categories of conflict of interest.[5] These common categories are listed below, and explained in a form more relevant to engineers and geoscientists.

- **Accepting secret commissions.** Accepting a secret payment or a significant gift from anyone with whom you have a business relationship generally creates a conflict of interest. The conflict exists, even if no obvious benefit is given in return. The most shameful example is accepting a bribe for ignoring shoddy work, or favouring certain suppliers or services. Secret commissions are illegal.
- **Misusing the employer's facilities.** Using the employer's computers, telephones, or supplies for private activities is theft, and needs no further explanation.
- **Secret employment or "moonlighting."** If you create a private business (even if you run it in your spare time), which is kept secret from your employer, you have a conflict of interest. The conflict is more severe if your business competes with your employer for clients, or if you spend time on your personal company that should be spent on your employer's projects. An employer pays a professional for knowledge, skill, and initiative, and it is unfair to divert these, secretly, to personal gain.
- **Self-serving decisions.** Using your position within an organization to hire relatives, or to divert business to a favoured company, is a conflict of interest. This is also called "self-dealing" or "abuse of privilege."
- **Influence peddling.** Using your position within an organization to support a group, cause, or political party for which you have affinity is a conflict of interest. This is a self-serving decision, as in the case above, but the benefit to the professional is less tangible.
- **Abusing confidential information.** As a professional, you have a duty to keep the employer's information confidential. Acting on confidential information for personal gain is an abuse of privilege and a clear conflict of interest.

- **Arranging future employment.** As a professional, you are free to resign and join a new company, or start your own practice. However, if the new company competes with (or even provides service to) your former employer, you may have a conflict of interest. Does the information that you obtained on the job belong to your former employer, or is it your professional experience? The interpretation depends on the facts of the case. If you have taken any of your former employer's trade secrets or confidential information to the new company, you have a serious conflict of interest, and the former employer may be able to sue you. Also, if the employment agreement with your former employer included a non-competition clause, you may have breached that contract. These points should be clarified before making a change, because a lawsuit is expensive, even if you may win.

Subcategories of Conflict of Interest

The conflicts of interest above may be further subdivided into at least three categories:

CLEAR (OR ACTUAL) CONFLICT In this case, the professional's service to the client or employer is clearly compromised. For example:

- A professional employee is responsible for monitoring the quantity and quality of concrete delivered to a runway construction site. The professional has a secret agreement with the owner of the concrete company: The professional ignores the fact that some concrete is made with cheap pit-run gravel, rather than crushed stone (as required by the contract). In return, the concrete company secretly supplies the concrete to build a dock at the professional's summer cottage.

POTENTIAL (OR LATENT) CONFLICT In this case, the professional does not have a conflict of interest at present, but the situation is such that a reasonable person would predict a conflict to exist in future. In other words, the potential for conflict exists, and a probable event could trigger it. For example:

- A professional engineer or geoscientist intends to run in an election for mayor of a small town. A childhood friend, who operates the sand and gravel company that spreads sand on the town streets in winter, offers to contribute $25,000 to the election campaign. The friend wishes to remain anonymous. (Municipal donors may remain anonymous in this province, and no tax receipts are given.) The professional accepts the money. Until the election is held, no conflict of interest exists. However, if the professional is elected as mayor, then the situation changes. One of the mayor's duties is to negotiate the annual winter contract for spreading sand on town streets. Therefore, a potential conflict of interest exists: the professional, if elected, must negotiate a sand-spreading contract with a person who has made a large secret payment to the professional's campaign. Obviously, secrecy is the problem. The potential conflict should be disclosed before the election. The voters can judge whether the conflict is serious, and can ensure that the professional, if elected, does not favour supporters.

PERCEIVED CONFLICT   In this case, the professional does not have a conflict of interest, but observers believe (or might believe) that a conflict of interest does exist. For example:

- A professional engineer or geoscientist is assigned to hire an administrative assistant. The job opening is advertised, candidates are interviewed, references are checked, and the best candidate is hired. However, by coincidence, the new administrative assistant has the same last name as the professional. Colleagues may believe that the assistant was hired because of a family relationship. To avoid this perceived conflict of interest, it is usually sufficient to explain publicly that no relationship exists.

Avoiding Conflicts of Interest

You can avoid conflicts very simply by refusing to accept gifts or bribes, refusing to misuse your authority for personal gain, and refusing to favour specific people. When a conflict arises accidentally or unavoidably, secrecy is the problem. Full disclosure will eliminate (or reduce) it. When a conflict is disclosed, the client or employer is able to ensure that no benefit or favouritism is involved. For example:

- Professional people, working for the same company, meet on the job and marry one another. This is fairly common; these links are emotional and rarely explainable. However, such arrangements create potential conflicts of interest, and must be disclosed, at least to superiors. The superior can ensure that no conflict of interest exists.

In summary, professional people are often in a position of authority or privilege, and must resist the temptation to give or receive favoured treatment. Conflicts of interest are unethical, contrary to the Code of Ethics, and, in many cases, illegal. Never accept a secret commission. Always disclose a conflict of interest.

INTRODUCTION TO CASE STUDIES

The case studies below (and in the following chapters) involve ethics, justice, and professional practice. Each case study describes a practical problem, and asks for a decision that is supported by the Code of Ethics and/or basic concepts of ethics and justice. Try to solve each case before you read the suggested solution.

Notes on Case Studies

- **Methodical strategy.** The first case (12.1) illustrates the strategy for solving ethical problems, in detail. The second case (12.2) applies the strategy less formally. The remaining cases compress the intermediate steps, and give only the final decision. Try the formal strategy a few times. After solving a few problems, readers usually understand the method and can develop practical ethical solutions quickly and intuitively.

- **Limitations.** Case studies are artificial, because readers are given only a summary of the facts, and cannot speak to the participants to gather the full information. Even so, case studies are useful for developing decision-making skills. In real cases, you would seek information assertively, in keeping with natural justice, which requires all viewpoints to be heard.

- **Similarity to real cases.** Case studies are based on real events, or reports of real events, but all names (and some details) have been altered for anonymity. Any similarity to real people in actual situations is entirely coincidental.

CASE STUDY 12.1

JOB APPLICATION DILEMMA

Statement of the Problem

Ralph X, a computer engineering student, is in his final university year, and about two months from graduation. He has applied for several jobs, but received no job offers. The university placement office tells Ralph of a job opportunity at a nearby tobacco company, and arranges an interview.

The job interviewer explains that the tobacco company manufactures cigarettes using recently built high-speed automated machinery, and the job involves digital control. In fact, Ralph's first assignment would be to optimize the control software to obtain maximum productivity from the new machinery. It is precisely the type of engineering job that Ralph wants, and would be a good entry to the digital control field. Ralph asks about alternative jobs, such as research work that is unrelated to manufacturing or promoting cigarettes, but the company has no other vacancies. The interviewer offers Ralph the job at an attractive salary.

Ethically and emotionally, Ralph does not want to work for a tobacco company. Ralph's father was a heavy smoker, who died of lung cancer about 10 years earlier, when Ralph was 12. This was a tragic episode in Ralph's life, and his father's death was attributed, at least in part, to smoking. Cigarettes have been proven to be addictive and harmful. Also, how could he face his mother? She suffered after his father's death, and she would be very sad to see Ralph working for a tobacco company.

Financially, Ralph has enough money to survive for another few months, but he wants to repay loans from his mother. His large student debt is currently interest-free. The banks are competing to offer him a line of credit as soon as he gets his engineering degree. According to the placement office, job placements continue into the summer, and other job opportunities are possible, but uncertain.

QUESTION

Should Ralph accept the engineering job offer at the cigarette factory?

AUTHOR'S SUGGESTED SOLUTION

This case was selected to illustrate the most difficult type of ethical dilemma—those that fall outside the Code of Ethics. This case concerns ethics and conscience, but tobacco production is legal, and the Code of Ethics gives us no guidance. However, the ethical theories and the problem-solving strategy discussed in Chapter 11 give us more insight:

Recognize the problem and gather information. Ralph recognizes that he faces an ethical dilemma: he wants a job, but his conscience rebels against manufacturing cigarettes. He has already gathered relevant information from the job interview and from the placement office, and he has examined his financial situation. He is still in his final year of university, so he is not really unemployed yet.

Define the problem. The problem may, at first, appear to be simply accepting or rejecting a job that has a good salary but a distasteful purpose. However, this is not the precise problem. The true question is whether to accept an immediate job offer, with the security that it offers, versus the uncertainty and work of applying for better jobs and attending interviews while struggling to pay bills and loans.

Generate alternative solutions. Ralph should not make a final decision without considering all the alternatives.

- Ralph could reject the job offer, and look more aggressively for a better job. If necessary, he could get a line of credit.
- Ralph could reject the offer, but accept a part-time job to avoid borrowing more money. The part-time job might interfere with his job search.
- Ralph could accept the offer, and try to ignore his conscience.
- Ralph could accept the offer, and keep looking for a better job. (Note: This option is dishonest, and is therefore ethically unacceptable.)

Evaluate the alternatives. Evaluate the four alternatives above, using the ethical theories discussed in the previous chapter:

- Mill's utilitarianism tells Ralph to balance the benefit of a secure job (and his ethical distaste for the job) against the benefit of a better job (and the uncertainty and financial struggle that goes with it). Duration and intensity of the benefits should be considered; a first job typically lasts from three to five years, and sets the tone for the rest of a career. Each reader would balance these factors differently, but most would likely conclude that three to five years in a distasteful job is too high a price for security.

- Kant reminds Ralph that human beings should always be treated as an end or as a goal, and never as a means of achieving some other goal. In accepting, Ralph would sacrifice his self-respect for security. A job that requires him to deny his true conscience would be demoralizing, and is unlikely to lead to a productive career.
- Locke's rights-based ethics would likely contribute little to resolving Ralph's dilemma. Cigarettes are harmful, but not illegal, so the company has the right to manufacture them. Unfortunately, no one has a right to a good job, although it may be an admirable concept.
- Aristotle's concept of seeking virtue is very relevant, since there is little virtue in manufacturing a harmful product.

Decision making. Given the above analysis, the author concludes that most theories show that Ralph should reject the job offer and continue to look, aggressively, for a better job. Some uncertainty is involved, but the risk is worth it.

Fairness check. We must make a final check for fairness. This case does not appear to have any unfair side effects, but the situation could be different. For example, if Ralph's mother were losing her home because of an unpaid mortgage, it would likely be unfair for Ralph to reject the tobacco job offer and delay repaying his debt to her.

Discussion of the decision-making strategy. This case illustrates an ethical dilemma that is not covered by laws or Codes of Ethics. Some readers might prefer to seek religious guidance in such cases. The ethical theories are consistent with most religions, and are intended to span cultural differences.

In closing, what conclusion would you reach? You may weigh the utilitarian benefits differently, or you may be able to suggest more alternatives. In questions of conscience, there is no right answer; however, a decision made in haste, or strictly for personal gain, is almost certainly wrong. The author respects other decisions, provided that they are based on gathering information, considering alternatives, and weighing the ethical nature of the alternatives.

CASE STUDY 12.2

ACCEPTING A JOB OFFER

STATEMENT OF THE PROBLEM

At a time of economic recession, an electrical engineering student, Joan Furlong, is nearing graduation. She is seeking a permanent position with an electronics company in digital circuit design and analysis, but she accepts interviews from both electronics and power companies. Her résumé clearly states her qualifications, job objective, and interests, which are in digital circuit design. With her graduation day approaching, she receives an offer from the Algonquin Power Company for a job working on scheduling substation maintenance. The salary is good, so she writes back immediately and accepts. About two weeks later, she receives a letter from Ace Microelectronics offering her a position on a new project in digital circuit design. The salary is roughly equal to the Algonquin offer, although the employment may end when the project does. Furlong is uncertain what to do. She sincerely wants to work in circuit design and not in scheduling maintenance. She identifies three possible courses of action.

Honour the Algonquin agreement. She could honour the agreement with Algonquin and decline the offer from Ace. She would likely thank Ace, and tell them she has already accepted an offer, although she might be able to join them on a later project in a few years' time, once she has satisfied her obligation to Algonquin.

Revoke the Algonquin agreement. She could write to Algonquin, tell them her plans have changed, revoke her earlier agreement, and apologize for the inconvenience. She is aware of the Code of Ethics of her provincial Association, but she is not yet a member of the Association and does not feel bound to follow the code. Although she is a student member of the Institute of Electrical and Electronics Engineers (IEEE), the IEEE code does not have any clause that pertains to this particular situation.

Revoke the Algonquin agreement but offer reimbursement. She could write to Algonquin as above, but also offer to reimburse them for the recruitment expenses they paid on her behalf.

QUESTION

From the ethical perspective, which of the above three options is best?

AUTHOR'S SUGGESTED SOLUTION

This problem is not specifically addressed in Codes of Ethics, although every code states or implies that a professional has an obligation to act in "good faith" or with "good will" toward clients, employees, and employers. However, the NSPE guidelines discuss this problem specifically: "Having accepted an employment offer, applicants are ethically obligated to honor the commitment unless and until they give adequate notice of intent to terminate."[6] Also, when we apply the ethical theories to this case, all of them clearly support the NSPE directive. To illustrate, consider the three alternatives.

Honour the Algonquin agreement. Clearly, this decision is most ethical. All of the ethical theories discussed in the previous chapter support this conclusion. Honouring agreements is a virtue, and utilitarians would argue that honouring agreements—even when they are not ideal—permits society to operate smoothly, benefiting everyone. Kant's philosophy would specifically tell Furlong that promises must be honoured: she made a promise to Algonquin and has a duty to fulfill it. Moreover, she may grow to enjoy the Algonquin job and end up feeling pleased that she followed her conscience. Furlong's personal rights are not an issue, because she freely accepted the offer. However, Algonquin has a legal right to require her to honour her agreement.

Revoke the Algonquin agreement. Revoking or ignoring the obligation to Algonquin is clearly unethical. As explained earlier, Furlong has an obligation to Algonquin that cannot be erased with a simple apology. The company has probably sent rejection letters to the other applicants for the position and may stand to lose more than the recruitment costs if the maintenance program is delayed. (People outside the company rarely understand the grave consequences of failing to recruit personnel.) The argument that the Code of Ethics does not yet apply to Furlong is spurious, legalistic, and unacceptable as a justification for her actions.

Revoke the Algonquin agreement, but offer reimbursement. Sometimes in life we make serious mistakes, and the only way to remedy these mistakes is to admit them, apologize, and offer restitution. Even marriage—a social contract that is at least as sacred as accepting a job offer—has provisions for divorce. If Furlong genuinely believes that accepting the Algonquin offer was a very serious error in judgment, she should admit her mistake and offer restitution. Algonquin will likely request the return of expenses paid during the recruitment (at least), and of course Furlong will never receive an offer from them again. So Furlong will pay a price for realigning her career path. This is not an ideal course of action. However, restitution acknowledges her ethical duty and ensures that the person who benefits most from this course of action (Furlong) compensates Algonquin for at least part of its losses.

In summary, the author leans strongly in favour of the first alternative—an offer of employment (like any legal agreement) should never be accepted unless it can be honoured. Furlong has a duty to fulfill it. Moreover, she may grow to enjoy the job at Algonquin Power, and may be pleased that she followed her conscience. This is clearly the most ethical decision.

CASE STUDY 12.3

PART-TIME EMPLOYMENT (MOONLIGHTING)

STATEMENT OF THE PROBLEM

Philip Forte is a licensed professional engineer who has worked for Federal Structural Design for 10 years. Unfortunately (for reasons that are not clear to either Forte or his employer), the company has not had many large contracts, and Forte's salary is very low. For the past 10 years of his employment, his pay raises have rarely exceeded cost-of-living increases. As a result, he has been forced to take on extra work in his spare time. He secretly brings the work to his office in the evening, where he uses the design and analysis software on his computer workstation. He is careful to pay for any office supplies or photocopying out of his own pocket, and he argues that the computer would be sitting idle in the evening anyway, so his employer is suffering no loss. In fact, Forte contends that his evening work is benefiting his employer, since it enables him to keep working for Federal Structural Design in spite of his low salary.

QUESTION

Is it ethical for Forte to carry on his part-time employment in this manner?

AUTHOR'S SUGGESTED SOLUTION

This case involves a clear conflict of interest, which is contrary to every Code of Ethics. Many engineers "moonlight" (that is, work part-time on evenings and weekends). It is not unethical for an employee to work for more than one employer, but employers must be fully informed of the situation in order to verify that the employee is not competing with the employer, and that the employee is not abusing or diverting the employer's resources.

Forte is not acting ethically in this situation, since he did not inform his employer of his part-time employment. Secrecy is the problem. It is irrelevant to argue that the employer is exploiting Forte by paying a poor salary, thus forcing Forte to work two jobs. Forte has an obligation to himself and to the profession (in every Code of Ethics) to insist on adequate pay for professional work.

Moreover, Forte may be placing his career at more risk than he realizes. If Forte is offering services to the general public when he moonlights, then in some provinces (such as Ontario) he must obtain a Certificate of Authorization, and the question of liability insurance must be addressed.

CASE STUDY 12.4

FALSE OR MISLEADING ENGINEERING DATA IN ADVERTISING

STATEMENT OF THE PROBLEM

Audrey Adams is a licensed mechanical engineer with marine experience, working for a manufacturer of fibreglass pleasure boats. She has conducted buoyancy tests on all the boats manufactured by the company and has rated the hull capacity of each, according to the procedure specified by Transport Canada. She notices that the company's sales brochures show photographs of six people on board a boat that is rated for a maximum of five people. In the printed specifications, the sales brochures state that the boat can hold a maximum of "five adults." She believes that the sales brochures are misleading and possibly hazardous. The error would be costly and perhaps embarrassing to change, since thousands of the brochures have been printed and distributed. Adams speaks briefly to a sales representative, who replies, "The boat was floating okay when we took the picture." Adams knows that the boat is safe in still water with six people on board, but could flood and sink in rough water.

QUESTION

What action (if any) should Adams take?

AUTHOR'S SUGGESTED SOLUTION

The Code of Ethics for every Association states that the safety and welfare of the general public must be considered paramount (most important). In this situation, the brochures create a false idea of the boat capacity. This is a potential hazard to the purchasers who use the boats, so Adams has an ethical duty to take immediate action to reduce or eliminate the hazard.

Adams's first step should be to inform the engineering manager about the problem. Most likely, she would do this by writing an internal memorandum describing the errors in the sales brochures. Most companies are honest and would take immediate action to correct the sales brochures, regardless of the cost. The new brochures should state that the old brochures are incorrect, and direct them to be discarded.

In rare instances, the brochures may indicate deliberate dishonesty. For example, if Adams discovered that incorrect capacities were printed in the brochures, and/or stamped on the boats' serial nameplates, the problem would be much more serious. The company's management could be guilty of misrepresentation, which is illegal, and possibly criminal. If a serious accident were to occur (such as an overloaded boat sinking, resulting in loss of life), the erroneous brochures and incorrect capacities would become public knowledge, and the engineer could be charged with incompetence or even with collusion in the misrepresentation.

A professional engineer or geoscientist who discovers that his or her employer is dishonest must quickly dissociate from any illegal activity. The professional must also consider whether to blow the whistle on a dishonest employer, or resign in protest (or both). Otherwise, the professional risks being labelled a participant in the illegal activity.

CASE STUDY 12.5

ALTERED PLANS AND INADEQUATE SUPERVISION

STATEMENT OF THE PROBLEM

Assume that you are a licensed civil (structural) engineer employed by a large retail company. You work directly for the company architect, who designs the new retail stores, including renovations or additions to existing stores. As an architect, your boss concentrates on creating stores that are eye-catching, colourful, and attractive to the customer—concepts that will get maximum sales. Your job is to show that the architect's concepts are safe by inserting the structural details, preparing the structural drawings, obtaining the building permits, and supervising the construction. Large construction projects are tendered out for contract. However, a company crew completes small jobs internally. The crew manager was trained as a frame carpenter, and was later certified as a technologist, but is neither an architect nor an engineer.

In the past seven years, you have successfully supervised the design of two large new stores and their construction by contractors. You have also planned about 20 small projects, which have been carried out by the company crew. Recently, however, you discovered that the crew manager has sometimes been deviating from your plans on the small jobs. In your technical discussions, he has always agreed to comply with your instructions. Now you realize that he has consistently been altering your plans—for example, by substituting different structural steel shapes, changing joint connection designs, changing column locations, and using salvaged structural steel instead of new sections. You are genuinely uncertain of the extent of his changes; you are also uncertain about the factors of safety in the "as built" structures.

You confront the crew manager, somewhat angrily. He responds that he was simply "using cost-saving measures" and that "the changes didn't affect strength." Moreover, he insists that he has "25 years of experience in project management" and that he has had "very few problems in the past." You report these facts to your boss, the architect, who is "simply too busy with project deadlines to worry about personnel problems." The boss instructs you to "sort it out yourself." You realize that you are partly to blame for this problem because you were responsible for "supervising," which includes ensuring that your plans were followed. The crew manager is employed by your company, so you trusted him more than you would have trusted an external contractor, and you failed to inspect these small projects and prepare as-built drawings.

QUESTION

What action should you take?

AUTHOR'S SUGGESTED SOLUTION

Under every engineering Code of Ethics, the engineer's primary responsibility is to ensure the safety of the public. The manager's assertion that "the changes did not affect strength" is not adequate, since he is not qualified to make that judgment. Clearly, you are uncertain about the factors of safety in the final design, so you must take some action on this problem. Regardless of the pressure of deadlines or the costs involved, you must convince your boss, the architect, that this problem must be remedied. Under the Code of Ethics for architects, she or he has an obligation to respect your expertise in determining structural strength.

In this case, the ethical analysis is easy. The Code of Ethics requires every professional to put public safety ahead of personal gain or inconvenience, and the main decision is how to do it. In order to guarantee safety, you must review all 20 of the projects as quickly as possible to determine the actual factors of safety in the as-built structures. Strengths will have to be recalculated where necessary, and will have to be filed with the original design calculations to document the as-built strength. If the as-built strength is marginal or inadequate, immediate structural repairs will have to be taken to increase strength. In the future, you will have to monitor the small projects more closely. Two incidents involving American structural projects are very relevant to this case study.

Hyatt Regency hotel. In 1981 in Kansas City, Missouri, two concrete walkways spanning the lobby of the Hyatt Regency hotel suffered a collapse that killed 114 people and injured more than 200 others. This was the deadliest structural failure in North America since the collapse of the Quebec Bridge in 1907 (discussed in Chapter 1). The cause of the Hyatt Regency collapse was eventually traced to a minor change in the design of the fittings supporting the upper walkway. Both walkways were to be hung from steel rods roughly 18 m (60 ft.) long. However, such rod lengths are difficult to transport, so the fabricator suggested that two shorter (9 m) steel rods be used, instead of the longer (18 m) rod. The shorter rods would both be bolted to the fitting that supported the upper walkway. This change seemed minor; in fact, however, it doubled the load on the fitting—a point that would have been obvious, if anyone had drawn a free-body diagram of the fitting. The design engineer did not recall seeing the change request, but the engineer's seal appeared on the revised drawings. The collapse was a tragedy for the victims. It was also costly for the insurance companies and for the engineers who lost their licences and were forced into bankruptcy.[7]

LeMessurier and the Citicorp Tower. William LeMessurier was hired as a structural consultant to the Citicorp Tower in New York City, a 59-storey building with a structural steel frame. After the building was completed, LeMessurier had occasion to recheck his strength calculations, at which point he realized that when strong winds blew from a certain direction, the forces on the bolted joints would be significantly greater than he had earlier calculated. He chose to face his error directly. Wind tunnel tests confirmed his fears, and he revealed his concerns to the building designers and to the client, Citicorp. Citicorp agreed to immediate action, and repairs were made in record time before the building risked being demolished by a hurricane. Surprisingly, even though the repairs cost millions of dollars, LeMessurier was shielded from most of the financial loss and was highly praised for his prompt, ethical actions. (Note: This important case is discussed in detail in Chapter 14.)

The above two cases illustrate that minor changes or errors can sometimes be critical. When faced with the possibility of structural flaws, the ethical response is to address the problem directly, determine the strength accurately, and take all necessary actions to ensure safety.

CASE HISTORY 12.1

THE *CHALLENGER* SPACE SHUTTLE DISASTER

The *Challenger* space shuttle explosion in 1986 is probably the most infamous engineering tragedy of all time. Millions of people were watching the televised launch when *Challenger* exploded, resulting in the loss of seven lives, immense costs, and severe problems for the American space program. (More viewers saw the collapse of the World Trade Center on September 11, 2001, but it resulted from a terrorist attack, and was not an engineering fault.)

Photo 12.1 — The* Challenger *Space Shuttle Explosion. *The Challenger space shuttle mission, unofficially called the "Teacher in Space Project," was launched on January 28, 1986. At 73 seconds after launch, a series of structural failures caused a fuel tank to explode. The shuttle and its crew of seven were lost, in the most heavily televised engineering failure in history. The subsequent inquiry revealed that managers over-ruled engineers who tried to delay the launch for safety reasons.*

Source: AP/Bruce Weaver.

Introduction

On January 28, 1986, the U.S. National Aeronautics and Space Administration (NASA) launched the space shuttle *Challenger* at Cape Canaveral, Florida. The launch had been delayed by bad weather, and the weather overnight had been exceptionally cold (for Florida). At 11:38 a.m., the rockets were finally ignited.

At first, the shuttle rose according to the flight plan; however, at 59 seconds into the flight a plume of flame was evident near the booster rockets. By 64 seconds, the flame had burned a hole in the booster; at 72 seconds, the booster's strut detached from the external tank. At 73 seconds into the flight, the booster struck *Challenger*'s right wing and then struck the fuel tank. The tank exploded. The shuttle was at an altitude of 14,600 m and travelling at about Mach 2 when the explosion occurred. The explosion may have killed the crew members; the crew module separated from the rocket during the explosion and was in free fall for 2 minutes and 45 seconds. It hit the ocean at a speed of about 320 km/h. There were no survivors. Fragments of the shuttle continued to rain down on the rescue team for about an hour after the explosion.[8]

The *Challenger* explosion caused the first deaths of American astronauts during a mission (although there had been three deaths in a ground test for the first Apollo mission, and three Soviet deaths when parachutes failed to deploy at the end of the first Soyuz mission). The *Challenger* disaster was a serious setback for the U.S. space program.

Investigation

American President Ronald Reagan convened a commission to investigate the *Challenger* explosion. The investigation involved over 6,000 people, and the resulting 256-page report was issued in June 1986.[9] After extensive delibera-tion, the commission concluded that the *Challenger* explosion was caused by the failure of a rubber O-ring seal between sections of a rocket booster. Hot gas from the rocket motor escaped past the O-ring (and past a secondary O-ring intended to double the factor of safety). This generated a lateral thrust that eventually broke a supporting strut. The strut failure permitted the booster rocket to swivel, puncturing the central hydrogen fuel tank, which led, in turn, to an explosion of the shuttle's hydrogen fuel.

The investigators also learned that on the eve of the launch, engineers tried to delay the launch, because they were uncertain how the O-ring seals would perform in such cold weather. The investigation then focused on NASA's management style; the commission concluded that the *Challenger* launch decision was flawed.

The O-Ring Problem

When the rockets fire, they create enormous stress in the rocket casing in all three dimensions. The joint between the rocket sections distorts, and the gap between the sections widens under this stress. The O-rings keep the joint sealed, preventing the hot gas from escaping. The O-rings, which are com-pressed in a groove, must be resilient enough to "spring back" to fill the gap and keep the joint sealed as it widens due to joint stress and distortion. The temperature is important, because lower temperatures increase the hardness of the O-rings and decrease their resilience.

Roger Boisjoly was the Morton-Thiokol engineer most familiar with the O-ring design. He had conducted temperature tests on O-rings, and as early as July 31, 1985, he had recommended in writing that the problem of O-ring erosion (burned by hot gas) be studied. Furthermore, he had warned that failure to "solve the problem with the field joint" could result in loss of a shuttle, probably on the launch pad.[10] He was authorized to set up an O-ring team, and in October 1985 he sought advice from 130 vendors and other seal experts. However, no help was forthcoming from these sources.[11]

The Evening Teleconference—Boisjoly Is Over-Ruled

The evening before the launch, weather forecasters predicted that the launch site temperature would drop to 20°F (–6°C) overnight. NASA engineering managers were worried about the effect of this unusually low temperature on the rocket boosters. A late-night teleconference was held, involving 34 people. Engineers and managers from Morton-Thiokol (the manufacturers of the rocket boosters) presented their concerns and recommendations to the NASA managers at the launch site. This critical evening conference is described in detail in a definitive book, *The Challenger Launch Decision*, written more than a decade after the tragedy.[12]

The teleconference focused on the performance of the O-ring seals at the circumferential joint in the rocket. Boisjoly stated that no previous shuttle had been launched at temperatures below 53°F (12°C), and that the rocket boosters recovered from that flight showed extensive damage to the primary O-ring, indicating that the O-ring had failed to seal properly. Fortunately, the joint has two O-rings, and the secondary O-ring had contained the hot gas. The engineering managers at Morton-Thiokol advised NASA's launch staff that the low temperature could cause failure of both O-rings; they then recommended that the launch be delayed until the ambient temperature reached at least 53°F (12°C). Boisjoly's data also included disconcerting results from a launch where the primary O-ring had failed when the launch temperature was 73°F (23°C).

At NASA, Lawrence Mulloy, an engineering manager at the next level (of a four-level launch approval protocol), questioned the recommendation to delay the launch. Mulloy pointed out the discrepancy in the data presented by Boisjoly from the previous boosters. One O-ring had failed during a fairly low-temperature launch, but one had failed during a fairly high-temperature launch. This might indicate that temperature is not the key factor in the joint failure. The Morton-Thiokol group asked for a brief delay so that they could discuss the question among themselves.

In the closed discussion with Morton-Thiokol engineers and managers, Boisjoly and the other engineers remained convinced that, in spite of the apparent discrepancy in the data, the cold temperature would seriously affect the O-ring performance. However, they could not explain the discrepancy. Their knowledge of O-ring performance at low temperatures was obviously inadequate. At this point, Morton-Thiokol's vice president, Joe Kilminster, intervened to prepare the formal recommendation to NASA. After some prodding, the four Morton-Thiokol engineering managers agreed to reverse the initial recommendation and approve the launch, effectively over-ruling Boisjoly. The teleconference with NASA resumed; Kilminster announced the change in opinion and recommended that the shuttle launch go ahead. The shuttle was launched the next morning at 11:38 a.m. and exploded 73 seconds later.

Discussion of the Ethics

In the aftermath of the explosion, there was plenty of blame to go around. Boisjoly was appointed to the investigation team and was initially involved in redesigning the seal. He provided information freely to the President's commission, which led to severe friction with his colleagues and superiors.

Eventually, he began to feel isolated. He drifted out of contact with his colleagues—especially the NASA management—and finally resigned. In 1987 he filed lawsuits against Morton-Thiokol and NASA for personal damages.

Boisjoly was seen as an ethical whistle-blower—his recommendation to delay the launch had been over-ruled. One unanswered question is whether he could have done more to obtain cold-temperature O-ring data and present a more convincing case for launch delay at the crucial late-evening teleconference. Boisjoly insists that he made the proper ethical choices during his engineering career, often at the risk of his job. In 1988 he received the American Association for the Advancement of Science (AAAS) award for Scientific Freedom and Responsibility for his efforts to act ethically in the events leading to the *Challenger* shuttle disaster.[13]

The key decision to over-rule the recommendations of the engineers, made by Kilminster on the eve of the launch, was clearly an ethical and management error. The best that can be said for the decision is that it was made under duress. The engineering managers were under intense pressure to meet schedules driven by political and financial priorities, with a space shuttle that was still experimental, and not a tested, proven vehicle. The pressure had changed the management philosophy from "launch only when engineers can prove it is safe to do so," to "launch unless engineers can prove it is unsafe to do so." It took several years for NASA to redesign and recertify the rocket boosters, and to get the space shuttle flying again.

The *Columbia* Disaster

On February 1, 2003, a second shuttle disaster happened: the *Columbia* broke up on re-entry from orbit. This disaster took place at a very high altitude, and observers on the ground saw the debris as several bright meteors streaking across the sky. An accident investigation board was convened. After intensive investigations, the board issued its final report on August 26, 2003.[14]

In simple terms, this second disaster happened because a piece of insulating foam broke away from the fuel tanks and struck the wing of the shuttle. The damage went unnoticed during flight, but the heat of re-entry was able to penetrate the left wing, weakening the internal structure. The shuttle disintegrated, and the pieces that did not burn at high altitude fell to the ground in a swath 1,000 km (622 miles) long. The disaster was a pointed reminder that the space shuttle is not a tested airliner, but a vehicle still under development.

DISCUSSION TOPICS AND ASSIGNMENTS

1. A professional employee is assigned to purchase computer hardware and software for a new design office to house 10 designers, engineers, and geoscientists. The professional calls for tenders and competitive tests for cutting-edge design software, but purchases the computer hardware from a local supplier, without competition. The hardware price is high, but not outrageous. The employer does not know that the supplier is the

professional's cousin. Does the professional person have a conflict of interest? If so, is it an actual, potential, or perceived conflict? What penalty (if any) is the professional subject to?

2. A professional engineer or geoscientist who accepts an offer of employment is creating a contract. In return for a salary and other benefits, the professional agrees to use his or her ability to help the employer achieve legitimate goals. Assume that you have been hired to design electrical or mechanical components for a machinery manufacturer. During a recession, the employer decides to diversify into new areas to attract more business. Would any of the following tasks be unacceptable, according to your conscience? That is, if the employer asked you to participate in designing the following machinery, which (if any) would you refuse to do, on ethical grounds?

- Bottling equipment for the beer and liquor industry.
- Medical equipment to make abortions safer and more convenient.
- Pill-making machines for the birth control or pharmaceutical industries.
- Security locks for the prison system.
- Equipment for nuclear power plants.
- Roulette wheels for casinos.
- Rifles or handguns for the Canadian armed forces.
- Rifles for hunters.
- Printing equipment for lottery tickets.

3. You are working as a professional engineer in a small consulting company that gives its engineering employees considerable latitude in scheduling tasks, meeting deadlines, and reporting expenses. The company president contacts you, and states that senior management has recognized your professional attitude and attention to high standards. The president also expresses concern about the lax attitudes of your colleagues, who seem to be abusing the freedom extended to them. Discuss, explain, and justify whether it would it be ethical for the president to offer, and for you to accept

- a secret assignment to monitor the behaviour of your colleagues and your immediate superior and report back to the president.
- a promotion to head engineer to replace your immediate superior, on the basis that the head engineer is not competent as a manager and should be replaced.

Additional assignments can be found in Appendix E.

NOTES

[1] Engineers Canada (formerly the Canadian Council of Professional Engineers—CCPE), *National Survey of the Canadian Engineering Profession*, 2002, Ottawa, available at <www.engineerscanada.ca/e/files/surveysummary2002.pdf> (June 15, 2009).

[2]   The Association of Professional Engineers, Geologists and Geophysicists of Alberta (APEGGA), *Guideline for Ethical Practice*, 2005, p. 17, available at <www.apegga.com> (May 23, 2008).

[3]   B.M. Samuels and D.R. Sanders, "Employment Law," Chapter 21 in *Practical Law of Architecture, Engineering and Geoscience*, Pearson Prentice Hall, 2007.

[4]   National Society of Professional Engineers (NSPE), *Guidelines to Employment for Professional Engineers*, 4th ed., Alexandria, VA, June 10, 2006, available at <www.nspe.org/resources/documents/pei/guidelines_rev4.doc> (May 23, 2008).

[5]   K. Kernaghan and J.W. Langford, *The Responsible Public Servant*, The Institute for Research on Public Policy, Halifax, 1990, quoted in M. Macdonald, "Ethics and Conflict of Interest," Centre for Applied Ethics (CAE), available at <www.armsdeal-vpo.co.za/special_items/reading/ethics.html> (June 15, 2009).

[6]   National Society of Professional Engineers (NSPE), *Guidelines to Employment for Professional Engineers*, Part 1—Recruitment (employee clause 2).

[7]   G. Voland, *Engineering by Design*, Addison Wesley, Reading, MA, 1999, p. 433.

[8]   L.C. Bruno, "Challenger Explosion," from *When Technology Fails: Significant Technological Disasters, Accidents, and Failures of the Twentieth Century* (ed. N. Schlager), Gale Research, Detroit, 1994, p. 613.

[9]   *Report of the Presidential Commission on the Space Shuttle Challenger Accident*, Washington, DC, 1986, available at <http://science.ksc.nasa.gov/shuttle/missions/51-l/docs/rogers-commission/table-of-contents.html> (May 23, 2008).

[10]   R.M. Boisjoly, Interoffice Memo to R.K. Lund, Vice President, Engineering, Wasatch Division, Morton Thiokol, Inc., July 31, 1985, in D. Vaughan, *The Challenger Launch Decision*, Appendix B, University of Chicago Press, Chicago, 1996, p. 447. Also in *Roger Boisjoly: The Challenger Disaster*, the Online Ethics Center for Engineering & Science, available at <www.onlineethics.org/cms/7123.aspx> (May 23, 2008).

[11]   R.M. Boisjoly, "Pre-Disaster Background," the Online Ethics Center for Engineering & Science, available at <www.onlineethics.org/CMS/profpractice/ppessays/thiokolshuttle/shuttle_pre.aspx> (May 23, 2008).

[12]   D. Vaughan, *The Challenger Launch Decision*, University of Chicago Press, Chicago, 1996.

[13]   Roger Boisjoly: *The Challenger Disaster*, the Online Ethics Center.

[14]   H.W. Gehman, Jr., et al., *Columbia Accident Investigation Board Report*, Volume 1, Columbia Accident Investigation Board, Arlington, VA, August 2003, p. 25, available at <http://caib.nasa.gov> (May 23, 2008).

Chapter 13
Ethics in Management

According to statistics, most professional engineers and scientists become technical managers during their careers. Promotion to management means more salary and authority, but also more responsibility. When ethical problems arise, such as negligence, conflicts of interest, or corruption, the manager must deal with them. Other employees look to the manager as a role model, so decisive, ethical leadership is important.

This chapter discusses some of the ethical issues faced by managers, particularly when hiring professionals, reviewing their work, and, where necessary, dismissing them. Six case studies illustrate practical ethical problems (and more case studies are located in Appendix F). The chapter ends with a case history of the collapse of the Vancouver Second Narrows Bridge, which shows that negligent professional practice can lead to tragedy.

ADHERING TO THE LICENSING ACT

The most obvious responsibility of a manager is to ensure that the professional engineering and/or geoscience Act is being obeyed. Two common infringements are allowing unlicensed personnel to work as professionals, and misusing titles. These practices are contrary to the Act in every province and territory.

Unlicensed Personnel

Every province and territory requires professional engineering and geoscience to be performed (or supervised) by a licensed professional. Infractions usually occur only in small companies where non-technical owners, unaware of the law, may hire unlicensed personnel to perform this work. However, if unlicensed people are working as engineers or geoscientists in your company, the manager must correct this practice, even if it has endured for years, and even if resentment and antagonism might result. Managers must insist that any employee practising engineering or geoscience be licensed. When this is not possible, the employee must be put under the supervision of a licensed professional.

Misuse of Engineering and Geoscience Titles

The misuse of engineering and geoscience titles is also an offence under every Act. Many companies have positions with the word "engineer" or "geoscientist" in the title, such as "Project Engineer" or "Chief Geoscientist." These imply that the person holding the position is a licensed professional. Two situations may arise, depending on the person's qualifications:

- **Unqualified personnel.** If an unlicensed person is using a professional title, and is indeed, practising engineering or geoscience, the problem is extremely serious—this is an offence punishable under the Act. Such situations must be rectified immediately. If the person is eligible for a licence, then an application should be made immediately; if the person is not qualified, then the person must be replaced by a licensed professional.
- **Erroneous title.** If an unlicensed person is using an engineering or geoscience title, but is performing tasks that do not require a licence, then a new job title is needed. The title may be elegant, but it must not contravene the Act. Assigning a new job title is usually a simple solution, although tact and diplomacy may be needed.

Using a misleading engineering or geoscience title contravenes the Act, so the manager's failure to change the title might be interpreted as contributing to the illegal practice of the profession, which is itself contrary to the Act.

HIRING AND DISMISSAL

The manager usually hires and dismisses technical staff. The manager should therefore be familiar with the provincial regulations for hiring and dismissal. Some key points follow.

Employment Contracts and Policies

The best method for employing professional engineers is through clear-cut employment contracts that specify duties, contract duration (either fixed length or indefinite), remuneration, pay raises, vacation entitlement, statutory holidays, and so forth. Some contracts even include definitions of just cause for termination, along with terms and amounts of severance pay.

However, a large company may have so many professional employees that personal contracts cannot be negotiated. In this case, the company management should establish and publish employment policies that apply to all employees. The professional then receives a letter of appointment that refers to these policies. However, the policies must be fair, and they must be negotiated and administered fairly, or the employees may resort to unionization. The NSPE *Guidelines to Employment for Professional Engineers* (discussed in Chapter 12, and reproduced in Appendix D) describes topics that should be included in professional employment policies.

Terminating Employment for Just Cause

A manager must take responsibility for terminating or discharging employees when their services are no longer required. These terminations must be in accordance with the employment contract or published company policies. In addition, employees may be discharged for *just cause*, which is defined as a "fundamental breach of the employment contract, justifying termination."[1]

Those matters that would allow an employer to terminate an employee, without notice or severance pay, are as follows:

1. Serious misconduct;
2. Habitual neglect of duty;
3. Serious incompetence, not just management dissatisfaction with performance;
4. Conduct incompatible with his or her duties, or prejudicial to the company's business;
5. Wilful disobedience to a lawful and reasonable order of a superior, in a matter of substance;
6. Theft, fraud or dishonesty;
7. Continual insolence and insubordination;
8. Excessive absenteeism, despite corrective counselling;
9. Permanent illness; and
10. Inadequate job performance over an extended period as a result of drug or alcohol abuse, and failure to accept or respond to the company's attempt to rehabilitate.

If one of these elements of misconduct exists, and is ascertained even after the employee has been discharged, the company can rely on that misconduct and not pay the employee any severance allowance.[2]

Wrongful Dismissal

When an employee is dismissed, and the reason is not just cause, as defined above, the dismissal may be wrongful. Wrongful dismissal cases sometimes end up in the courts, so both the employee and the manager may need legal advice.

In a comprehensive article on wrongful dismissal, lawyer Howard Levitt described six situations that could also be considered wrongful dismissal even if, technically, the employee was not dismissed:

- Forced resignation;
- Demotion;
- A downward change in reporting function;
- A unilateral change in responsibilities;
- A forced transfer; and
- Serious misconduct of the employer toward the employee.[3]

In summary, managers must be alert to the complexities of supervising the work of others. A manager needs leadership ability, sensitivity, and a professional attitude. Knowledge of the law (or access to legal advice) is also useful—advice should always be sought before the hard decisions are made, not after the fact.

REVIEWING WORK AND EVALUATING COMPETENCE

The law requires professionals, whether they are employees or managers, to practise only within their limits of competence. Professional employees should not undertake—and managers should not assign—work that is not within the competence of the professional.

Reviewing Work for Accuracy

In engineering and geoscience, key decisions are never made on the basis of a single unchecked calculation. Important analyses or calculations, and the assumptions on which they are based, are always double-checked for errors. Professionals expect their work to be routinely reviewed by a second person, because errors could be extremely costly. For example, structural design, aerospace, and nuclear industries have standard procedures for double-checking all calculations, and the decisions resulting from them.

This routine check greatly increases the accuracy of the work, improves confidence and safety, and lowers the risk of failure and liability. The person who performed the original calculations is always informed prior to the review, given the opportunity to clarify any doubtful points, and shown the results.

Reviewing Work to Assess Performance

It is common practice for the manager to evaluate the performance of all employees annually. These annual reviews must be discussed with the professional, and typically, the professional is given an opportunity to challenge or explain an unsatisfactory evaluation and to include these comments with the manager's evaluation.

Additional performance reviews may be necessary if performance shows an inadequate level of competence. However, a manager should never ask a professional to review secretly the work of another professional. This precept, which is included in most Codes of Ethics, is simple common courtesy and should apply to any professional employee. A secret review is like a trial in absentia, and this is contrary to natural justice. Professional work is based on specialized knowledge, and the person who checks it must be able to understand the opinions and conclusions expressed. A careless or cavalier review could inadvertently damage someone's professional reputation, which is a valuable asset, created through years of study and experience.

In summary, frequent reviews of a professional's work for accuracy, and an annual review for performance, are routine and should be expected. However, secret reviews of one employee's work by another employee contravene the

Code of Ethics. Managers should be sensitive to this professional courtesy. This rule does not apply, of course, when the professional has left the place of employment, and duties have been reassigned. Obviously, any new employee must review the departed employee's work for continuity.

Maintaining Competence

Most provincial Acts specifically require professional engineers and geoscientists to maintain competence, as explained in Chapter 3, but managers have a duty to encourage and assist. The employee must show initiative, but the manager must support it.

For example, when a company takes on a new project or installs a new computer system, professional employees must upgrade their skills and/or knowledge. The manager should arrange orientation, review courses or workshops, and so forth to assist employees to learn the new skills needed. This is a win–win arrangement for employee and employer.

When a professional fails to keep skills up-to-date, drifts into incompetence, and refuses to exert the effort to become more effective, the manager must make a difficult decision. An incompetent professional should not be practising; in fact, every licensing Act cites incompetence as a basis for removing a licence. Dismissal may be necessary, but the task of the manager, as in any problem-solving exercise, is to gather information, generate alternatives, examine those alternatives, and seek the optimum course of action. The outcome will depend on the facts of the individual case, but the manager must deal with the professional employee fairly and ethically.

CONFLICT OF INTEREST IN MANAGEMENT

Chapter 12 gives a detailed overview of conflict of interest, and lists seven common forms of conflict (or "abuse of privilege," as it is sometimes called). Managers must set an example by avoiding conflicts of interest. Otherwise, they will not be able to monitor employees for this problem. Managers should take the initiative by establishing clear policies on conflict of interest.

- **Policy on unsolicited gifts.** Some companies have a strict policy that no gifts or benefits (meals, entertainment, or travel) may be accepted from clients, suppliers, or colleagues. This is simple, easily understood, and easily applied. Other companies set a limit for the value of small gifts, such as $20, but this is always subjective; the value of a gift or benefit may not be known.
- **Disclosing conflicts of interest.** All employees are expected to disclose conflicts of interest, and this should be stated clearly in a company policy.

DISCRIMINATION IN PROFESSIONAL EMPLOYMENT

The manager plays a key role in hiring professionals, evaluating their performance, promoting them, and (when necessary) dismissing them. The manager is in the front line of the battle against discrimination in employment. In Canada, the Charter of Rights and Freedoms prohibits discrimination. Hiring must be on merit, promotion must be based on performance, and dismissal must be based on *just cause*, as discussed above. Although much progress has been made in overcoming discrimination in recent years, certain groups, such as women, Aboriginal peoples, and people with disabilities, are still under-represented in engineering and geoscience. The problem is especially obvious where women are concerned—they are a slight majority in the general population, yet they are definitely a minority in our professions.

These groups have a legal right to be treated fairly. They do not expect preferential treatment, but artificial obstacles must not be created for them. Discrimination is discussed in detail in Chapter 10, "Fairness and Equity in the Professional Workplace."

CASE STUDY 13.1

UNLICENSED PROFESSIONAL

STATEMENT OF THE PROBLEM

You are the manager of the engineering design department for a fairly large consulting engineering firm. As part of your job, you hire and dismiss department staff members, including engineers, designers, computer technicians, and clerical workers. Six months ago you hired Jorge Xavier, who had recently moved to your area from another province. During the employment interview, you emphasized that it was essential that he be licensed. The letter of appointment sent to him stipulated that he was being hired as a professional engineer. After Xavier started work, you had a sign placed on his door and had business cards printed, both of which had the "P.Eng." designation after his name.

You are startled to receive a complaint from a client who claims that Xavier is not a licensed professional engineer. The client is furious that you and your company would send unqualified people to work on her project. You contact the provincial Association, which confirms that Xavier does not hold a licence. Now, *you* are furious.

QUESTIONS

Who is responsible for this problem? Can you fire Xavier for just cause? Would it make any difference if

- Xavier is licensed in another province, but has neglected to apply for a licence in your province?
- Xavier has applied for a licence in your province, but the provincial Association is still processing the licence?
- Xavier has never been licensed in any province?

AUTHOR'S SUGGESTED SOLUTION

This case involves a breach of a professional Act. A professional licence is valid only for the province in which it was issued, although a licence can be obtained in the new location fairly easily. When a person moves to a different province or territory, a licence application must be submitted to the Association in that jurisdiction. The process is routine, and additional licensing conditions are rarely required. A new licence will generally be issued with a minimum of inconvenience.

There can be little doubt that Xavier is guilty of practising professional engineering without a licence. He has used the business cards that clearly say "P.Eng." without protest or correction, and he is not licensed in the province where he is working. Consequently, he is committing an infraction of the Act, although the fact that you, as manager, had the business cards prepared could be considered a mitigating factor. You will be guilty of a breach of the Code of Ethics if you permit Xavier to continue to practise engineering.

More importantly, you must determine what work Xavier has done for the client. If Xavier was in a junior or training position during his first few months with the firm, and if his work has been supervised by another engineer, as would usually be done initially, then the risk to the client or to the public may be minimal. Damages may be limited to the embarrassment and the possible over-billing of fees.

However, if Xavier has been making independent decisions on engineering projects, then your engineering firm would be liable for any problems that arise from those decisions. You must review the work that Xavier has done, and discuss this liability problem with the company lawyer. You may also have some liability, as manager, since you are responsible for verifying the qualifications of those who work for you.

As to whether Xavier should be dismissed, and the grounds for that dismissal, it depends on which of the three following situations applies:

- If Xavier failed to apply for a licence after six months of employment, but has a valid licence from another province, then he has committed professional misconduct (or neglect of duty), which is just cause for dismissal.
- If Xavier has applied for a licence, but the application is still being processed, and if he has a valid licence from another province, then he has probably complied with your requirements, and dismissal would probably be unjust.
- If Xavier has never been licensed in another province, and if he has been unable or unwilling to obtain a licence in your province, then he has been dishonest in his employment interview with you, and such fundamental dishonesty would be just cause for dismissal.

Xavier clearly contravened the Act when he used the "P.Eng." title in his work without a valid licence. Therefore, he may also be subject to a charge under the Act, even if he does have a licence from another province. The provincial Association would decide whether to charge him, and the charge would be heard in the provincial court.

As the manager, you must bear some responsibility for any embarrassment or liability the firm suffers. You stated the requirement for a licence clearly, but you did not follow up to verify that, in fact, Xavier had a valid licence, and that he had the legal right to use the "P.Eng." designation. A company that offers engineering services to the public has a duty to verify that its engineers keep their licences up-to-date.

CASE STUDY 13.2

CONCEALING A CONFLICT OF INTEREST

STATEMENT OF THE PROBLEM

Assume that you are an engineering manager in a fairly large automotive parts manufacturing company. You are a member of a standards committee for automotive equipment. The committee comprises ten people: three industry representatives (including you), three government representatives, and three engineering professors. The chair is a staff member from a technical society.

During the meeting, Mr. X, an industry representative, proposes a change to the standard for an automotive component. The change would improve the quality slightly, but it requires special manufacturing equipment. Your company manufactures this component, and you realize that if the change were approved, your company would benefit greatly. Your company has a patented process for the component, which could be adapted to make the change. However, competing parts manufacturers would suffer, since they would have to develop new technology, or license yours.

You believe Mr. X's company would have the same advantage, since his company has a different patented process, but yields the same result. You are uncertain whether you should mention all this to the committee. You did not propose this change, but it would improve the automotive component, slightly, and your company would benefit, strictly by chance.

QUESTIONS

Is this a conflict of interest? Do you have an ethical obligation to inform the committee that your company will benefit from this change? Do you have an obligation to point out that Mr. X (who proposed the change) may also benefit?

AUTHOR'S SUGGESTED SOLUTION

You have a clear conflict of interest, which you must disclose to the committee. The Code of Ethics states that an engineer must place the welfare of society above narrow personal interest. The main function of a standards committee is to serve the public welfare, not the financial interests of its members. After you disclose your conflict of interest, it might be acceptable for you to answer questions and/or express your opinion of the change. However, you definitely should not participate in the formal vote on changing the standard.

Mr. X (who proposed the change) has a similar obligation to disclose that he has a conflict of interest. Your disclosure will likely encourage him to do so. If not, his silence might be an unethical attempt to benefit from this conflict of interest. You should not accuse him without more evidence, but you could ask all members of the committee to declare whether they have a conflict of interest similar to yours.

Conflicts of interest are fairly common on standards committees, for the simple reason that the best-informed people are those involved in the design and manufacture of the components concerned. However, conflicts of interest must be disclosed. We must not tolerate unfair or unethical abuse of such positions of trust.

CASE STUDY 13.3

DISCLOSING ERRORS IN PLANS AND SPECIFICATIONS

STATEMENT OF THE PROBLEM

You are the engineering manager for Acme Assembly, which designs, fabricates, and assembles machinery. You have received a contract to construct 20 large gearboxes that have been designed by Delta Designs, a company that sometimes competes with Acme. However, Delta is extremely busy, and does not have the capacity for this work at the present time.

One of your engineers notices that the sizes of shafts and gears on the drawings seem rather small for the torque and power ratings of the gearboxes; your rough calculations confirm that opinion. You call the chief engineer at Delta, who tells you he is too busy to double-check the drawings. He has full confidence in his designers and says you should "get on with the job." He points out that you are employed in this contract as the fabricator, not as the designer, and should not be reviewing his work.

QUESTIONS

Do you have an ethical obligation to pursue this apparent discrepancy? Would it make any difference if failure of the gearboxes would cause injury or death, rather than just inconvenience and repair?

AUTHOR'S SUGGESTED SOLUTION

Under the Code of Ethics, an engineer has an obligation to ensure that the client is fully aware of the consequences of failing to follow the engineer's advice (as explained in Chapter 12). In this case, a single telephone call probably would not satisfy this requirement, either ethically or legally. You should follow up the telephone call with a letter or e-mail that describes the reasons for your concerns and requests written instructions to proceed. The chief engineer at Delta would likely realize the gravity of the situation, and thank you for avoiding wasteful expense.

However, if the chief engineer instructs you, in writing, to proceed with the fabrication, then you should follow his instructions. In this contract, your job is fabrication, not design, and the designer is usually the only person who fully understands the expected loads, operating conditions, and intended use of the gearboxes.

The only exception would be the case where you have convincing proof that the design is clearly inadequate, and that failure of a gearbox would be likely to cause injury, death, or serious risk to public welfare. The manufacturing contract obviously does not include funds for you to review the design, so further advice would be needed. You would consult your employer about negotiating a new contract, or declining (breaching) the existing contract. Since the problem involves the Code of Ethics, in many provinces you could also call your Association for confidential advice on how to proceed. (See the discussion later in this textbook on whistle-blowing.)

The probability is very low that the chief engineer at Delta is negligent, incompetent, or involved in fraud. However, several well-known tragedies have occurred, in which whistle-blowing would have saved many lives. For example:

- **Ford Pinto.** The Ford Pinto, built in the 1970s, was prone to gas-tank rupture when the vehicle was hit from behind. A design modification to prevent the rupture (and the fire that often ensued) would have cost only about US$11 per vehicle. An engineer resigned from Ford in protest, and disclosed the faulty design.[4]
- **DC-10 cargo door latch.** Convair had a subcontract for the design of the cargo door on the DC-10 passenger airplane. The door latch failed during a cabin pressurization test, and a senior engineer wrote a memo to the management at Convair, itemizing the dangers in the design. Convair managers kept the information confidential, even though a cargo door blew out in flight over Windsor in 1972. The information was revealed only after a DC-10 crash at Orly Field in Paris killed 346 people in 1974.[5]

In serious cases such as these, the duty to public safety must take priority over the duty to an employer, client, or colleague.

CASE STUDY 13.4

DISCLOSING PRELIMINARY MINING DATA

STATEMENT OF THE PROBLEM

You are a professional geologist responsible for all of the ore assays in a mine. You report directly to the mine's chief executive officer, who is an accountant by training. You have just finished evaluating initial ore assays for a newly opened part of the mine. These show much lower ore content than anticipated. The CEO is very disappointed at the news. You reassure him that the results are preliminary and that more thorough results will be available in a week or so. The CEO had hoped to present good news about the exploration to shareholders at a meeting to be held in the next few days.

The CEO tells you to keep the results confidential and not to report or discuss them—not even with the company's employees—until after the shareholders' meeting.

QUESTION

Is it ethical to hide this information from the shareholders, who are the owners of the company?

AUTHOR'S SUGGESTED SOLUTION

This question is important, but it reveals a misunderstanding about corporate management structure. The shareholders are indeed the ultimate owners of the corporation, but they do not run it. The shareholders elect directors, who form a board of directors. In turn, the board appoints the officers of the company—the president, CEO, treasurer, and so on—and these people are responsible for the day-to-day operations of the company. Employees, including the geologist, take direction from these company officers. So the simple answer to this question is that the geologist has no legal or ethical duty to report directly to the shareholders.

In the mining, oil, and gas industries, geological data are extremely sensitive information and can be the basis for important financial decisions. Unauthorized disclosures can lead to abusive stock market tactics. Most boards authorize only the CEO to issue public statements.

In fact, all public disclosures from mineral companies must follow the strict guidelines set out by the Canadian Securities Administrators (CSA). The CSA guidelines are a fairly recent set of rules introduced after the Bre-X mining fraud (discussed in Chapter 2). Every geoscientist involved in preparing mineral studies should be familiar with the CSA guidelines, which regulate all public statements relating to mineral projects, be they oral or written (including news releases, prospectuses, and annual reports). The guidelines also require all disclosures to be based on a technical report prepared by a "qualified person" (as defined in the document and discussed in Chapter 2). Furthermore, they require this report to adhere to a particular format.[6] So the answer to the question in this case study is very clear—any disclosure of the ore assay results by the geologist would be unethical and probably illegal.

CASE STUDY 13.5

PROFESSIONAL ACCOUNTABILITY IN MANAGEMENT[7]

STATEMENT OF THE PROBLEM

Ethel Eager, P.Eng., is a mechanical engineer at a well-known specialty chemicals company. The company makes consumer products in Canada for the North American market. It also has plants in the United States, which compete with Canadian plants for North American production mandates.

Eager started out five years ago in a junior production position, reporting to the production supervisor Cam Complacent, P.Eng. When Eager started at the Canadian plant, it was highly successful. However, over the five years of her employment, the plant has become steadily less competitive relative to other firms and its sister plants in the United States. When Complacent retired recently, Eager was promoted to fill his job.

Eager passed her Professional Practice Exam while working at the company, and she is aware of the importance of professional ethics in engineering. Over the past five years she has noticed several unusual practices and events in the plant and in the office. For example, supplies often run out before forecast, inventory is invariably balanced by assuming losses, and there are frequent shortages in customer shipments. In the human resources area, she has noticed a tendency to "horseplay" on the graveyard shift, as well as what she would consider to be instances of racial and sexual harassment. Also, procedures for recording the hours that employees actually work are very casual, and overtime is high.

These discrepancies disturbed her, and Eager had approached her boss, Cam Complacent, about them several times. Each time, he played down her concerns and said being "easy" on these subjects helped keep morale and productivity up. Although Eager was personally convinced that some employees were cheating their employer by taking products home and misrepresenting their hours of work, as a junior employee she had decided to take her manager's advice to keep quiet.

However, shortly after she replaced Complacent as supervisor, Eager was informed, early one Monday morning, that there had been a major theft at the plant on the weekend. A truck had pulled up to the warehouse without being challenged, loaded up, and disappeared. Fortunately, the police soon caught the two thieves, who turned out to be employees, one of them a relative of a senior employee. Indeed, the police soon found that a network of employees was involved. They now want to interview Eager about further investigations.

Meanwhile, Eager has just received a fax from the company's vice president for North American manufacturing, who wants to investigate why the Toronto plant's costs have been so high and why productivity has been so low relative to the company's other plants. The fax concludes: "Understand major theft has occurred. Will be in Toronto tomorrow to review your situation." The future of Eager and her plant looks grim.

QUESTIONS

Should Eager be held accountable for the employees' actions? What lessons, if any, can be learned from this case?

AUTHOR'S SUGGESTED SOLUTION

As a middle manager and a professional engineer, Eager is accountable to her superiors, possibly to the police, and to her profession, because she knowingly allowed a dishonest environment to flourish. All of the stakeholders involved— Eager, her superiors, her peers, her employees, and even her suppliers—have suffered, or will suffer. Because she is a professional engineer, Eager has a duty under the provincial Code of Ethics to all of these stakeholders to act at all times with devotion to high ideals of personal honour and professional integrity. She also has a duty to expose, before the proper tribunals, unprofessional or unethical conduct by another engineer.

Although there are mitigating circumstances in this case (e.g., Eager's relative inexperience and her employer's lack of an ethics program), Eager has learned two valuable lessons:

- The meaning of accountability;
- That there are no small ethics problems.

In hindsight, Eager now knows that turning a blind eye to the problems at her plant was wrong. She must also realize that there would have been benefits to dealing with her concerns as they arose, and that now there are consequences to her having ignored them. She should have explained to Complacent that, as a professional engineer, she was duty bound to act on her concerns. She should have suggested to him that, together, they discuss the subject with senior management. Had Complacent been unwilling to consider this approach, as a last resort, Eager could have considered going alone to senior management or obtaining advice from the provincial Association.

CASE STUDY 13.6

STUDENT PLAGIARISM

STATEMENT OF THE PROBLEM

Oliver T. is an engineering student in the last week of his co-op summer work term at a large manufacturing company. Oliver has enjoyed the job, worked hard, learned much, and been well paid. He is on very good personal terms with his boss, and the boss will likely give Oliver an excellent work evaluation. However, Oliver has not yet written a work report. When he returns to university next week, he must submit a work report, describing a professional project he undertook during his work term. A professor marks the report, university standards are applied, the grade is recorded, and Oliver's graduation could be delayed if the report is unsatisfactory. (This is a routine requirement at most Canadian universities.) Oliver had several interesting jobs that were suitable report topics, but he was "too busy" to write anything down. He is beginning to worry—he has no plausible excuse for failing to write a report.

Oliver mentions his problem to his boss. Surprisingly, the boss suggests that Oliver could save a lot of time by simply submitting an older work report as his own. The boss has a file cabinet containing copies of work reports written by former work-term students. Most of the reports are also stored in computer memory or on CDs.

QUESTIONS

Should Oliver submit a work report written by a former student as his own? If he does so, and is caught, what penalties does he face? Is he less guilty because his boss suggested the idea? Is the boss guilty of any unethical action?

AUTHOR'S SUGGESTED SOLUTION

A person who submits a report written by someone else is committing plagiarism. Every Code of Ethics or ethical theory condemns plagiarism as unethical, either directly or indirectly. (In common parlance, this is an ethical "no-brainer.") Obviously, Oliver should resist the temptation to submit a plagiarized work report, get busy, and use the remaining week to write the report.

We all have a duty to prevent plagiarism, yet plagiarism has increased dramatically in the past decade (especially in schools and universities). This increase is probably caused by the easy availability of written material on the Internet. Unfortunately (for some), the Internet also provides tools for detecting plagiarism, and universities have severe penalties for it.

Plagiarizing a work report undermines both the educational process and the cooperative program. Co-op employment is not just a job. Co-op integrates work experience with academic study, and the work report tests this integration. If Oliver's plagiarism were discovered, the consequences would typically be as follows:

- **Academic penalties.** Plagiarism is a serious academic offence that usually results in suspension for a first offence and expulsion for a second offence. Both suspensions and expulsions are usually shown on grade transcripts. Either will delay graduation far more than a late report.
- **Future recommendations.** Obviously, professors and co-op employers will hesitate to recommend students for job openings or for graduate admission if the students have been suspended or expelled for plagiarism. Character references are also needed to obtain a professional licence.

The fact that the boss was willing to help Oliver commit plagiarism would not lessen the penalties for Oliver; in fact, it might extend the disciplinary action to include the boss (assuming that the boss is a licensed engineer or geoscientist). The boss has breached the Code of Ethics and could be reported to the Association. Also, the university would almost certainly bring such collusion to the attention of the boss's employer, and could prevent the employer from hiring co-op students in future.

- **How to avoid plagiarism.** If your work includes any material (including sentences, photos, drawings, or figures) from any other source, cite the complete source—it is easy to do. Failure to cite sources is plagiarism. In particular, any material cut and pasted from websites must be fully identified with a proper reference that cites the URL and the date. Authors are guilty of plagiarism if they submit reports containing Internet material that is not fully cited.

- **How to detect plagiarism.** Written material plagiarized from the Internet is easy to detect. A key word search, using words from the plagiarized report, on any search engine will generally detect the source very quickly. In addition, a Web service for detecting plagiarism has been developed and is available to professors for a nominal fee.[8] Many universities are also developing in-house solutions for plagiarism that involve scanning parts of submitted reports to create a database for searching. Even one passage in a report could be proof of plagiarism if the source has not been cited.

- **Where to learn more.** Several excellent guides to avoiding plagiarism can be found by a simple Internet search using "plagiarism" as the search term.

CASE HISTORY 13.1

THE VANCOUVER SECOND NARROWS BRIDGE COLLAPSE

This case history reminds engineers and geoscientists that serious failures can occur during construction. In fact, the risk of failure may be higher for temporary supports, forms, and scaffolds because they are rarely analyzed as thoroughly as the main, permanent structure. Managers must double-check calculations for critical components, especially when less experienced professionals conduct the work.

On June 17, 1958, two spans of the Vancouver Second Narrows Bridge (now known as the Iron Workers Memorial Bridge) collapsed during construction. Eighteen workers were killed. The lateral buckling of beam-webs in a temporary tower caused the tragic accident. A fairly simple calculation would have shown that the beam-webs were unsafe. The following description of the tragedy is reprinted, with permission, from W.N. Marianos, Jr., "Vancouver Second Narrows Bridge Collapse."

Background

The Second Narrows Bridge connects Vancouver, British Columbia, with its northern suburbs across Burrard Inlet, the city's harbor. The structure was built for the British Columbia Toll Highways and Bridges Authority. The main bridge, a steel cantilever truss structure, has a total length of over two

Photo 13.1 — Vancouver Second Narrows Bridge Collapse. *Eighteen people were killed when failure of temporary construction supports caused the Vancouver Second Narrows Bridge spans to collapse on June 17, 1958. The bridge was rebuilt and opened in 1960, and was later re-named the Iron Workers Memorial Bridge.*

Source: © Bettmann/CORBIS.

thousand feet (610 m). Unlike older, simpler, and shorter bridges whose spans or sections rest independently on their piers or abutments, those of a cantilever bridge run continuously over or extend beyond the piers. The main bridge has three spans: a 1100 foot-long (335 m) centre span balanced by two side spans, one 465 feet (142 m) and one 466 feet (142 m) long. Four steel truss spans, each 276 feet (84 m) long, make up the northern approach to the main bridge. The structure was designed by Swan, Wooster and Partners, a Vancouver engineering firm. Dominion Bridge Company was the contractor for the construction of the steel spans. The foundations and bridge piers were constructed by Peter Kiewit Sons and Raymond International.

Details of the Collapse

By mid-June 1958, the approach spans were in place and erection of the northern side span of the main bridge was in progress. The length of the span required the use of two temporary supports for construction, since the side span would not be self-supporting until its full length was in place. Each temporary support, called a falsework bent, consisted of two columns, one under each side of the span. The columns were built on temporary piers in the harbor. These piers were supported by a group of foundation piles. The load from each column was distributed to the foundation piles by a grillage—a two-layer grid of steel beams. The lower set of beams sat on top of the foundation piles. The upper layer, a set of four beams set side by side, supported the column bases.

On June 17, the first side span was supported on a permanent concrete pier at one end, and was overhanging the first falsework bent, designated "bent N4," at the other. At 3:40 P.M. that afternoon, bent N4 collapsed, plunging the partially completed span into Burrard Inlet. The falling metalwork pulled the permanent pier it was resting on out of line, which caused the adjacent approach truss to collapse as well.

Immediately after the accident, the government of British Columbia appointed a royal commissioner, Sherman Lett, chief justice of the provincial supreme court, to determine the cause of the collapse. The commissioner selected five leading engineers to investigate and report on the matter: FM. Masters and J.R. Giese of the United States; J.R.H. Otter and Ralph Freeman of Britain; and A.B. Sanderson of Canada. Materials testing and special investigations were conducted at the University of British Columbia and testing laboratories in Vancouver.

The commissioner's report concluded that the collapse was caused by failure of the four upper grillage beams. The webs (the vertical portion) of the beams buckled laterally, causing the collapse of the falsework bent columns.

Faulty design of the falsework or temporary columns led to the grillage failure. The commission discovered two major errors in the Dominion Bridge Company's grillage design calculations. The first mistake was in checking the grillage beam shear strength (the capacity of a beam to carry a load in its vertical plane; shear stress tends to tear a beam vertically, usually at supports or at points of concentrated load). The cross-sectional area of the entire beam was used in the calculation rather than just the areas carrying the load. This mistake would lead the grillage designer to believe the beam strength was about twice as much as it actually was.

A second calculation, which checked the need for web stiffeners, was also incorrect. Stiffeners are metal plates welded to beam webs to give them additional stiffness and resistance to buckling. The contractor's engineer had used the one-inch (2.5 cm) thickness of the beam flanges (the horizontal elements) rather than the actual 0.65-inch (1.6 cm) web thickness. This led to the erroneous conclusion that no stiffeners were needed.

A separate investigation by Dominion Bridge Company came to the same conclusion—that incorrect calculations led to a fatally inadequate grillage design. One of the errors in calculation was even discovered before the accident, but no corrective action was pursued. The two engineers responsible for the calculations were both killed in the collapse.

Wood blocks and plywood pads had been included in the grillage to provide some bracing of the beams. Laboratory tests indicated that these blocks were only marginally effective at best. Most of the wooden blocks were not even located at the most effective bracing points. The investigation performed at the

University of British Columbia also indicated that the ability of the beam webs to resist buckling was not adequately predicted by the usual design formulas for column buckling.

In his report, the royal commissioner laid the blame for the collapse on the Dominion Bridge Company. The commission found the contractor negligent for "(a) failing properly to design and substantially construct false bent N4 for the loads which would come upon it . . . (b) failing to submit to the engineers plans showing the falsework the contractor proposed to use in the erection . . . and (c) leaving the design of the upper grillage of false bent N4 to a comparatively inexperienced engineer, and failing to provide for adequate or effective checking of the design and the calculations made in connection with the design."

The commissioner also found that a failure in the construction process had contributed to the accident. His report pointed out that the bridge design engineers, Swan, Wooster and Partners, had a responsibility to make sure the contractor submitted the falsework plans and calculations for their approval, as required by the project contract. The engineers certainly knew that the bridge was under construction, and they had prepared the section of the project specifications that required engineer's approval of the temporary falsework structures. Commissioner Lett concluded that "there was a lack of care on the part of the engineers in not requiring the contractor to submit plans of the falsework." Ironically, the satisfactory performance of Dominion Bridge on earlier projects may have contributed to the design engineer's laxness in pursuing the falsework plans and calculations for review.

The commissioner recommended that on future large bridge projects the consulting engineers recommend allowable stresses for temporary construction support structures, and that the contractor be required to submit all construction plans and calculations for approval prior to construction. The contractor, however, would always remain legally responsible for the adequacy of construction methods and temporary structures.

Impact

After the inquiry, construction resumed on the bridge. Two concrete bridge piers damaged in the collapse had to be rebuilt. This required the careful removal of two thousand cubic yards (1529 m^3) of reinforced concrete. The collapsed superstructure spans were salvaged, and some undamaged members were reused. Erection of the bridge continued according to the original plan, with the notable addition of careful checking and review of all construction calculations and plans. The additional time and materials required to reconstruct the damaged portions of the bridge added four million dollars to the original contract price of sixteen million dollars.

The editors of *Civil Engineering* magazine noted that the collapse "illustrates the ever-present risks that are inherent in construction, due to human error. The failure emphasizes the need for utilization of all possible checks on construction procedures."

Today, the leading bridge design firms continue to carefully review and check the contractor's construction plans and calculations. The collapse of the Vancouver Second Narrows Bridge was neither the first nor the last incident of mistaken temporary construction calculations leading to disastrous consequences. The accident vividly highlights the importance of independent checking of critical aspects of the construction process.[9]

DISCUSSION TOPICS AND ASSIGNMENTS

1. Assume that you are a project manager, and one of your key responsibilities is to make time and cost estimates for the project. The project is fairly complex, so your calculations result in very high estimates—so high that you fear the project may be cancelled. Older colleagues on your team tell you: "Many earlier projects would have been cancelled if the true costs had been known this early in the game." Moreover, they argue that no one can ever be really sure of what something is going to cost: "After all, these are only estimates!" You know from experience that several earlier projects exceeded the cost estimates, but the funds were found, and the projects were successful.

   The older colleagues urge you to reduce your estimates so that the project will not be cancelled. You have put a lot of careful work into your estimates, and believe your figures are as correct as any estimate of the future can ever be. Therefore, if you reduce the estimates, you know you will be lying. Furthermore, you know your own reputation in the company will be damaged if it becomes apparent that you shaved your estimates. However, if your project is cancelled, some of the people in your project team may be laid off. You are caught in a dilemma, and as a manager, you must decide one way or the other. Explain how you would try to solve this ethical dilemma. What additional information would you need? Assume that you have this information (and state it), then write a brief summary of your decision and your reasons for it.

2. Assume that you are a licensed professional engineer in a fairly large engineering and geoscience consulting company. You have recently been promoted to manager and transferred to manage a branch in a western city. The branch is swamped with work, and you rely heavily on an older assistant, who has been with the company for about three decades. You notice that the assistant has the sign: "Associate Geoscientist" on his door, and the same title on his business cards. You check the personnel files, and realize that the assistant has extensive personal experience and is an incredibly valuable employee, but obtained his expertise in the field and has neither a licence as a geoscientist, nor certification as a technologist or technician. Is the "Associate

Geoscientist" contravening the professional geoscience Act in your province or territory? If so, what (if anything) would you do about it, as the manager?

3. Renée Langlois is a professional geoscientist who has recently been appointed president of a large dredging company. She is approached by senior executives of three competing dredging companies and asked to cooperate in bidding on federal government dredging contracts. If she submits high bids on the next three contracts, the other companies will submit high bids on the fourth contract and she will be assured of getting it. This proposal sounds good to Langlois, since she will be able to plan more effectively if she is assured of receiving the fourth contract. Is it ethical for Langlois to agree to this suggestion? If not, what action should be taken? If she agrees to this suggestion, does she run any greater risk than the other executives, assuming that only Langlois is a licensed professional?

Additional assignments can be found in Appendix E.

NOTES

[1]   B.M Samuels and D.R. Sanders, *Practical Law of Architecture, Engineering, and Geoscience,* Canadian Edition, Pearson Prentice Hall, Upper Saddle River, NJ, 2007, p. 238.
[2]   Howard A. Levitt, "The Law of Dismissal in Canada," as quoted in *CSPEAKER,* Canadian Society of Professional Engineers (CSPE), September 1981, pp. 1–4. Reprinted with permission of Howard A. Levitt.
[3]   Levitt, "The Law of Dismissal in Canada."
[4]   C.E. Harris, M.S. Pritchard, and M.J. Rabins, *Engineering Ethics: Concepts and Cases,* Thomson Wadsworth, 2005, p. 142.
[5]   Harris, Pritchard, and Rabins, *Engineering Ethics: Concepts and Cases.*
[6]   Canadian Securities Administrators (CSA), *National Instrument 43-101: Standards of Disclosure for Mineral Projects.* Document NI-43-101 can be found on several securities websites, such as those in British Columbia <www.bcsc.bc.ca>; Ontario <www.osc.gov.on.ca>; Quebec <www.lautorite.qc.ca>; or Alberta <www.albertasecurities.com> (May 10, 2008).
[7]   Case Study 13.5 is adapted from James G. Ridler, P.Eng., "Accountability: At the Core of Professional Engineering," *Engineering Dimensions,* vol. 18, no. 1, January–February 1997, pp. 40–41. Used with permission of James G. Ridler and PEO.
[8]   *Turnitin* (plagiarism prevention) website at <www.turnitin.com/static/home.html> (May 23, 2008).
[9]   W.N. Marianos, Jr., "Vancouver Second Narrows Bridge Collapse," from *When Technology Fails: Significant Technological Disasters, Accidents, and Failures of the Twentieth Century,* 1st edition by Schlager, Neil (Editor). 1994. Reprinted with permission of Gale, a division of Thomson Learning: www.thomsonrights.com. Fax 800 730-2215.

Chapter 14
Ethics in Private Practice and Consulting

Professional engineers and geoscientists in private practice have the same basic duties as employees or managers, but usually have more variety in their work and more flexibility in their schedules. They also make more money, providing that they can cope with the added responsibilities and stress of running a private enterprise. In this chapter we discuss several important aspects of private practice: the consultant's relationship with clients, the ethical aspects of advertising, the competition for contracts, and the review of another professional's work. Several case studies illustrate ethical pitfalls, and a case history recounts how an engineering consultant overcame an immense ethical challenge.

THE CLIENT–CONSULTANT RELATIONSHIP

The engineer or geoscientist in private practice is usually a consultant, advising a client. Typically, the client hires the consultant to monitor a project—for example, the design of a building, or the development of an ore deposit. The client needs the consultant's advice to confirm that a contractor's work is properly performed. This creates a three-way relationship between

- **the client** (or owner, who may or may not have any technical knowledge),
- **the contractor** (designer, builder, or developer, skilled in the activity, but who may or may not be a licensed professional), and
- **the consultant** (professional engineer or geoscientist, whose role is to watch over the interests of the owner, technically, but also assists the project to move along quickly, safely, honestly, and fairly).

The client–consultant relationship may take many forms, depending on the knowledge, skill, and personalities of those involved. The result is a wide range of client–consultant relationships. D.G. Johnson describes three typical relationships along this range:

The "independent" model: The client explains the problem and then turns over decision-making power to the consultant, who takes charge of the problem and makes decisions for the client. The consultant does not provide technical knowledge to the client, but acts in place of the client, keeping the client's interests in mind, in a paternalistic but independent way. This is one end of the client–consultant spectrum, and it is generally unacceptable, since it robs the client of the ability to make any choices.

The "balanced" model: The consultant interacts with the client, by providing engineering advice and evaluating the risks and benefits of various alternatives, but the client makes the choice of the action to follow. This relationship is similar to the ideal patient–physician relationship, where the professional may have the knowledge and expertise to solve the client's problem, but the client must be informed of the possible choices, and their benefits and risks, before making a decision to proceed with treatment. In a balanced relationship, the client and consultant must treat each other as equals. The consultant has a responsibility to provide engineering expertise to the client, but the client retains power to make the key decisions. The balanced relationship is the approximate mid-point of the spectrum, and is generally the optimum client–consultant relationship.

The "agent" model: The consultant is simply an agent or "order-taker" for the client, and contacts the client for instructions before acting. This is the other end of the client–consultant spectrum, and it is also generally unacceptable, since the client does not make full use of the engineer's knowledge. This relationship may also be seen as demeaning by the consultant.[1]

In any project, the consultant and the client should agree on the working relationship, which would usually be near the middle of the spectrum described above. In a good relationship, information flows both ways between client and consultant, both are adequately aware of key areas of the work, and well-informed decisions are made in a timely manner, leading to a successful result.

ADVERTISING FOR NEW PROJECTS

A professional in private practice may need to advertise. This is a thorny issue that plagues all the professions. Advertising is important to attract clients and to ensure that they obtain accurate information about professional qualifications and experience. Unfortunately, mediocre, self-serving advertising dominates all our communications media, from newspapers to the Internet, and it is demeaning to promote professional services as if they were soap, soup, or chewing gum. Every province and territory therefore restricts advertising of professional services. The key goals are to ensure honesty, fairness, and respect.

General restrictions are usually found in the Act or the regulations, and more detailed advice is usually in the Code of Ethics or in professional practice guidelines. Advertising a professional's availability, experience, and areas of expertise is fair, acceptable, and expected. "Calling card" or "business card" advertising, in the back pages of most technical publications, is very professional.

In Alberta, the APEGGA *Guideline for Ethical Practice* requires all advertisements, proposals, presentations, and solicitations for professional engagement to be "factual, clear and dignified."[2] In Quebec, Regulation 10 under the Act gives very precise rules concerning the information that may be conveyed on business cards, stationery, newspapers, magazines, directories, and signs on work premises, offices, and vehicles.

Ontario regulations permit advertising, provided that it is done in a professional and dignified manner; that it is factual and does not exaggerate; and that it does not directly or indirectly criticize another licensed engineer or the employer of another licensed engineer. The regulation also expressly forbids the use of the engineer's seal or the Association's seal in any form of advertising.[3] Licensed professionals may use the Association's name and logo on business cards and letterhead to signify licensing by the Association, but a professional seal cannot be used. The seal has a legal significance (as explained in Chapter 6) that is totally incompatible with advertising. In its *Guideline to Professional Practice*, PEO (Ontario) publishes several advertising rules, which are reproduced below with permission.

Advertising may be considered inappropriate if it:

i. Claims a greater degree or extent of responsibility for a specified project or projects than is the fact;
ii. Fails to give appropriate indications of cooperation by associated firms or individuals involved in specified projects;
iii. Implies, by word or picture, engineering responsibility for proprietary product or equipment design;
iv. Denigrates or belittles another professional's projects, firms or individuals;
v. Exaggerates claims as to the performance of the project; or
vi. Illustrates portions of the project for which the advertiser has no responsibility, without appropriate disclaimer, thus implying greater responsibility than is factual.[4]

In summary, advertising is acceptable if it is factual, truthful, and communicates accurate information about qualifications, experience, location, or availability in a dignified manner.

COMPETITIVE BIDDING FOR NEW PROJECTS

A detailed procedure for selecting an engineer in private practice, Quality-Based Selection (QBS), was described in Chapter 7. The procedure involves three stages and separates the process of selecting the best-qualified engineer (or firm) from the process of negotiating the fees. This prevents many problems that commonly arise when engineers are selected on a competitive basis by lowest bid.

However, it should be emphasized that seeking professional services by lowest bid is legal and ethical; in fact, fair competition is beneficial. Ingenuity thrives on healthy competition. However, there is a danger in competitive bidding, as explained in Ontario's *Guideline to Professional Practice:*

> With professional services there are ultimately only two elements that a client is retaining, i.e. the engineer's knowledge and time. Short-changing on a professional engineering fee will result in the substitution of less skilled engineers or less time put into the assignment, thus potentially short-changing the project.[5]

However, some competitive activities in obtaining contracts are considered unfair and unethical. For example, any agreement to pay a kickback, gift, commission, or consideration, either openly or secretly, would be an unfair and unethical (and likely illegal) method of obtaining contracts. Many codes also describe supplanting a colleague as unethical, where *supplanting* is defined as intervening in the client–engineer relationship of a colleague and, through inducements or persuasion, convincing the client to fire the engineer and hire the intruding engineer.

ENSURING COMPETENCE

Professional competence, gained through education or experience, is a valuable asset. The client is paying for that competence. Any professional who accepts an assignment that is beyond his or her level of competence could be guilty of either unprofessional conduct or incompetence. Either of these could lead to disciplinary action.

This does not mean that a professional must be an expert in every phase of a proposed project before accepting it. However, the person must be confident that he or she can become competent, through study or research, in a reasonable period of time. Alternatively, the professional must be able to hire a colleague or consultant with the needed expertise, without delaying the project. The essential point is that the client's project must not be placed at risk (or become needlessly expensive) because of the professional's lack of competence.

You are expected to know your level of competence; you are also expected to expand your knowledge and experience and maintain your competence (as discussed in Chapter 3). You must be realistic about your abilities—a difficult task at the best of times. However, no one knows the limits of your knowledge better than you do yourself.

CONFIDENTIALITY IN PRIVATE PRACTICE

Under the Code of Ethics, professionals are obliged to keep the client's affairs confidential. A client will sometimes ask a consultant to sign a confidentiality agreement. Since professionals intend to maintain confidentiality anyway, they are usually willing to sign these agreements. The requirement for confidentiality can create ethical problems, however. Consider the following two cases:

• If a consultant is hired by a new client who is a competitor of a former client, a conflict of interest may result. The consultant must not accept a contract that requires disclosure of a previous client's affairs, be they technical, business, or personal. This applies especially to proprietary information and to trade secrets that could result in financial loss to the former client, if disclosed. Even if the competitor does not expect the professional to disclose the information, the appearance of a conflict of interest may remain. In this situation, any confidentiality agreement signed with the former client should be reviewed. Undertaking work for the competitor is obviously risky.

- Another problem with confidentiality agreements arises in environmental projects. Where danger to the public is involved, the Code of Ethics (or environmental regulations) may require the consultant to reveal information. Consider a case where a consultant advises a client to remedy an environmental hazard. What happens if the client refuses to do so? If the consultant has signed a confidentiality agreement and later blows the whistle to the authorities, this could be interpreted as a breach of contract. Clearly, the consultant is facing a serious ethical dilemma: breach the contract, or obey the law. Ontario's *Guideline to Professional Practice* suggests a compromise: include a clause in the confidentiality agreement stating that if the client should fail to act on certain hazards within a specified period of time, the consultant is entitled to fulfill any reporting requirements that are specified in law, after first notifying the client.[6] Any consultant practising in the environmental area should get legal advice on the proper wording of such agreements.

CONFLICT OF INTEREST IN PRIVATE PRACTICE

As noted earlier, a conflict of interest arises in a professional relationship when the professional has an interest that interferes with the service owed to the client. For example, an engineer who recommends that a client purchase goods or services from a company in which the engineer has partial ownership (of which the client is not aware) has created a serious conflict of interest that is contrary to the Code of Ethics.

Conflicts can be much simpler than this. A consultant may be tempted to suggest that the client adopt a course of action where the main benefit is to reduce the consultant's workload. Unless there is a similar reduction in fee, the consultant has a conflict of interest that must be disclosed fully to the client. In every instance of conflict (or potential conflict) of interest, the consultant must make a full disclosure to the client of any personal interest, whatever that may be. If the client agrees that the conflict is insignificant, then work can proceed, but the client is making a fully informed choice.

A client who later learns that a consultant benefited personally and secretly from a decision that was ostensibly based on technical factors would be justified in contacting the provincial Association to lodge a complaint of professional misconduct.

REVIEWING THE WORK OF ANOTHER PROFESSIONAL

The issue of reviewing another professional's work is especially sensitive when they are consultants in private practice. As a general rule, a professional must be informed when his or her work is to be reviewed, but it is not necessary to obtain the person's permission for that review. In all instances, the welfare of the client or the general public must come before the professional's personal wishes. Ontario's *Guideline to Professional Practice* summarizes the situation as follows:

> The Code of Ethics permits engineers to be engaged to review the work of another professional engineer when the connection of that engineer with the project has been terminated. Before undertaking the review, reviewers should know how the information will be used. Even when satisfied that the connection between the parties has been terminated, reviewers should, with the agreement of the client, inform the other engineer that a review is contemplated. They should recognize that the client has the right to withhold approval to inform the engineer, but [should] satisfy themselves that the reasons for the owner's decision are valid before proceeding with the review.
>
> If a client asks an engineer to review the work of another engineer who is still engaged on a project, either through an employment contract or an agreement to provide professional services, the reviewer should undertake the assignment only with the knowledge of the other engineer. Failure to notify the engineer under this circumstance constitutes a breach of the Code of Ethics. On the other hand, should a second engineer be engaged by another person (say, a building department) to provide professional engineering services on the same project, he or she would have no obligation to advise the original client of the commission.
>
> Senior engineers are often asked to review a design prepared by another engineer. (Most engineers are expected to have their work routinely reviewed as part of an ongoing quality control and professional development process.) If reviewers find that design changes are necessary, they should inform the design engineer of these findings and the reasons for the recommended changes. During the design stage, reviewers (who are acting as the client's agent in this case) and engineers may agree on changes to the engineers' proposal. However, design engineers must not agree to any change or alternative suggested by reviewers that could result in an unworkable installation, be in conflict with the relevant codes, or create a risk of damage or injury.
>
> Reviewers must administer the design contract and evaluate engineers' work at arm's-length, so that the engineer of record maintains full responsibility for the design. . . .
>
> It is emphasized that the acceptance of mutually agreed-upon changes does not relieve the original design engineer of responsibility for the design or work under review.
>
> Once the review has been completed, there is no obligation or right for the reviewers to disclose their findings to the other engineer. In fact, in most cases, disclosure of the findings would not be permitted by the client. Reviewers' contractual obligations are to the client. However, reviewers should seek the client's approval to inform the engineer of the general nature of the findings, and if appropriate, should try to resolve any technical differences.[7]

OWNERSHIP OF DESIGN CALCULATIONS

A client may ask a professional to submit calculations that support a recommendation. This amounts to a review of the professional's work, but obviously it is done with the full knowledge and cooperation of the professional. The client has an ethical right to review these calculations and to make a copy for permanent record. However, the time necessary to prepare the calculations in a format understandable to the client should be included as part of the contracted service.

Occasionally the computation techniques or the data on which the computation is made may be proprietary, and the professional may not wish to divulge them. In this situation, the conditions for reviewing the calculations should be negotiated beforehand, and the extent of disclosure should be understood in advance. The usual procedure is to provide the proprietary data to the client with the clear understanding that they will be kept confidential.

NEGLIGENCE AND CIVIL LIABILITY

The professional in private practice must avoid the two main sources of liability: breach of contract and negligence. These are usually inadvertent and unfortunate, because they can lead to massive personal financial loss. Some protection against loss is available for these risks.

- A *breach of contract* is a failure to complete the obligations in a contract. By incorporating a practice, the individual may gain some protection against massive personal financial loss due to lawsuits resulting from breach of contract.
- *Negligence* is a failure to exercise due care in the performance of professional duties. Liability insurance protects the individual against massive personal financial loss due to lawsuits resulting from negligence.

However, incorporation and insurance do not protect the individual from disciplinary action for negligence, incompetence, or professional misconduct (which are discussed in Chapter 4). Ontario's *Guideline to Professional Practice* summarizes the situation as follows:

> An individual engineer can protect personal assets against an action for damages for breach of contract by incorporating the practice. After incorporation, it is the company that is the contracting party and not the individual. As far as protection from liability for negligence there is nothing available to an engineer other than careful, thorough engineering and insurance.[8]

CODES OF ETHICS FOR CONSULTANTS

Consulting engineers and geoscientists must be licensed in Canada, and are governed by the same Code of Ethics as employee engineers and geoscientists. However, two advocacy groups have been created, mainly by consulting firms, to work on behalf of members. Each group publishes guidelines, specifications, and fee schedules to assist consulting practice. Both groups publish a voluntary Code of Ethics that applies to their members. The codes are very similar to the Codes of Ethics of the licensing Associations, but lack the enforcement provisions of the provincial and territorial Acts, of course. A few clauses apply specifically to private practice, as mentioned below:

ACEC   The Association of Canadian Engineering Companies (ACEC) is an advocacy body working on behalf of Canadian engineering companies and engineers in private practice. Although it is not a licensing body, ACEC publishes a voluntary Code of Consulting Practice that applies to member firms of ACEC and requests them to fulfill their duties with honesty, justice, and courtesy toward society, clients, other consulting engineers, and employees.[9] The ACEC Code specifically forbids supplanting a colleague. This action has always been considered unethical. As explained earlier, *supplanting* means using inducements or persuasion (or derogatory comments) to convince a client to fire a professional and to hire the intervenor. Association codes also forbid supplanting, even when it is not specifically mentioned, because it is unfair to colleagues. The ACEC Code of Consulting Practice is reproduced in Appendix C.

FIDIC   The International Federation of Consulting Engineers (FIDIC) is an international advocacy group for consultants. FIDIC publishes a voluntary Code of Ethics that guides the conduct of its members, who are mainly firms of consulting engineers.[10] The FIDIC code is similar to the ACEC code and to the Association codes, and requests members to fulfill their duties with responsibility, competence, integrity, impartiality, and fairness, and to avoid corruption. The FIDIC code encourages sustainable development (discussed in Chapter 17) and Quality-Based Selection (QBS), discussed in Chapter 7. The FIDIC Code of Ethics is reproduced in Appendix C.

ETHICS IN FOREIGN CONSULTING

A consultant engaged on a project in a foreign country must obey that country's laws. But what if the laws are much harsher and the working conditions are much lower than the standards we would expect in Canada? Which laws and working conditions should apply? Very few Codes of Ethics deal with this particular situation.

Uniquely, New Brunswick's Code of Ethics[11] has a clause that advises professionals to "observe the rules of professional conduct which apply in the country in which they practise and, if there are no such rules, observe those established by this Code of Ethics." However, in an extreme case, this rule may lead to a moral dilemma with no useful outcome.

Consider, for example, an engineer or geoscientist who is working in a developing country, wracked by civil war, where no local professional rules apply. The country is likely to have extremely low wage scales, and a local militia or police force that demands extortion to permit the company to operate. Both of these acts—exploiting the poor by low wages and paying the extortion—would be contrary to any Canadian Code of Ethics. It is impossible for a single consulting firm to impose Canadian standards on an entirely lawless state, so the New Brunswick code could not be applied. However, abandoning developing countries will not advance them, either to the higher standard of living or the more civilized ethics that we enjoy in Canada. Another solution must be found.

In a case such as this, Aristotle's concept of the golden mean (discussed in Chapter 11) may be useful. The golden mean seeks some intermediate level of virtue between the extremes of excess and deficiency. In this case, the extremes are accepting the anarchic or self-serving practices in a lawless country, or seeking to impose impossibly high Canadian standards and eventually abandoning the country to its fate. Each situation requires an individual analysis, and the ethical problem-solving strategy proposed in Chapter 11 will likely be of value. The end result should satisfy the contract, benefit the country as a whole, and yet agree as closely as possible with Canadian practices.

The text by Harris, Pritchard, and Rabins contains an informative chapter on this problem and suggests nine factors (or "culture-transcending" norms) that should be considered as guides when deciding on a course of action.[12] Harris et al. suggest that fairness can be determined easily for each of these factors by applying the well-known Golden Rule: If a foreign corporation were involved in a project in Canada, how much should the corporation adjust to our practices and traditions? (This situation is not difficult for most Canadians to imagine.) The nine factors are listed below, but explained in a more laconic style:

- **Avoid exploitation.** In Canadian and international law, a contract must be entered into freely, and must provide some benefit to both parties. Exploitation occurs when one party sets all the terms of the contract, and the other party receives inadequate benefit. This form of exploitation is unethical and violates all of the ethical theories described earlier.
- **Avoid paternalism.** Paternalism is the process of controlling the behaviour of citizens and making personal decisions for their best interests without their involvement. For example, suppose that a large corporation decides to pay all workers by bank transfers when the workers are used to cash payments, and many do not have access to banks. Such payments may be more secure, may help to track income tax, and may be convenient for the company, but if the process is contrary to the country's tradition, it may be paternalistic and unfair to workers who are impeded from receiving their pay.
- **Avoid bribery.** Bribery is a very common problem in international consulting. Bribes or "secret commissions" are illegal under Canadian law, and every Code of Ethics forbids them, either directly or indirectly. Bribery distorts the fair trade of goods and services, and corrupts the reputation of all involved. The process of giving "gifts" is common in many countries, so some discretion may be necessary to avoid giving offence. This is a matter that should be investigated before undertaking a contract in a foreign country. In recent years, the term "grease money" has entered the Canadian vocabulary, and it is only subtly different from a bribe or extortion. To clarify: a *gift* is freely given, to enhance the personal

relationship, with no expectation of benefit. However, a *bribe* is given in expectation of favourable treatment, *grease money* is intended to "open doors" so that favourable treatment will arrive more quickly, and *extortion* is a demand from a person in authority for a payment before providing an action that should be given freely.

- **Respect human rights.** Human rights are guaranteed in Canada under the Charter of Rights and Freedoms,[13] and similar rights apply to almost all societies under the United Nations' Universal Declaration of Human Rights,[14] which is older than our Charter. Even in a wealthy country such as Canada, not all citizens enjoy all of the rights, and our goal must be to extend these rights. However, we cannot improve the rights in a foreign country if we abandon it, so some compromise may occasionally be needed. Each situation must be analyzed to determine when the acceptance of an infringement on rights will improve the country's social conditions, so that human rights will improve in future. We may, for example, accept that equality before the law has many interpretations (even in Canada), but clearly there are some limits that cannot be crossed. The UN Declaration specifically forbids slavery and torture, and such behaviour cannot be condoned in any civilized society.

- **Respect cultural norms and laws.** Consultants in foreign countries must recognize the laws and customs of these countries. This is simple courtesy. However, some practices may conflict with Canadian customs. For example, many countries forbid alcohol, and some countries forbid women to drive vehicles. Canadians must respect these laws when working in those countries, even if they are contrary to Canadian laws.

- **Promote the country's welfare.** As mentioned above, any contract requires both parties to benefit. The public welfare should also benefit. That is, consultants should not collaborate in any activity that would be harmful to the general public, or would be a criminal activity in that country or ours. In particular, consultants should not agree to conceal the adverse effects of their activities on the general public.

- **Protect health and safety.** Professional consultants must seek to protect the public from harm, whether in Canada or elsewhere. This is particularly important when the employees' work environment does not meet Canadian Occupational Health and Safety standards. A consultant must decide whether the employer is adequately protecting the health and safety of employees. If the situation is unacceptable, and your continued involvement will not lead to improvement, then it is more ethical for you to abandon the project than to damage the health of citizens and leave them even more vulnerable.

- **Protect the environment.** Of course, in foreign countries, as in Canada, all professionals must protect the environment. The problem is complex, and involves more than simple efficiency, because global warming is changing our climate. This topic is discussed in more detail in the next three chapters of this textbook.

- **Promote the society's legitimate institutions.** Professional consultants in underdeveloped countries should support the laws, customs, and institutions that will guide the country to stability and financial well-being. In general, this means supporting (providing they are operating honestly) the elected government, the health system, the established police and justice system, the banks and other financial institutions that are essential to the free trade of goods and labour, all of which can improve the life of the average person.

In summary, a Canadian professional may face many ethical problems when working in a foreign country (particularly an underdeveloped country). The professional must seek the intermediate level of virtue between the extremes of excess and deficiency, and should not accept either the extremes of anarchic or self-serving practices or, conversely, try to impose impossibly high Canadian standards. Each situation requires an individual analysis, and the ethical problem-solving strategy proposed in Chapter 11 should be applied. The nine factors listed above are common issues that must be addressed. The end result should satisfy the contract, benefit the country as a whole, and yet agree as closely as possible with Canadian practices.

CASE STUDY 14.1

BENEFITING FROM A CONFLICT OF INTEREST

STATEMENT OF THE PROBLEM

Edward Beck is a consulting engineer in a small town. He also sits on the town council as an elected councillor, a part-time job that he considers a form of public service. Beck is hired by a developer to draw up plans for the street layout, water supply, and sewage facilities for a new residential subdivision in the town. The developer's submission to town council includes Beck's drawings and specifications. The town council discusses the subdivision at a regular meeting. During the discussion, Beck does not publicly state his relationship with the developer, nor does he conceal it. His signature and seal are on some of the plans submitted to council. When the town council comes to the vote, Beck votes to approve the subdivision. Everyone knows that he is the only engineer in town who does this type of work, and he is certain that they would prefer to see local people hired for this project.

QUESTION

Has Beck acted unethically by voting to approve this project in town council?

AUTHOR'S SUGGESTED SOLUTION

This situation sometimes occurs in small towns with few engineers, where a conflict of interest cannot be avoided. Professional engineers and geoscientists certainly should not be disqualified from projects because they are serving the public as members of town councils. However, in this case it is not enough that "everyone knows" that Beck has a business relationship with the developer. Beck created a serious conflict of interest when he voted to approve plans that he himself prepared. He should have made a clear, unequivocal statement of his involvement in the project and his relationship with the developer, then withdrawn from the debate and abstained from the vote. By participating in a formal vote without declaring the conflict of interest, Beck has acted unethically. More importantly, he has exposed himself to the possibility of a complaint to the Association, and possible disciplinary action.

CASE STUDY 14.2

CONSULTING IN A FOREIGN COUNTRY

STATEMENT OF THE PROBLEM[15]

Professional Engineer Epsilon was an employee of the Canadian consulting firm ABC International Inc., and was sent as a resident geological engineer to Pradonia, South America. ABC International had been hired by the Pradonian government to oversee a major project being undertaken by another Canadian company, XYZ Overseas Inc. The project involved the construction of a 400-km highway across a mountainous region. Although relatively new to ABC International, Epsilon, with more than 25 years of experience, was given the key assignment of ensuring that contract agreements between the Pradonian government and XYZ Overseas were met. Epsilon's signature on the payroll certifies that the interests of the Pradonian government were being served.

Almost immediately, Epsilon began to experience doubts about the project. The design for the highway, which, as it turned out, was originally done by ABC International, called for cutting deep channels—some of them more than one hundred metres deep—through the mountains, with cliffs rising sharply on both sides of the road. Epsilon was concerned that, with the instability of the mountains, it did not appear as if enough geological borings had been taken to identify potential slide areas. Epsilon's fears were confirmed, unfortunately, when several landslides and other construction accidents occurred, killing some workers. XYZ Overseas asked Epsilon to add [a sum] to the payroll to cover the substantial costs for slide removals.

Epsilon viewed the request as "padding" and, therefore, not justified by anything in the contract. At first Epsilon's position was supported by Epsilon's firm [ABC International]; however, with mounting pressure from XYZ Overseas Inc., ABC International ordered Epsilon to add the slide-removal costs to the payroll. Epsilon refused to do so, insisting that it would be a violation of the Pradonian government's interests, which ABC International was charged to protect. Epsilon was relieved of his resident engineer's responsibility and was subsequently fired by ABC International.

QUESTIONS

Discuss Epsilon's actions, as well as those of ABC International and XYZ Overseas, in terms of your Association's Code of Ethics. Is there a recommended recourse that Epsilon might pursue in view of his dismissal?

AUTHOR'S SUGGESTED SOLUTION

The typical client/contractor/consultant relationship exists in this case. The client is the Pradonian government; the contractor is XYZ Overseas, hired to build the highway, and the consultant is ABC International, hired to "oversee" the highway construction. Epsilon's role, as an employee of the consultant (ABC), is to act on behalf of the client (the Pradonian government). Epsilon's employer (ABC) also designed the highway, but this fact should not influence Epsilon's judgment.

Epsilon's precise duties and responsibilities are not fully stated in the question. (In an exam situation, we must make assumptions, but, in real life, these duties would be specified in Epsilon's employment contract.) Apparently, Epsilon's main task was monitoring the quantities and quality of materials and labour provided to this massive project by the contractor (XYZ).

This case is curious (or ironic) because the direct employer (ABC) dismissed Epsilon for being too vigilant in defending the client (the Pradonian government). Four ethical issues are involved: workplace safety, design safety, the additional payments to remove landslide debris, and Epsilon's dismissal. Let us look at these separately.

Workplace safety. The most critical events were the landslides, which caused deaths and debris. Workplace safety is the responsibility of the prime contractor performing the work (in this case, XYZ), who would likely employ a site engineer on such a major project. Mountainous terrain is dangerous, and only the contractor knows what is achievable with its labour and resources. When a workplace accident causes death in Canada, the workplace is closed for a safety inspection under Occupational Health and Safety (OHS) legislation (as explained in Chapter 8). Work does not resume until the cause of the accident is understood and safety restored. Unsafe practices can lead to charges under OHS laws. The question does not mention OHS; presumably, such legislation does not exist in Pradonia.

Design safety. The question hints that the highway design may be faulty. Epsilon had doubts about the design because the geological borings appeared to be inadequate to identify potential slide areas. However, we must separate doubts from evidence. If Epsilon had the power to investigate his concerns, he should have done so, even though his employer (ABC) was also the designer. Failure to act on evidence of unsafe conditions contravenes the Code of Ethics and Canadian OHS legislation. In this case, it is unlikely that Epsilon had the resources or the authority to investigate the design. Therefore, we cannot say for sure, but since the contractor (XYZ) did not question the design, we must conclude that XYZ did not believe the workplace was unsafe.

Appropriateness of payments. The payments are a minor issue compared to the workplace deaths. However, if we assume that the landslides were the result of XYZ's negligence or poor workplace safety, and if XYZ asked Epsilon to add the landslide clearing costs to the payroll to conceal the costs (or XYZ's poor construction practices) from the Pradonian government, then these hidden costs are indeed padding. It would be contrary to the Code of Ethics (and likely illegal) for Epsilon to approve such payments.

Dismissal. Although ABC was Epsilon's direct employer, Epsilon had an obligation to act in good faith on behalf of the client (the Pradonian government). The disputed payments for the landslide costs were not in the contract, and ABC's instructions to pay XYZ were motivated by external pressure (and perhaps to cover up ABC's design flaws). As mentioned earlier (in Chapter 13), an employee cannot be directed to perform an act that is illegal or clearly contrary to the Code of Ethics. Although the work took place in Pradonia, Epsilon could contact a lawyer in Canada to discuss the possibility of a wrongful dismissal suit.

In summary, the real tragedy lies in the workplace deaths. An OHS investigation should have been held to determine the cause of the landslides. Regrettably, safety and ethical issues were not addressed before the death and damage occurred, so this case will likely be resolved in the law courts. (Note: The statement of the problem is very brief; many assumptions were needed. More information or different assumptions might lead to other conclusions.)

CASE STUDY 14.3

CONTINGENCY FEE ARRANGEMENTS

STATEMENT OF THE PROBLEM

As an engineer in private practice, you are considering whether to offer your services on a contingency basis, an arrangement whereby you would be paid a percentage of the outcome. Two clients wish to retain you.

- Client A wants to retain you to act as an expert witness in a lawsuit against a third party. The lawsuit, if successful, should result in the award of a very large sum as a settlement.
- Client B has shown a tentative interest in retaining you to recommend changes to the energy usage of a manufacturing process. You believe Client B would be more responsive if fees were contingent on the savings. After an initial study of the problem, you believe that the energy savings could be immense.

QUESTION

Would it be ethical to offer your services on a contingency basis to either of these clients, with the understanding that you would be paid a percentage of the legal settlement (Client A) or a percentage of the value of the energy savings (Client B)?

AUTHOR'S SUGGESTED SOLUTION

These two cases seem similar, but are distinctly different.

Client A. An expert witness is permitted to express opinions, whereas a non-expert witness must confine his or her testimony to known facts. Therefore, an engineer testifying as an expert witness must have an impartial attitude toward the outcome of the case. However, as a recipient of a percentage of the potential settlement from Client A, you would have a conflict of interest and your testimony would be suspect. Therefore, it would be unethical to accept this case on a contingency basis. You should bill Client A for your time and expenses so that the reimbursement is independent of the outcome of the case.

Client B. The case of Client B is somewhat different, since there is no need for impartiality. In fact, your bias toward reducing energy consumption could be very beneficial to the client. Also, you have a duty to yourself and to your colleagues to charge an adequate fee. From your study, you evidently believe that this fee will be adequate. Therefore, the proposal to base the fee on a contingency is not unethical. However, a word of warning: there might be a perception of unethical behaviour unless the results can be measured accurately and impartially and can be achieved without degrading the client's product or facilities. Therefore, although this method of setting a fee is not unethical, some risks are associated with it. You would be well advised to use one of the more common billing methods (as described in Chapter 7) unless the client expresses a preference for the contingency method and the savings can be clearly and unequivocally measured.

CASE STUDY 14.4

ADHERENCE TO PLANS

STATEMENT OF THE PROBLEM

A professional engineer in private practice is engaged by a building contractor to prepare drawings for the forms and scaffolding needed to construct a reinforced concrete bridge. The forms and scaffolding must sustain the weight of about 1,400 tonnes of concrete until the concrete is cured. The engineer prepares the drawings and signs and seals the originals, which he gives to the contractor. The contractor later engages the engineer to inspect the completed structure. The engineer finds that the contractor has made several major deviations from the plans. He is not sure whether the structure is safe or unsafe. The contractor has stated that time is of the essence, and concrete is to be poured in the next 48 hours. The engineer feels an obligation to the contractor because of their previous professional relationship and hopes that it will continue.

QUESTION

What should the engineer do?

AUTHOR'S SUGGESTED SOLUTION

Two issues are at stake here. Once the engineer passed the sealed original drawings to the contractor, control was lost. It is possible that changes were made to the originals that, if unsafe, could create serious problems for the engineer. As a general rule, only prints should be signed and sealed so that modifications will be evident. In this case, apparently no changes were made. However, the contractor did not construct the forms and scaffolds according to the plans, and the engineer is now faced with the unpleasant task of informing the contractor that the deviations from the plans must be evaluated to ensure that they are safe. This will undoubtedly require some calculations and perhaps a second inspection.

The engineer should notify the contractor in writing that concrete must not be poured until the review and re-inspection is complete and that the structure could constitute a hazard to workers and the general public. The strength analysis should be carried out as quickly as possible. If the 48-hour deadline cannot be met, the project must not proceed until all safety concerns have been satisfied. It is perhaps useful to point out that the contractor could have consulted the engineer about the changes earlier in the construction, so that the delay could have been avoided. Failures occur more often in temporary structures, because they do not receive sufficient analysis. The collapse of the temporary supports for the Vancouver Second Narrows Bridge illustrates this point. (Read the case history in Chapter 13.)

CASE STUDY 14.5

FEE REDUCTION FOR SIMILAR WORK

STATEMENT OF THE PROBLEM

Susan Johnson is a professional engineer in private practice. She is hired by Client A to design a small, explosion-proof building for storing flammable paints, chemicals, and explosives. The work is carried out in her design office, and copies of the plans are provided to her client. After construction is complete, Client B contacts her. Client B needs a similar building, but suggests that the fee should be substantially reduced, since the design is already finished and only minor changes would be required.

QUESTION

Would it be ethical for Johnson to reduce her fees as suggested? Would it be good business practice?

AUTHOR'S SUGGESTED SOLUTION

First, Johnson should clarify whether she or Client A owns the copyright for the drawings. (This should have been specified in the original agreement with Client A.)

Second, establishing fair and reasonable fees depends on five factors:

- Level of knowledge and qualifications required;
- Difficulty and scope of the assignment;
- Responsibility that the engineer must assume;
- Urgency with which the work must be accomplished (will overtime or extra personnel be required?); and
- Time required (number of people hours).

Of these five factors, only the last two (urgency and time required) are likely to be reduced because of the earlier project. If Client A were requesting a second building of the same design, then it might be appropriate to pass on some of the savings in time. However, Client B benefits from receiving a design that has been tested and is likely to be more dependable and easier to construct. The level of knowledge, the qualifications, and, most importantly, the responsibility that the engineer must assume are unchanged. In summary, it would be unfair to Johnson and poor business practice to accept a substantial fee reduction for providing the drawings for this structure.

CASE HISTORY 14.1

LEMESSURIER AND THE CITICORP TOWER

This case shows an exceptional engineer resolving a serious ethical dilemma. It therefore differs from most of the case histories in this text, in which negligent or incompetent decisions led to disaster.

Photo 14.1 — Citicorp Tower.
William LeMessurier designed the 59-storey Citicorp Tower in New York City. Because of lot restrictions, four massive pillars support the building, as shown in the photo. In May 1978, a few months after the building's completion, LeMessurier learned that a design fault could cause the building to collapse under certain severe wind conditions. The gripping story of his successful efforts to strengthen the building before the start of the storm season is a tale of professional ethics at its best.

Source: Kathleen Voege/Getty Images.

It is also instructive to note that the designer, LeMessurier, applied a problem-solving process several times in this case. Early on, before the public became aware of the design flaw, LeMessurier faced an ethical dilemma. He pondered several alternatives, including concealing the flaw. He decided that the most ethical route was to face the problem squarely. He then went through the problem-solving process three more times (at least) to resolve technical, implementation, and financial problems.

Introduction

Structural engineer William LeMessurier was hired as a consultant to advise the designers of the 59-storey Citicorp Tower in New York City. The design was severely constrained by the building site, because a church occupied one corner of the lot. It was agreed that the Citicorp Tower could occupy the space above the church, but it could not touch the ground on the church's corner of the site. To satisfy the constraint, four massive pillars were constructed, each rising 35 m (114 ft.) from the ground. These four pillars support the building in the middle of each of the four walls, instead of the corners, and thus avoid the church. The building rises 59 storeys above the pillars, with a crown structure that peaks at 279 m (914 ft.).

LeMessurier created a simple, innovative design for supporting the corners: a steel frame with six sets of structural steel braces running diagonally upward from the columns at the middle of the walls to the corners, each brace welded to the floor-beams of nine stories. The diagonal braces support the corners and give the finished building a distinctive appearance. In addition, the building was the first major skyscraper to incorporate an active damper to reduce sway during high winds. This was a 363-tonne (400-ton) concrete block on the top of the building. Sensors control the dampers by measuring the building's lateral acceleration and, using hydraulic pressure, drive the massive block to counteract the sway. In this way, the dampers cancel some of the structural stress.

LeMessurier's analysis of the loads and design of the main supporting structure were essential to guarantee the strength of the building. His analysis was given to the design engineers, who determined the structural details. The design engineers were not required to inform LeMessurier of all of the subsequent design details, provided that the structural strength appropriately exceeded the predicted loads. The building was completed in 1977.

First Problem: Recognizing and Resolving the Ethical Dilemma

In May 1978, a few months after the building's completion, LeMessurier learned that the design engineers had actually used less expensive bolted joints in the construction, instead of the welded joints he had recommended. This was not a serious change, provided the bolted joints were as strong as the welded joints. But in June 1978, motivated by a question from a graduate student studying the Citicorp Tower design, LeMessurier reviewed the design to prepare a lecture explaining the analysis. He came to a shocking realization: when strong winds blew from a quartering direction (that is, at an angle to the wall, hitting two sides simultaneously), the stresses in the structural members and the forces on the bolted joints would be significantly greater than he had earlier calculated. This quartering load condition had not been part of the New York building code when he calculated the stresses during the design phase, and the design engineers had not considered it either. Although the diagonal braces would withstand quartering loads, the bolted joints specified by the design engineers likely would not.

LeMessurier now suspected that the finished building was understrength and might pose a serious hazard. He faced an ethical dilemma. By revealing the design flaw, he would be risking almost certain humiliation and financial ruin. Yet by concealing his knowledge, he would be placing tenants and neighbours of the building at risk of disaster. According to one source, LeMessurier "contemplated, ever so briefly, destroying his notes or even killing himself. . . ."[16] LeMessurier evaluated the alternatives and decided that the ethical choice was clearly to investigate the design flaw directly, remedy it if necessary, and accept the consequences, however cruel they might be.

Second Problem: Determining the Hazard and the Best Technical Remedy

LeMessurier consulted Alan Davenport, an award-winning Canadian civil engineer. (Among his many achievements and commissions, Davenport carried out the wind-tunnel tests for the CN Tower.) Davenport had run earlier wind-tunnel tests for the Citicorp Tower; he ran the tests again using winds at an angle to the wall surfaces. Davenport's results reinforced LeMessurier's concerns. A severe storm could cause one of the mid-level joints to fail; if this happened, the entire structure would come cascading down. Statistically, such a storm could be expected to occur about every 16 years. Moreover, although the active damper might help reduce the loads and stresses, the damper required electric power, which might be disrupted during a severe storm.

The best technical solution was fairly clear—the joints could be reinforced by heavy steel strapping, which would be welded across the joints from the inside. But who would pay for such repairs, and how could they be completed without panicking the tenants and neighbours of the huge building? Moreover, time was of the essence: the design flaw became clear to LeMessurier in July 1978; the hurricane season typically reaches New York in September.

Third Problem: Correcting the Problem without Inciting Panic

LeMessurier acted on the last day of July 1978. His initial contacts were through the lawyers for the company that had hired him as a consultant and that company's insurance company. Many meetings followed, and Leslie Robertson, also a prominent structural engineer, was retained to review LeMessurier's calculations and conclusions. (Ironically, Robertson was the structural designer for the twin towers of the World Trade Center, which collapsed minutes after the terrorist attacks of September 11, 2001.)

Citicorp, the tenants of the building, were next to be informed, and their response was critical. Fortunately, Citicorp realized the value of LeMessurier's advice, and although questions of cost and inconvenience arose, the gravity of the problem took precedence for everyone concerned. Citicorp authorized repairs to start almost immediately. At the same time, weather experts were hired to monitor and predict wind conditions, and New York City officials were contacted to work out an evacuation plan for the area around the Citicorp Tower. A press release was issued. Worries were diminished somewhat by explaining that the repairs were intended to cope with higher winds. This was partially true, since slightly higher winds were predicted for the autumn of 1978.

Robertson became a key figure in designing and supervising the repairs. The welding was done at night to minimize the inconvenience and allay the concerns of the tenants. Standby generators were installed to ensure that the electrical supply to the active damper could not be interrupted by a storm. Plywood shelters concealed the work areas while repairs were made.

Repairs to the Citicorp Tower had not been completed by September 1978, when the weather office observed Hurricane Ella moving toward New York. Fortunately, the hurricane moved back out to sea without incident. By October, the repairs were complete. The building's strength now significantly exceeds the original design objectives. It is believed that the tower will be able to withstand a windstorm that, statistically, can be expected to occur only once every seven hundred years! It is now one of the safest structures in the world.[17]

Final Problem: Paying the Bill

Citicorp willingly paid for the repairs but also informed LeMessurier that it expected to be reimbursed. After negotiation, LeMessurier's liability insurance company agreed to pay US$2 million. Citicorp eventually agreed to accept that amount as the settlement, and also to exonerate LeMessurier. Citicorp did not reveal the total costs, but a rough estimate would be at least double the insurance coverage.

LeMessurier fully expected his liability insurance premiums to rise steeply; after all, his design error had resulted in a large claim and a great deal of anguish. However, the insurance company agreed that LeMessurier had acted promptly and ethically; it also agreed that if he hadn't done so and if the building had collapsed, the company would have been liable for a much greater amount in death and injury claims. In fact, perhaps not surprisingly, LeMessurier's liability insurance premiums were reduced.[18]

Conclusion

Obviously, LeMessurier could have ignored any responsibility for this problem. Initially, he had no firm knowledge that the structure was deficient, and with good luck (no serious windstorms) the building might have survived indefinitely. However, as an ethical engineer, his perspective was simple and clear: if you have a licence from the state to hold yourself out as a professional, you have a corresponding responsibility. If your structure poses a risk to the lives of others, you must do something about it.

DISCUSSION TOPICS AND ASSIGNMENTS

1. Assume that you and a colleague are starting a consulting engineering partnership. Your partner suggests that the business cards and stationery should include:

   • a stylish logo;
   • your engineering seal, reduced in size to fit; and
   • the slogan "The best in the business!"

   Which (if any) of these advertising components would be acceptable under the advertising rules in your provincial or territorial Code of Ethics (and/or professional practice guidelines)? Cite a reference to justify each answer, and comment whether you agree with the rule, and why.

2. You have been hired as a machine design consultant to a soap manufacturer to suggest methods of speeding up a liquid detergent production line. In the course of your work, you inadvertently gain access to confidential company documents and discover that the company is adding minute quantities of a known carcinogen to the detergent without listing it as an ingredient. You know this substance has been banned. This confidential information is totally irrelevant to the job you were hired to perform, and you have discovered it entirely by chance. You have a duty to the client to maintain confidentiality, but do you have a duty to the public to act on this information? Does it matter that the information is unrelated to your work, and was discovered accidentally? What action would you take? Explain your answer citing your Act or your Code of Ethics, or by referring to the basic ethical theories.

3. Assume that you are a consultant to a large Canadian manufacturing corporation, hired to assist the chief engineer to establish a branch plant in a developing country. Your task is to supervise the installation and

commissioning of the manufacturing equipment. Local people, with little or no education, will operate the equipment.

As soon as you arrive on the site and familiarize yourself with the plan, which is well under way, you realize that the manufacturing line to be installed was removed from service in Canada because it created toxic waste. The waste can be disposed only by special incineration equipment that was too expensive to install in Canada, and was one of the reasons the line was removed. The manufacturing line would not be permitted in Canada, but the developing country does not have environmental laws that would prevent its installation and operation.

You have some ethical concerns about this project and discuss them with the chief engineer. He is sympathetic, but points out that the manufacturing line ran in Canada for more than 10 years before pollution laws stopped it, and no illnesses or deaths were attributed to it. Moreover, the local people will be much better off when the line is running, giving employment, and producing manufactured products for domestic use or export.

What guidance does your provincial Code of Ethics provide for this problem? (Obtain the code from your Association's website in Appendix A, or as reproduced in Appendix B.) Does the code apply to activities conducted in a foreign country? What alternative courses of action are open to you? Which course is best from the ethical standpoint?

Additional assignments are in Appendix E. Additional Case Studies are in Appendix F. Sample questions from the Professional Practice Examinations are in Chapter 18.

NOTES

[1]   Excerpts from D.G. Johnson, "Engineering Ethics," in *The New Engineer's Guide to Career Growth and Professional Awareness* (ed. I.J. Gabelman), IEEE, Piscataway, NJ, 1996, p. 173. Copyright © 1996 IEEE. Reprinted with permission of IEEE.

[2]   Association of Professional Engineers, Geologists and Geophysicists of Alberta (APEGGA), *Guideline for Ethical Practice,* Edmonton, March, 2003, p. 22, available at <www.apegga.org> (May 26, 2008).

[3]   Ontario Regulation 941, Section 75, made under the *Professional Engineers Act,* R.S.O. 1990, c. P.28, available at <www.canlii.org/en/on/laws/regu/rro-1990-reg-941/latest> (June 15, 2009).

[4]   Professional Engineers Ontario (PEO), "Advertising," *Guideline to Professional Practice,* Toronto, 1988 (revised 1998), p. 21, available at <www.peo.on.ca> (May 27, 2008). Excerpt reprinted with permission. Please note guidelines will be revised in the near future.

[5]   Professional Engineers Ontario (PEO), "Selection of an Engineer," *Guideline to Professional Practice,* p. 9. Excerpt reprinted with permission. Please note guidelines will be revised in the near future.

[6]   Professional Engineers Ontario (PEO), "Recommended Confidentiality Agreement," *Guideline to Professional Practice,* p. 15.

[7]   Professional Engineers Ontario (PEO), "Rules of Practice," *Guideline to Professional Practice,* p. 6. Excerpt reprinted with permission. Please note guidelines will be revised in the near future.

[8]   Professional Engineers Ontario (PEO), "Contractual Liability," *Guideline to Professional Practice*, p. 19. Excerpt reprinted with permission. Please note guidelines will be revised in the near future.

[9]   Association of Canadian Engineering Companies, (ACEC), *Code of Consulting Practice*, Ottawa, available at <www.acec.ca/en/who/membership/code.asp> (June 15, 2009).

[10]   International Federation of Consulting Engineers (FIDIC), *Code of Ethics*, Geneva, Switzerland, available at <http://www1.fidic.org/about/ethics.asp> (May 27, 2008).

[11]   "By-Laws and Code of Ethics—2005," *Engineering and Geoscience Professions Act*, Section 2, Province of New Brunswick, available at <www.apegnb.ca> (May 27, 2008).

[12]   C.E. Harris, M.S. Pritchard, and M.J. Rabins, "International Engineering Professionalism," Chapter 10 in *Engineering Ethics: Concepts and Cases*, Thomson Wadsworth, 2005.

[13]   Government of Canada, *Canadian Charter of Rights and Freedoms*, Schedule B to the Canada Act 1982, Ottawa, Canada, available at <http://www.laws.justice.gc.ca/en/charter/> (May 27, 2008).

[14]   United Nations, *Universal Declaration of Human Rights*, Office of the High Commissioner for Human Rights, Geneva, Switzerland, available at <www.un.org/Overview/rights.html> (December 19, 2007).

[15]   Professional Engineers Ontario (PEO), *Professional Practice Exam, Part A—Ethics*, April 2006. Excerpt reprinted with permission.

[16]   J.R. Chiles, *Inviting Disaster: Lessons from the Edge of Technology*, HarperCollins, New York, NY, 2002, p. 196.

[17]   G. Voland, *Engineering by Design*, Addison Wesley, Reading, MA, 1999, p. 398.

[18]   C. Whitbeck, *Ethics in Engineering Practice and Research*, Cambridge University Press, Cambridge, UK, 1998, p. 146.

Chapter 15
Environmental Ethics

Our environment is a public resource, and protecting it is everyone's duty, but professional engineers and geoscientists have a special responsibility because their decisions sometimes can cause great harm. Professionals must avoid unsafe, unethical, or illegal environmental practices, even when clients or employers request them. In some cases, assertive action may be needed—the professional engineer or geoscientist may be forced to report unethical practices to the appropriate authorities.

This chapter discusses the professional's duty to protect the environment, and lists some of the laws regulating environmental hazards. The environmental guidelines published by Engineers Canada (and several provincial Associations) are important additions to the Code of Ethics. This chapter reviews these guidelines, as well as corporate guides for environmental responsibility. Although reporting (or "whistle-blowing") is rarely necessary, this chapter reviews when it might be justified, and describes how it should be done. The chapter concludes with a history of the 1982 Lodgepole blowout—a well-known Alberta case of sour gas (hydrogen sulphide) emission. The next chapter (Chapter 16) reinforces the importance of ethical treatment of the environment by describing several environmental threats and disasters.

THE DUTY TO SOCIETY

Engineers and geoscientists are bound by the Code of Ethics to protect the public welfare, which includes the environment. In general, they do a good job. Most projects involve low to moderate risks, so established methods usually ensure safety. Perfection is not attainable, and some risk always exists. One judge in a tort liability case remarked: "Engineers are expected to be possessed of reasonably competent skill in the exercise of their particular calling, but not infallible, nor is perfection expected, and the most that can be required of them is an exercise of reasonable care and prudence in the light of scientific knowledge at the time, of which they should be aware."[1]

The duty to society therefore does not require perfection, but it does require reasonable care, so we may ask, "What is reasonable?" In environmental terms, reasonable care, prudence, and scientific knowledge mean the following:

KNOWLEDGE OF ENVIRONMENTAL LAW   Professional engineers and geoscientists should seek advice before taking any action that might contravene an environmental law, regulation, or bylaw. Compliance with environmental law is essential. General environmental laws are listed later in this chapter.

ADEQUATE TECHNICAL KNOWLEDGE   Before releasing any substance into the environment, professional engineers and geoscientists must have an adequate knowledge of the effects of the release, even when the substances are not toxic. This information is too voluminous for inclusion here, but is immediately available through the Internet, as explained later in this chapter.

THOROUGH ANALYSIS   In any new process, and in large or dangerous projects such as a chemical plant or nuclear facility, a "cradle to grave" systems approach is necessary to ensure that hazards are controlled. The designers must foresee the problems of handling, storing, and disposing of hazardous substances. They must also consider the decommissioning and disposing of the plant itself, even though this may be 50 years in the future. Designers must also foresee how plant operations might go astray. Sophisticated failure analyses such as failure modes and effects analysis (FMEA), event tree analysis, and fault tree analysis are useful for estimating operating hazards. The designers must try to find every conceivable mode of failure, evaluate the probability that it will occur, and devise a remedy. This topic is discussed further in Appendix G-1.

INSISTENCE ON HIGH ETHICAL STANDARDS   Obviously, high ethical standards are essential, and this textbook is devoted to encouraging these standards. Engineers and geoscientists must follow their Codes of Ethics (as discussed in Chapter 11) as well as the environmental guidelines (discussed later in this chapter).

THE DUTY TO THE EMPLOYER (AND ITS LIMITS)

If you are an employee, you have an obligation to your employer. In rare cases, however, an employer may instruct an employee to carry out acts that are contrary to the welfare of society. For example:

- An engineer may be asked to design a factory cooling system that takes in water from a nearby stream and dispels polluted waste water into the city sewer system, without the knowledge of city authorities; or
- A geoscientist may be asked to falsify ore records for a mine to suggest that lower than actual quantities of toxic chemicals are being disposed in the tailings.

Fortunately, the situations suggested by these examples do not arise often. As a professional, you must refuse to carry out such unethical activities. However, you may have to defend or explain your refusal. The following advice is adapted from a previous chapter:

ILLEGAL ACTIONS Any activity that contravenes an environmental law or regulation is illegal. No employer has the authority to direct an employee to break the law. A professional engineer or geoscientist must refuse to perform any activity that is clearly illegal, and must take action if other employees are observed in illegal activities. By participating in (or merely by ignoring) illegal acts, the professional employee becomes liable for the penalties prescribed by the law; in addition, he or she may face disciplinary action by the provincial Association. In extreme cases, it may be appropriate to report the illegal acts to the appropriate authority.

ACTIONS CONTRARY TO THE CODE OF ETHICS OR ENVIRONMENTAL GUIDELINES A professional engineer or geoscientist must refuse to carry out any activity that, while not clearly illegal, is a breach of the Association's Code of Ethics or is a breach of the environmental guidelines developed by the Association. In some situations, the employer may simply be unaware of the legal significance of the Code of Ethics and may need to be informed. If the employer is an engineer or geoscientist, he or she has a similar obligation to obey the code. The employee has a legal basis for insisting on ethical behaviour—an employer cannot direct a professional engineer or geoscientist to take an action that would result in a loss of licence.

ACTIONS CONTRARY TO THE CONSCIENCE OF THE EMPLOYEE An employee may be asked to perform an act that, while not illegal and while not clearly a violation of the Code of Ethics or the environmental guidelines, nevertheless contravenes the employee's conscience or personal moral code. For difficult situations such as these, the decision-making procedure described earlier (in Chapter 11) may be useful.

Ethical problems sometimes have unfair personal consequences. Obeying your conscience and refusing to follow the employer's directive may result in disciplinary action or dismissal. That is, you may be fired, or pay raises or promotions may be delayed. You must consider the possibility of dismissal, the consequences of unemployment, and the remedies for wrongful dismissal (as discussed in Chapter 13). Decisions such as these must be made carefully.

CANADIAN ENVIRONMENTAL LAW

If your actions affect the environment, your first responsibility is to know the law and to follow it. The federal, provincial, and territorial governments (and many municipalities) have environmental laws and regulations that may limit or regulate your activities. Fortunately, these laws are now very easy to find on

Photo 15.1 — Stelco's Hilton Works. *Large industrial plants such as the Stelco steel works, shown here, are essential to Canada's prosperity. But they also emit carbon dioxide, a greenhouse gas that causes global warming. In the interests of efficiency, engineers and geoscientists must endeavour to reduce energy consumption and the consequent emission of greenhouse gases.*

Source: CP/Kevin Frayer.

the Internet. The most practical information is often found in the Regulations, made under the authority of the Act (and usually available on the same websites). Some environmental websites also include guidelines, advisory notices, information on climate change, and information explaining the legislation.

- **Federal, provincial, and territorial laws.** These laws are available from the government websites listed below.
- **International summary.** A comprehensive summary of environmental law, including equivalent American and Mexican law, is published by the North American Council for Environmental Cooperation (CEC). It is also available on the Internet.[2]
- **Provincial summaries.** Summary publications are available in most provinces to explain the myriad of provincial environmental laws. Two examples are: *Handbook of Environmental Compliance in Ontario*,[3] and *Guide to Environmental Compliance in Alberta*.[4] A simple Internet search may reveal a similar guide for your province.
- **Local laws.** Your municipal government may have specific laws that apply only locally. Contact your city hall or regional government for information.

A sample of the most relevant Canadian environmental laws is listed below, along with a Web address (as of June 2008) for the legislation, or for the department that administers the legislation.

Federal Government Acts

The Environment Canada website <www.ec.gc.ca> monitors the environment (wind, water, climate change, ozone layer, and so forth). It also contains links to *EnviroZine,* the informative online environmental news magazine; to the Canadian Environmental Assessment Agency; and to the federal environmental laws (although some laws are more easily accessed from the Department of Justice website at <http://laws.justice.gc.ca/en>). The key Acts or laws are as follows:

- **Canadian Environmental Protection Act.** This is the main federal law regulating the environment. It is administered by Environment Canada and is aimed mainly at preventing pollution as the best way to protect the public. Its 12 sections deal with an extensive range of topics, such as public participation, information gathering, codes of practice, pollution prevention, controlling toxic substances, biotechnology, waste management, and enforcement. The Act is administered by Environment Canada, and provides enforcement officers with powers similar to those awarded to police under the Criminal Code.
- **Fisheries Act.** This Act also protects the environment, and forbids activities that might degrade any fish habitat. The Act provides severe penalties to prevent people from dumping toxic materials into water that contains fish. The Department of Fisheries and Oceans administer the law. However, the sections most relevant to this chapter—those that concern placing deleterious substances in "water frequented by fish"—are administered jointly with Environment Canada.
- **Canadian Environmental Assessment Act.** The goal of this Act is to encourage sustainable development. The Act applies to projects for which the Government of Canada has decision-making authority, be it as proponent, land manager, source of funding, or regulator. All projects must receive an environmental assessment before they can proceed. The Environmental Protection Section of Environment Canada administers the Act.

Provincial and Territorial Acts

Every province and territory has environmental laws and regulations. Table 15.1 lists the Web address of each provincial and territorial environmental department and the key laws or Acts administered by the department (as of June 2008). The regulations are usually available on the same websites as the Acts. (If necessary, click on "Legislation" or search the website.) In addition, the websites usually contain guidelines, advisory notices, information on climate change, and information explaining the legislation.

TABLE 15.1 — Summary of Provincial and Territorial Environmental Departments and Legislation

Alberta
<www.environment.alberta.ca>
Environmental Protection and
 Enhancement Act
Water Act
Climate Change and Emissions
 Management Act

British Columbia
<www.env.gov.bc.ca/epd>
For Acts and Regulations:
<www.leg.bc.ca/legislation/index.htm>
Environmental Management Act
Water Act
Water Protection Act

Manitoba
<www.gov.mb.ca/conservation/
 environmental.html>
For Acts and Regulations:
<http://web2.gov.mb.ca/laws/statutes/
 index. php>
The Environment Act
The Water Protection Act
The Waste Reduction and Prevention Act

New Brunswick
<www.gov.nb.ca>
For Acts and Regulations:
<www.gnb.ca/0062/acts/index-e.asp>
Clean Air Act
Clean Environment Act
Clean Water Act

Newfoundland and Labrador
<www.env.gov.nl.ca/env>
Environmental Protection Act
Water Resources Act

Northwest Territories
<www.enr.gov.nt.ca>
Environmental Protection Act
Environmental Rights Act
Pesticide Act
Waste Reduction and Recovery Act

Nova Scotia
<www.gov.ns.ca/nse>
Environment Act
Environmental Goals and Sustainable
 Prosperity Act
Water Resources Protection Act

Nunavut
<www.justice.gov.nu.ca>
Nunavut legislation is similar to that of
 the Northwest Territories.

Ontario
<www.ene.gov.on.ca>
For Acts and Regulations:
<http://www.e-laws.gov.on.ca/index.html>
Clean Water Act, 2006
Environmental Protection Act
Environmental Bill of Rights Act
Environmental Assessment Act
Ontario Water Resources Act
Guideline for Use at Contaminated Sites
 in Ontario

Prince Edward Island
<www.gov.pe.ca/enveng/index.php3>
For Acts and Regulations:
<www.canlii.org/en/pe>
Environmental Protection Act
Natural Areas Protection Act
Pesticides Control Act

Quebec
<www.mddep.gouv.qc.ca/index_en.asp>
For Acts and Regulations:
<www.publicationsduquebec.gouv.
 qc.ca/accueil.en.html>
Quebec Environment Quality Act
Sustainable Development Act
Quebec Civil Code (Code civil du Québec)

Saskatchewan
<www.environment.gov.sk.ca>
The Clean Air Act
Environmental Assessment Act
Environmental Management and
 Protection Act, 2002
Fisheries Act (Saskatchewan)

Yukon
<www.environmentyukon.gov.yk.ca>
For Acts and Regulations:
<www.canlii.org/en/yk>
Environment Act
Waters Act
Solid Waste Regulations
Air Emissions Regulations

All websites are valid as of June 15, 2009.

ENVIRONMENTAL GUIDELINES FOR ENGINEERS AND GEOSCIENTISTS

Although the care of our environment is everyone's responsibility, professional engineers and geoscientists must assume a key role because their decisions often have a great impact on the environment. Environmental guidelines are published by several Associations and by Engineers Canada.

Historical Development

The Association of Professional Engineers, Geologists, and Geophysicists of Alberta (APEGGA) developed the *Guideline for Environmental Practice* in 1994, which was revised in 2004.[5] The 1994 guideline was also adopted by Professional Engineers Ontario (PEO) in 1998, with some changes to the explanatory instructions.[6]

The Association of Professional Engineers and Geoscientists of British Columbia (APEGBC) also developed their *Guidelines for Sustainability*, in 1995.[7] The APEGBC guide has many clauses that are similar to the APEGGA guideline, but emphasizes the importance of sustainability—a philosophy that links a viable economy to protection of the environment and to social well-being. These APEGBC guidelines are advisory and are intended to help members maintain a state in which these features flourish indefinitely. The APEGBC guidelines on sustainability specifically do not create any legal duty or obligation by any member to any person.

In 2006, the Canadian Engineering Qualifications Board, a standing committee of Engineers Canada, prepared a *National Guideline on the Environment and Sustainability*,[8] based on the 2004 APEGGA guideline, and incorporating additional material from the APEGBC guidelines. In the past few years, many provincial Associations have endorsed the *National Guideline*.

These guidelines are complementary to the Code of Ethics. They commit professional engineers and geoscientists to protecting the environment and safeguarding the public's well-being. The original documents contain a wealth of explanatory information. The nine precepts of the APEGGA and the *National Guidelines* are almost identical, although the APEGGA guideline is addressed to both engineers and geoscientists and is reproduced, with permission, as follows:

Guideline Summary

Professional Members:

1. Should develop and maintain a reasonable level of understanding, awareness, and a system of monitoring environmental issues related to their field of expertise;
2. Shall use appropriate expertise of specialists in areas where the Member's knowledge alone is not adequate to address environmental issues;
3. Shall apply professional and responsible judgment in their environmental considerations;

4. Should ensure that environmental planning and management is integrated into all their activities which are likely to have any Adverse Effects;

5. Should include the costs of environmental protection among the essential factors used for evaluating the economic viability of projects for which they are responsible;

6. Should recognize the value of environmental efficiency and Sustainability, consider full Life-Cycle Assessment to determine the benefits and costs of additional environmental stewardship, and endeavour to implement efficient, sustainable solutions;

7. Should engage and solicit input from Stakeholders in an open manner, and strive to respond to environmental concerns in a timely fashion;

8. Shall comply with regulatory requirements and endeavour to exceed or better them by striving toward the application of best available, cost-effective technologies and procedures; they shall disclose information necessary to protect public safety to appropriate authorities; and

9. Should actively work with others to improve environmental understanding and practices.[9]

Implementing Environmental Guidelines

Legally, the above guidelines appear to be voluntary guidance, except in Alberta, where the mandatory wording ("shall") in precepts 2, 3, and 8 appears to make these three clauses more forceful, and therefore possibly enforceable under the APEGGA Code of Ethics.

Because the guidelines apply to every discipline, they are very general, and do not give discipline-specific guidance. Therefore, the professional must:

- find and comply with the appropriate regulations for the professional's discipline,
- apply professional and responsible judgement,
- call for specialist guidance when it is needed, and must
- disclose information when necessary, to protect public safety.

ENVIRONMENTAL GUIDELINES FOR CORPORATIONS

The Ceres Principles

In addition to the above guidelines, which apply to individuals, a set of environmental principles has evolved for corporations. These principles have been developed over the past 13 years by the Coalition for Environmentally Responsible Economies (Ceres), an American coalition of institutions—mainly environmental, public interest, and community groups, as well as investors, advisors, and analysts.

The Ceres environmental principles were developed in the wake of the Exxon Valdez oil spill, an environmental disaster that polluted the Alaska shoreline in March 1989. The Exxon Valdez ran aground in a bay on the Alaskan coast. The accident was attributed to navigational errors; also, it is possible that the captain was drunk. The resulting 38,800-tonne oil spill killed wildlife and coated the shoreline up to 750 km (almost 470 miles) from the accident site. It was one of the worst environmental disasters in North American history.[10]

In the autumn of 1989, Ceres published the Valdez Principles (later renamed the Ceres Principles), a 10-point code of corporate environmental conduct. Ceres asks industrial corporations to support the Ceres principles by adopting them as a corporate Code of Environmental Ethics. The 10 principles are reproduced in Appendix C, with permission of Ceres, and include protection of the biosphere; sustainable use of natural resources; reduction and disposal of wastes; energy conservation; environmental risk reduction; safe products and services; environmental restoration; informing the public; management commitment to environmental issues, and regular audits and reports.[11]

Registration under ISO 14001

Of course, when corporations are serious about demonstrating their commitment to responsible interaction with the environment, the most visible route to follow is registration under the ISO 14000 series of Environmental Management System Standards, established in 1996—in particular, ISO 14001 (discussed earlier in this textbook). ISO 14001 registration requires a commitment from senior management, a review of all of the applicable environmental laws, an audit of the environmental impact of the corporation's operations, development of environmental policies, establishment of measurement techniques and methods for recording the measurements, preparation of a procedures manual to define who does what, training of employees, full communication within the corporation, and regular audits to ensure that the system is working and achieving its goals. Obviously, registration indicates that the corporation has made a serious commitment to act responsibly in environmental matters. Moreover, since the ISO 14001 standard is recognized internationally, registration should be an aid in international trade.

THE DUTY TO REPORT—WHISTLE-BLOWING

An engineer or geoscientist who observes unsafe, unethical, or illegal practices must take action. Each case is different, of course. Although no procedure works every time, a direct, personal contact will resolve most problems. Some general rules apply.

- First, you must decide whether the situation is dangerous to human life, because a much more aggressive approach is necessary in this case. Many cases involve infractions of the Code of Ethics, but more serious cases may also be offences under the Criminal Code.
- Second, you must decide whether the problem is caused by the situation or by the individual, and what the simplest remedial action would be. Obviously, some personal judgment is required at this stage, and the problem-solving technique described earlier (in Chapter 11) may be useful.

- Third, a direct, but informal, personal conversation with the closest person involved (presumably a colleague or your boss), proposing a solution (not just the description of the problem), usually yields the best results. If this yields no results, then you would speak to someone further up the chain of authority. (For example, speak to your boss's boss, and so on.)
- Finally, if you can see no resolution in sight, it would be appropriate to consult your Association for further guidance.

For example, a concern about unethical billing practices is not usually urgent and can usually be resolved informally. A situation where workers' lives are at risk is clearly more urgent and might be referred immediately to the authorities (including the police) if a delay in acting might cause injury or death. Failure to take immediate action to protect human life is professional misconduct.

Provincial Associations have often served as mediators to help professionals who believe that clients, colleagues, employers, or employees are involved in unsafe, unethical, or illegal practices. The Association can play a useful role by helping define the ethical issues involved, advising the professional, communicating the concerns to the client or employer in an unbiased way, and generally mediating as informally as possible.

Professional Engineers Ontario (PEO) defines the procedure for reporting in their publication *A Professional Engineer's Duty to Report—Responsible Disclosure of Conditions Affecting Public Safety*. An excerpt follows:

The Reporting Process

Engineers are encouraged to raise their concerns internally with their employers or clients in an open and forthright manner before reporting the situation to PEO. Although there may be situations where this is not possible, engineers should first attempt to resolve problems themselves.

1. If resolution as above is not possible, engineers may report situations in writing or by telephone to the Office of the Registrar of PEO. In reporting the situation to PEO, engineers must be prepared to identify themselves and be prepared to stand openly behind their judgements if it becomes necessary.
2. The Office of the Registrar will expect the reporting party to provide the following information:
   a) The name of the engineer who is reporting the situation;
   b) The name(s) of the engineer's client/employer to whom the situation has been reported;
   c) A clear, detailed statement of the engineer's concerns, supported by evidence and the probable consequences if remedial action is not taken.
3. The Office of the Registrar will treat all information, including the reporting engineer's name, as confidential to the fullest extent possible.
4. The Office of the Registrar will confirm the factual nature of the situation and, where the reporting engineer has already contacted the client/employer, obtain an explanation of the situation from the client/employer's point of view.

5. Where the Office of the Registrar has reason to believe that a situation that may endanger the safety or welfare of the public does exist, the Office of the Registrar will take one or more of the following actions:

    a) Report the situation to the appropriate municipal, provincial and/or federal authorities;

    b) Where necessary, review the situation with one or more independent engineers, to obtain advice as to the potential danger to public safety or welfare and the remedial action to be taken;

    c) Request the client/employer to take steps necessary to avoid danger to the public safety or welfare;

    d) Take such other action as deemed appropriate under the circumstances;

    e) Follow up on the action taken by all parties to confirm that the problem has been resolved.

6. Wherever possible, the Office of the Registrar shall maintain accurate records of all communications with the reporting engineer, any authorities involved and the client/employer.

In Summary: The Office of the Registrar will cooperate with any engineer who reports a situation that the engineer believes may endanger the safety or welfare of the public. Wherever possible, the confidentiality of reporting engineers and the information they disclose will be maintained. The Office of the Registrar will emphasize in all dealings with the engineer's client/employer and the public the engineer's duty to report under the Act and Regulations, and will provide the reporting engineer with an endorsement of the performance of his/her duty, provided that the Registrar has determined that the engineer has acted properly and in good faith.[12]

The above PEO policy is orderly and impartial. It clearly places public welfare first. However, not all Associations have implemented a similar policy.

THE ETHICAL DILEMMA OF WHISTLE-BLOWING

Whistle-blowing always involves an ethical dilemma. Every Code of Ethics requires engineers and geoscientists to consider their duty to society as paramount. However, every code also stipulates duties to clients, employers, colleagues, and employees. At what point does the duty to society override these other duties? For example, the code forbids disclosing confidential information, but conversely, the code requires disclosure of any situation that may endanger the health or safety of the public. Obvious, these duties may conflict.

Whistle-blowing is a controversial act, so we must define the term clearly. Connie Mucklestone provides a good definition in her article "The Engineer as Public Defender":

Whistleblowers are people (usually employees) who believe an organization is engaged in unsafe, unethical or illegal practices and go public with their charge, having tried with no success to have the situation corrected through internal channels.[13]

True whistle-blowing is rare because, as the above quote emphasizes, a true whistle-blower must be concerned about "unsafe, unethical or illegal practices" and such valid complaints can usually be resolved simply by communicating the facts to the people in charge. Personal complaints or disputes are not a proper basis for whistle-blowing.

A whistle-blower is different from a troublemaker in two important ways: the motive of the person involved and the methods used to protect the public. These points are illustrated in the following quote:

> Engineers must act out of a sense of duty, with full knowledge of the effect of their actions, and accept responsibility for their judgement. For this reason any process which involves "leaking" information anonymously is discouraged. There is a basic difference between "leaking" information and "responsible disclosure." The former is essentially furtive, and selfish, with an apparent objective of revenge or embarrassment; the latter is open, personal, conducted with the interest of the public in mind and obviously requires that *engineers put their names on the action and sometimes their jobs on the line.*[14] (italics added)

The whistle-blower must be aware that the process may involve public exposure and scrutiny and may place his or her career in jeopardy. Obviously, whistle-blowing should not be done casually, unknowingly, or wantonly. The provincial Association should be contacted, and its reporting process should be followed.

In summary, before using the reporting process described above, an engineer or geoscientist should consider the following three points:

- **Informal resolution.** It is extremely important to try to resolve problems informally and internally, in an open and professional manner. In the vast majority of cases, clear communication is all that is required. A professional must be certain that an informal internal solution cannot be obtained before whistle-blowing, and must assume the responsibility and consequences of any harm that results from a frivolous accusation.
- **Confidentiality.** Professionals must always report unethical cases to the appropriate regulating body and not to the news media. The goal is to remedy a problem, not to embarrass individuals.
- **Retaliation.** When an employee reports an unethical, illegal, or unsafe act to public authorities, the employer may retaliate by firing the employee. Professional engineers and geoscientists should know that reporting (when justified) is not a basis for dismissal (as explained in Chapter 13). A wrongfully dismissed employee can sue to recover lost wages and costs.

A DISSENTING VIEW OF THE DUTY TO SOCIETY

The first clause of every Code of Ethics says that professionals must consider their duty to society to be paramount. This obligation seems clear and unequivocal. However, one well-known expert on engineering ethics, Samuel

Florman, spoke against this clause as a general guide, because it does not have a precise meaning. His comments are quoted, in part, below:

> If this appeal to conscience were to be followed literally, chaos would ensue. Ties of loyalty and discipline would dissolve, and organizations would shatter. Blowing the whistle on one's supervisors would become the norm, instead of a last and desperate resort. It is unthinkable that each engineer determine to his own satisfaction what criteria of safety, for example, should be observed in each problem he encounters. Any product can be made safer at greater cost, but absolute freedom from risk is an illusion. Thus, acceptable standards must be specifically established by code, by regulation, or by law, or where these do not exist, by management decision based upon standards of legal liability. Public-safety policies are determined by legislators, bureaucrats, judges, and juries, in response to facts presented by expert advisers. Many of our legal procedures seem disagreeable, particularly when lives are valued in dollars, but since an approximation of the public will does appear to prevail, I cannot think of a better way to proceed. . . .

> The regulations need not all be legislated, but they must be formally codified. If we are now discovering that there are tens of thousands of potentially dangerous substances in our midst, then they must be tested, the often-confusing results debated, and decisions made by democratically designated authorities—decisions that will be challenged and revised again and again. . . .

> This is an excruciatingly laborious business, but it cannot be avoided by appealing to the good instincts of engineers. If the multitude of new regulations and clumsy bureaucracies has made life difficult for corporate executives, the solution is not in promising to be good and eliminating the controls, but rather in consolidating the controls themselves and making them rational. The world's technological problems cannot even be formulated, much less solved, in terms of ethical rhetoric: especially in engineering, good intentions are a poor substitute for good sense, talent, and hard work.[15]

Florman's comments are thought-provoking and refreshing. A professional person should not have to be a martyr, or sacrifice his or her career, to protect the public welfare. The public (as represented by the government) must assist by providing the regulation and controls upon which a professional can depend when making a difficult decision. Florman's recommendation for developing standards and regulations based on solid research deserves support—such standards are essential to the practising professional, especially where dangerous chemicals are concerned. However, many well-known regulations and standards are already in print, and yet some companies and individuals still do not follow them because of ignorance, inertia, or unethical attitudes.

Instead of discarding the public safety clause in the Code of Ethics, as Florman suggests, it might be better to provide more mediation between whistle-blowers and their employers (as some provincial Associations are now doing), and to provide protection against retaliation for professionals who, after exhausting all other routes of action, report unethical practices.

Also, it must be remembered that Florman is referring to the United States, where licensing laws are much less restrictive than Canadian laws. As explained in Chapter 2, Canadian laws require a licence to practise engineering, whereas U.S. laws permit anyone to practise engineering; a U.S. licence is required only to use the title of Professional Engineer.

In spite of the apparent contradiction, both the Code of Ethics and Florman's suggestions are useful. The Code of Ethics expresses an ideal, but clear regulations and standards (as proposed by Florman) are a way of making the ideal more attainable.

CASE HISTORY 15.1

THE LODGEPOLE BLOWOUT

Introduction

Residents of the Drayton Valley area of Alberta will long remember the autumn of 1982. At 2:30 p.m. on October 17, the Amoco Lodgepole oil well being drilled near Drayton Valley encountered sour gas (that is, gas laden with hydrogen sulphide) and blew out of control. Over the next two months, while specialists fought to regain control of the well, residents living within a 20- to 30-km (12- to19-mile) radius were twice exposed to the rotten egg smell of hydrogen sulphide (H_2S) and the threat of H_2S poisoning. The first H_2S exposure period lasted 16 days; the second lasted 12 days. During attempts to cap the blowout, two employees were overcome by H_2S and died, and the well was twice engulfed in flames. About 28 people were voluntarily relocated to avoid the H_2S, and several homes were ordered evacuated during especially heavy H_2S concentrations on October 29 and from November 17 to 24. Even people living far from the well were subjected to noxious and unpleasant odours, depending on the prevailing winds.

The well was not capped successfully until December 23. In January 1983, a Lodgepole Blowout Inquiry Panel was convened to investigate the causes of the blowout, the actions taken to prevent it and to regain control, the hazard to human health, and the impact on the environment, and to recommend what should be done to avoid future blowouts at wells in Alberta. The Panel issued a comprehensive report in December 1984.[16]

Events Leading Up to the Blowout

The Amoco Lodgepole oil well, known officially as Amoco Dome Brazeau River 13-12-48-2, is located about 140 km (90 miles) west of Edmonton (about 40 km/25 miles west of Drayton Valley). The well was named after the nearby hamlet of Lodgepole, which is situated about halfway between the well and Drayton Valley. The Amoco Canada Petroleum Company obtained a licence to drill the Lodgepole oil well from the Alberta Energy Resources Conservation Board (also known for several years as the Alberta Energy and Utilities Board). The well was "spudded" (started) in August 1982. Drilling proceeded to a depth of about 3,000 m (10,000 ft.) without problems. An intermediate casing was then installed, and the drilling crew began coring operations to examine the strata prior to drilling into the oil-bearing formation. Two cores were obtained without apparent problems. On October 16, the crew was obtaining a third core when they realized that fluid was entering the well from the oil- and gas-bearing formation.

The drill crew stopped the coring operations to deal with this problem, which is known as a "kick," because reservoir fluids enter the wellbore and force the drilling mud out of the well. For the next 16 hours, the crew fought to regain control of the well, but eventually the drill pipe "hydraulicked" up the hole and the kelly hose was severed, at which point the well was out of control. The intense pressure caused a continuous, uncontrolled flow of mud and sour gas into the atmosphere. The exact flow rate is unknown; however, during the inquiry it was estimated at 1.4 million m^3 of gas per day. Later tests indicated that the flow could have been even greater.[17]

Emergency Measures

Amoco immediately implemented its Major Well-site Incident Response Plan, and key Amoco personnel were notified of the blowout. People and equipment were dispatched to the site, including safety personnel, paramedics, ambulances, helicopters, and firefighting equipment (including breathing apparatus). Hydrogen sulphide monitoring equipment was ordered for both on-site and off-site monitoring. The company immediately hired specialists to cap the well, and special equipment to do so was ordered.

Over the next two months, several plans for capping the well were tried without success. On November 1, 1982, a failed attempt resulted in a fire that engulfed the well. A new control plan was developed, and on November 16, the fire was extinguished with explosives, prior to implementing the plan. Two days later, while attempting to execute the plan, an accident occurred that resulted in the deaths of two employees who were overcome by H_2S. On November 25, the well was again on fire. It was later determined that this fire probably resulted from an undetected underground muskeg fire that had been smouldering for some time. Amoco decided to try to cap the well while it was still ablaze; however, the well specialists declined to attempt this procedure, which had seldom succeeded with other blowouts. On December 1, new well-capping specialists were hired, and by December 23, they had installed a blowout preventer (BOP) over the stub of the intermediate casing. Lines were then connected to flare off the gas and pump mud into the well. Over the next five days, 96 m^3 of mud were pumped into the well, the pressure was stabilized, and the crisis was brought to an end.

What Went Wrong at the Lodgepole Well?

The blowout occurred basically because Amoco personnel were unable to control the hydrostatic pressure in the well. This is a critically important and delicate balancing procedure. The control strategy—usually called the well control plan—sets out the basic principles and procedures that must be followed to ensure that a well will not blow out during drilling, completion, or production operations.

The well control plan is rarely a single document. Rather, it is the sum total of all drilling program documents, special instruction bulletins—including those posted at the site—company procedure manuals, and other books, manuals, and written and verbal instructions that guide drilling procedures. To help you understand the importance of the control plan and how it applies to the critical balancing of hydrostatic pressure in the well, the following explanatory note is reproduced from the Inquiry report, with permission:

> The drilling fluid (mud) system has a dominant position in the general well control plan for any well. The plan requires that the hydrostatic pressure in the wellbore be greater than the formation pressure. Hydrostatic pressure depends on the height of the column of drilling mud and the density of the mud. A reduction in either or both of these will reduce the pressure that results from the column or head of drilling mud. If the hydrostatic pressure is too low, a state of underbalance exists and fluids from the formation, such as gas, may flow into the wellbore. Unless this flow is properly controlled, a blowout will result. On the other hand, if the hydrostatic pressure is too great and a state of excessive overbalance exists, the drilling mud may flow into the formation. This is referred to as "lost circulation" and will result in a loss of hydrostatic pressure which can also lead to a blowout.
>
> ... In developing a drilling program, an operator must therefore consider formation pressures encountered at other wells in the general vicinity of the well being planned for and select a drilling mud density which will ensure a modest overbalance. To ensure that neither a state of underbalance nor of excessive overbalance develops as the well is drilled, close attention must be given to:
>
> (a) mud density,
> (b) any contamination of the drilling mud that will change its effective density, such as by drill cuttings (increase) or by air introduced during tripping (decrease),
> (c) maintaining a full mud column,
> (d) the rate of lowering the drill pipe,
> (e) the rate of hoisting the drill pipe, and
> (f) pumping rate and pressure.
>
> Should the hydrostatic pressure from the mud column prove to be insufficient, and as a consequence, formation fluids such as gas enter the wellbore, a kick condition would exist. Procedures have been developed such that those fluids may be controlled within the wellbore using the BOP [blowout preventer] system mounted on the well casing. The gas is flared at the surface and control operations are continued until the flow is progressively restricted and stopped. If the kick has occurred and efforts to control and contain the in-flowing formation fluids have failed, an uncontrolled flow or blowout results and re-establishing control may be both technically difficult and dangerous.

The general well control plan must include the procedures to be used to circulate out the kick, and it must also ensure that proper equipment will be available should a kick occur. This includes the BOP system but additionally, casing and drill pipe design and selection are important components of the plan.

In order to carry out the general well control plan, the detailed drilling plan must provide for the integrity of the drilling fluid system throughout the drilling operation regardless of the circumstances encountered. The operator should have on site, at critical times, experts in geology and mud properties. The operator must also ensure that the drilling and well equipment, particularly the BOP, is functioning properly. Finally, for the general well control plan to be implemented effectively, on-site supervisors and the drilling crew must be properly trained, regularly briefed, and always prepared to act promptly in carrying out prescribed kick control procedures.[18]

Evaluating Amoco's Actions

Amoco assisted the Inquiry Panel by providing complete documentation on its drilling plan, drilling mud program, rig, and well equipment, and on the qualifications of the well-site crew and supervisors. It also provided a detailed chronology of events leading to the blowout and expert witnesses to explain the events surrounding the blowout. An especially important point was the density of the drilling mud used. Obtaining the right mud density is a key part of the balancing act: it requires knowledge of the formation pressures and careful monitoring of the hydrostatic pressure. The problem is explained in the Inquiry report as follows:

> [I]t is necessary to use a mud density which is neither too low, thus allowing an influx from the formation, nor too high, which could result in lost circulation. This means there is a range of mud density within which operations must take place. The closer one is to the upper or lower limit, the more careful drilling procedures must be. If the mud density is within the range but towards the "high limit," care must be taken to avoid lost circulation. For example, the mud volume must be carefully monitored and the crew must be ready to add lost circulation material. If the mud is closer to the "low limit," drilling must proceed slowly, the potential for a kick must be carefully monitored, and plans to quickly weight up the mud must be in place.
>
> . . . In developing its drilling plan, Amoco reviewed information concerning formation pressures encountered at a large number of other wells in nearby areas. These indicated that the pressures in the formation of interest at the [Lodgepole] well would ordinarily be around 33 000 to 35 000 kPa. In isolated cases, pressures as low as 22 430 kPa and as high as 46 540 kPa had been reported. Amoco decided to design the drilling mud density to meet a pressure of 33 600 kPa with provision for increasing the density if higher pressure was encountered.

Amoco also indicated that its normal practice was to design mud density to provide a mud column overbalance of some 1500 kPa above the expected formation pressures. During drilling or coring operations, mud pump pressures would add further to this, resulting in an overbalance which should avoid the possibility of excess fluid head and lost circulation but at the same time prevent influx of reservoir fluid.

Calculations by the Panel indicate that, at the predicted Nisku reef depth of 3035 m, the planned mud density of 1150 kg/m3 would result in an overbalance of some 630 kPa relative to the expected pressure of 33 600 kPa. . . .

. . . The range of appropriate mud densities varies for each situation, and at the [Lodgepole] well, because of the high reservoir pressure with very good permeability, [the range] was likely relatively narrow. The planned pressure overbalance of 630 kPa plus or minus the effects of operations such as pumping or pulling out of the hole, was less than the 1500 kPa normally used by Amoco. . . .

. . . In summary, the Panel concludes that the planned mud density for the [Lodgepole] well was on the low side and therefore extra care should have been specified during the critical period of drilling into the Nisku zone. Although substantial seepage losses were reported and these might have been interpreted by on-site personnel as an indication that the mud was on the heavy side, an analysis of the situation indicates that the reported losses were likely due to errors. This is an indication that the drilling practices being used were less than satisfactory.[19]

Assigning Responsibility

Although the Lodgepole Inquiry Panel concluded that "no single element in the chain of events was the sole cause of the blowout," the Panel's examination of the events led it to conclude that the initial kick occurred mainly because "drilling practices during the taking of cores were deficient." When combined with the marginally adequate mud density being used, this deficiency permitted the entry of reservoir fluids into the wellbore.[20]

Amoco expected the Lodgepole well to find "sweet" oil, but the company's control plan definitely recognized and accounted for the possibility of encountering sour gas. The Panel accepted this testimony and did not believe that "the expectation of sweet oil played a direct role in the cause of the blowout. However, it may have influenced planning for the well and may have led to less caution in the drilling operations than might have been the case if the well was being drilled specifically for sour gas."[21]

The Panel also concluded that in all likelihood the kick was not controlled because the drilling crew did not immediately recognize the problem and therefore did not immediately apply and maintain standard kick-control procedures. With regard to contributing factors, the Panel noted that several pieces of vital equipment did not function properly, and that supplies of mixed drilling mud were not adequate during the kick-control operations. As the Panel wrote in its report:

The unexpected entry of reservoir fluids into the wellbore was probably due to a combination of Amoco not adhering to sound drilling practices and only marginally adequate mud density. If the degasser had operated effectively, the initial kick might have been circulated out of the system, and subsequent kicks would likely have been avoided. Even with the failure of the degasser, control might have been maintained if there had been sufficient and properly weighted mud on hand to pump into the well. Additionally, if the casing pressure instruments had been operational from the outset, the crew might have recognized the kick at an earlier stage and implemented standard kick-control procedures when fluid influx was still relatively small. If the decision had been made to use hydrogen sulphide (H_2S)-resistant pipe for the full drill string, the pipe might not have parted and the succession of kicks might still have been successfully circulated out. And finally, if the travelling block hook latch had not failed, it may have been possible to retain control of the well by "top kill" methods.[22]

The Panel was satisfied that Amoco applied reasonable judgment in selecting the type of drilling rig, the degasser, and the type of drill pipe even though, in retrospect, other choices might have been better. However, the Panel concluded that Amoco's actions were deficient with respect to

(a) drilling practices during coring operations (cores No. 2 and 3),
(b) implementation of standard kick control procedures,
(c) ensuring adequate mixed drilling mud was available at all times, and
(d) maintaining equipment in satisfactory operating condition (casing pressure instruments).

It appears to the Panel that the fundamental problem was that Amoco did not apply the necessary degree of caution while carrying out operations in the critical zone. Amoco did not appear to be sufficiently aware of the potential for problems that could occur when coring into the Porous zone and thus the need to be fully prepared in the event of a fluid influx. Consequently, when a kick developed, there were delays in responding to it. Then, when equipment problems occurred and supplies of mixed mud were inadequate, Amoco was forced into further delays of precious time in implementing kick-control actions.[23]

In the second phase of the Inquiry, the Panel made a series of recommendations for reducing the possibility of future blowouts.[24]

Lodgepole Blowout: Conclusion

Oil drilling is a demanding, uncertain, and dangerous job. The Lodgepole blowout, which resulted in two tragic deaths and millions of dollars in financial losses, illustrates this point. There are other losses that are harder to put a price on, such as the threat of H_2S poisoning, the disruption of life, and the inconvenience caused to almost all the residents within a 30-km (19-mile) radius of the well. It is also difficult to evaluate the magnitude of loss suffered by Amoco and its technical staff as a result of the negative publicity resulting from the blowout. Safe, standard procedures on drill sites are essential if these dangers are to be avoided.

Additional Environmental Case Studies

Chapter 16 discusses the major threats to the Canadian environment, and includes several environmental case histories.

DISCUSSION TOPICS AND ASSIGNMENTS

1. You have graduated from university and have been working for five years as a plant design and maintenance engineer for a pulp and paper company in northern Canada. The company is a wholly owned subsidiary of a large multinational conglomerate. When you received your P.Eng. licence, you were promoted to chief plant engineer. You work directly for the plant manager, François Bédard, who reports to the head office, which is not in Canada. The company employs about 150 people—most of the adult population of the nearby village—either directly as employees or indirectly as woodcutters.

    In the course of your work you have become aware that the plant effluent contains a very high concentration of a mercury compound that could be dangerous. In fact, since the plant has been discharging this material for 25 years, water in the nearby river downstream from the plant is thoroughly unfit for drinking or swimming. You suspect that a curious new illness in an Aboriginal village about 40 km (25 miles) downstream is really Minamata disease, the classic symptoms of which are spasticity, loss of coordination, and, eventually, death. You also suspect that the fish in the river have been contaminated with the mercury and have spread the contamination to all the downstream lakes.

    Remedying these problems would involve drastic changes to the plant that would cost at least $10 million. So far, you have told no one of your suspicions except Bédard, with whom you have discussed the problem at length. Bédard, who is not an engineer, has confided that the head office considers the plant only marginally profitable and that an expenditure of this magnitude is simply not possible. He has also told you that the head office would close down the plant, causing massive unemployment in the area and probably forcing the workers to abandon their homes to seek work elsewhere. What should you do?

2. Between 40,000 and 50,000 people are killed every year in car accidents in North America, yet people apparently consider driving a car to be worth the risk. Fewer than 10 people have been killed in North America in nuclear reactor accidents, yet many people are afraid of nuclear power. Many people die every year while producing food (even farming is dangerous), and many thousands of miners have been killed in the 20th century (coal mining is especially dangerous). Clearly, there is a discrepancy between perceptions of danger and probability of death (as reflected in safety statistics). Yet perceptions often have the stronger impact on the public when it comes to public support of engineering projects. Using the information resources available to you, complete the following assignments:

    a. Examine the risks associated with the various energy sources (solar, wind, wave, geothermal, fission, fusion, and so on), and develop a fair method for comparing the risks and benefits of each. That is, find the statistics for the probability of injury or death per unit of energy produced. Compare this with automobile travel on the basis of risk per unit of energy consumed.

b. Using the concepts provided in this chapter and in Chapter 6, state in one or two pages an ethical guideline for deciding when construction of a dangerous facility (such as a nuclear power plant) or production of a dangerous chemical (such as a pesticide) is morally justified. Include financial, engineering, and political arguments in your answer as well as ethical concepts.

3. We North Americans consume more energy and resources per capita than any other people on earth. By making even small lifestyle changes, we could reduce our consumption significantly; yet as a society we resist doing so for reasons that are unclear. The following is a brief list of simple ways that we could reduce our consumption of resources. Can you add to this list? How would you go about convincing the general public to "do the right thing" in each of the following cases?

a. Although many cities have Blue Box recycling programs, some residents insist on discarding bottles, cans, and plastics with garbage. The recyclable materials are dumped in landfills, which exceed their capacities more quickly. The value of the recyclable materials is lost, and costly new landfill sites must be found.

b. In many homes, the cellar drainage sump (which should be connected to the municipal storm drain system) is actually connected to the municipal sewage system pipe, which is usually closer. The sump typically collects rainwater from the house's perimeter drain, so this connection permits rainwater to flow into the sewage treatment plant, which must process the otherwise clean rainwater along with the sewage. After a heavy rain, the sewage plant may not be able to cope with the flow. The overflow, which is now polluted with sewage, is usually released into a stream or lake, fouling the environment.

c. Some car owners who change their own oil do not take the old oil to a gas station for recycling. Instead, they simply dump the used oil into a storm drain or septic sewer. Yet even a small amount of oil can seriously harm the environment—most obviously, by killing aquatic life. The oil may even enter the municipal drinking water system.

d. Many trailers and recreational vehicles have self-contained toilets that must be emptied regularly. Some people pollute the environment by dumping the toilet contents in parks, fields, or storm sewer systems, rather than into septic systems that would carry it to sewage treatment plants.

Additional assignments can be found in Appendix E.

NOTES

[1] D.L. Marston, *Law for Professional Engineers: Canadian and International Perspectives*, 3rd ed., McGraw-Hill Ryerson, Whitby, ON, 1996, p. 34.
[2] Commission for Environmental Cooperation (CEC), *Summary of Environmental Law in North America*, 2003, available at <www.cec.org/pubs_info_resources> (June 1, 2008).

[3] *The Handbook of Environmental Compliance in Ontario,* McGraw-Hill Ryerson, Toronto, ON, 2003. This book may be previewed at <http://books.google.com> (June 1, 2008).

[4] D. Buchanan and B. Berzins, *Guide to Environmental Compliance in Alberta,* Hazard Alert Training Inc., now published by Carswell, and available through <www.carswell.com> (June 15, 2009).

[5] Association of Professional Engineers, Geologists and Geophysicists of Alberta (APEGGA), *Guideline for Environmental Practice,* Edmonton, V 1.0, February 2004, available at <www.apegga.org/Members/Publications/guidelines.html> (June 1, 2008).

[6] Professional Engineers Ontario (PEO), *Guideline to Professional Practice,* Toronto, 1988, revised 1998, available at <www.peo.on.ca> (June 1, 2008).

[7] Association of Professional Engineers and Geoscientists of British Columbia (APEGBC), *APEGBC Guidelines for Sustainability,* Vancouver, BC, May 1995, available at <www.apeg.bc.ca/resource/publications/otherguidelines.html> (June 1, 2008).

[8] Canadian Engineering Qualifications Board, a committee of Engineers Canada, *National Guideline on Environment and Sustainability,* 2006, available at <www.engineerscanada.ca/e/pu_guidelines.cfm> (June 15, 2009).

[9] Association of Professional Engineers, Geologists and Geophysicists of Alberta (APEGGA), *Guideline for Environmental Practice,* p. 6. Excerpt reprinted with permission.

[10] "History & Facts," Valdez, Alaska, available at <www.valdezalaska.org/history/oilSpill.html> (June 1, 2008).

[11] Ceres (formerly the Coalition for Environmentally Responsible Economies), *The Ceres Principles,* available at <www.ceres.org/> (June 1, 2008). Reprinted with permission of Ceres.

[12] Professional Engineers Ontario (PEO), *A Professional Engineer's Duty to Report: Responsible Disclosure of Conditions Affecting Public Safety,* Toronto (undated brochure), pp. 3–5, available at <www.peo.on.ca> (June 15, 2009). Excerpt reprinted with permission of PEO.

[13] C. Mucklestone, "The Engineer as Public Defender," *Engineering Dimensions,* Professional Engineers Ontario (PEO), vol. 11, no. 2, March–April 1990, p. 29.

[14] Excerpt from PEO, *A Professional Engineer's Duty to Report: Responsible Disclosure of Conditions Affecting Public Safety,* PEO, Toronto, undated. <http://www.peo.on.ca/>, (January 5, 2008). p. 2. Excerpt reprinted with permission of PEO.

[15] Samuel C. Florman, "Moral Blueprints," *Harper's Magazine* (October 1978). Copyright © 1978 by Harper's Magazine. All rights reserved. Reproduced from the October issue by special permission.

[16] *Lodgepole Blowout Inquiry: Phase 1—Decision Report,* Report to the Lieutenant Governor in Council with Respect to an Inquiry Held into the Blowout of the Well: Amoco Dome Brazeau River 13-12-48-12, Energy Resources Conservation Board, Calgary, AB, December 1984.

[17] Ibid., p. 7–17.

[18] Ibid., p. 5–2. Excerpt reprinted with permission of Alberta Energy and Utilities Board.

[19] Ibid., p. 5–5. Excerpt reprinted with permission of Alberta Energy and Utilities Board.

[20] Ibid., p. 1–1.

[21] Ibid., p. 5–30.

[22] Ibid., p. 1–2. Excerpt reprinted with permission of Alberta Energy and Utilities Board.

[23] Ibid., p. 1–2. Excerpt reprinted with permission of Alberta Energy and Utilities Board.

[24] *Lodgepole Blowout Inquiry: Phase 2 Report—Sour Gas Well Blowouts in Alberta; Their Causes, and Actions Required to Minimize Their Future Occurrence (bound as Appendix 5 of Lodgepole Blowout Inquiry: Phase 1—Decision Report),* Energy Resources Conservation Board, Calgary, AB, April 1984.

Chapter 16

Environmental Threats and Disasters

The Canadian environment is our pride and wealth, but our water, land, sea, and sky are threatened by pollution, negligence, and abuse. This chapter discusses many disturbing environmental threats that could affect engineers and geoscientists.

This chapter has three general themes: the first few sections illustrate the wide range of environmental hazards. The chapter then discusses the ethical dilemma called the "tragedy of the commons," which is a common pattern for many of today's environmental problems, and points the way to overcoming them. The chapter closes with case histories on toxic pollution and nuclear safety, describing well-known disasters that resulted from environmental negligence.

CANADA'S ENVIRONMENTAL HEALTH

Engineering, geoscience, and technology have been of immense benefit to Canada and to humanity. Medicine gives us health, and the humanities give us pleasure, but technology gives us time to enjoy them both. However, industrialization brings problems, even while improving our lives. The lifestyle of North Americans involves high resource consumption and extremely high energy usage.

The National Pollutant Release Inventory (NPRI)

Millions of tonnes of pollutants are discharged in North America every year as a result of industrial activity. Canadians can obtain information about pollution in their communities through the National Pollutant Release Inventory (NPRI), established in 1992 to monitor the facilities that release or recycle pollutants across Canada.[1]

Over 9,000 "reporting facilities"—factories, processing plants, generating plants, oil and gas operations, and similar industrial plants that emit any of the 300 substances on the NPRI substance list—are required to report the disposition, release, or recycling of these substances to Environment Canada, under the Canadian Environmental Protection Act, 1999. The NPRI website lists all 300 pollutants, including:

- **Core:** over 200 substances from acetaldehyde to xylene and zinc compounds
- **Threshold:** heavy metal compounds—arsenic, cadmium, chromium, lead, and mercury compounds
- **Polycyclic Aromatic Hydrocarbons:** 18 organic compounds, from acenaphthene to pyrene
- **Dioxins, Furans, and Hexachlorobenzene**
- **Criteria Air Contaminants (CACs):** carbon monoxide, nitrogen oxides, sulphur dioxide, particulate matter

Reporting is required when quantities of compounds exceed a lower limit, which differs for each class of compound. This information is compiled and made available to the public through NPRI. The purpose of this reporting is explained as follows:

> The information collected by the NPRI can help to identify potential environmental and health risks from pollution, to examine the environmental performance of facilities and communities, and to assist in the development of environmental and emergency plans. The NPRI also includes information on pollution prevention practices. Public access to the NPRI motivates industry to prevent and reduce pollutant releases. NPRI data helps the Government of Canada evaluate releases and transfers of substances of concern, identify and take action on environmental priorities, and implement policy initiatives and risk management measures.[2]

The current data show mixed results: Over the period from 1996 to 2006, about 900 facilities showed an 8 percent decrease in pollutant releases to air, water, and land. However, because of recent production growth, and the expanding number of facilities that report to the NPRI, the total emissions reported have increased substantially since 1996.

The facilities that report pollutant releases to NPRI are easily found using the Internet. NPRI provides a "map layer," which, when implemented on Google Earth, shows the locations of all 9,000 reporting facilities on the Google Earth map of Canada. It is therefore easy to find which reporting facilities are located in your community.[3] The extensive data available from NPRI is illustrated in Figure 16.1, which plots the location of each of the 9,000 industrial facilities that reported to NRPI in 2006.

The development of the NPRI and several similar pollutant registers around the world was stimulated by the major industrial accident near Bhopal, India, in December 1984. The methyl isocyanate gas leak from a Union Carbide pesticide plant killed thousands of people in neighbouring communities. People around the world realized that they needed to know about industrial threats within their communities. The Bhopal tragedy is described in Case History 16.1, later in this chapter.

FIGURE 16.1 — Locations of Industrial Facilities Reporting Pollution Releases or Transfers, 2006

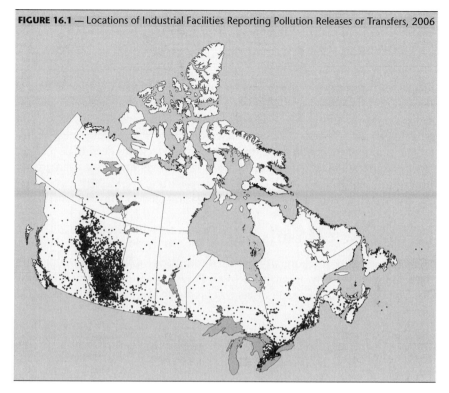

This map shows the locations of 9,000 industrial facilities that reported to National Pollutant Release Inventory (NPRI) in 2006. The NPRI data is available free through the Internet for analysis, research, comparisons, or general interest.

Source: Environment Canada, *2006 NPRI Summary,* Figure 3.1–2, available from the Environment Canada website at <www.ec.gc.ca/inrp-npri> (June 15, 2009). Figure reprinted with permission of Environment Canada.

Commission for Environmental Cooperation (CEC)

The NPRI data on the release of pollutants into the environment is also provided to the Commission for Environmental Cooperation (CEC), an agency set up by the governments of Canada, Mexico, and the United States as part of the North American Free Trade Agreement (NAFTA).

In September 2007, the CEC released its 11th annual report on industrial pollutants in North America. This 162-page report (based on data for the year 2004, the latest year for which such data could be compiled) found that

- facilities in Canada and the U.S. reported 3.12 million tonnes of chemicals released and transferred for 2004. Almost one-quarter of this total was released into the air (707,500 tonnes). Jurisdictions with the largest releases and transfers in 2004 were Texas, Ontario, Indiana, and Ohio.

- the 1998–2004 trend analysis reveals that facilities in Canada and the U.S. have reduced releases and transfers of chemicals by 9 percent over the seven-year period.[4]

The decrease was mainly the result of reductions in discharges by large facilities. Unfortunately, many smaller facilities are increasing their chemical releases at an alarming rate. "The evidence is clear that industry and government action to limit chemical releases is showing steady progress," said Adrián Vázquez-Gálvez, CEC's executive director. "It is equally clear that a large number of small and medium-size industrial facilities need to do a better job in reducing their waste and emissions if we are going to see even greater progress in North America."[5] The airborne pollution releases are particularly significant, since they cause global warming and lead to climate change.

GREENHOUSE GAS EMISSIONS

"Global warming" is a popular term used to indicate a change in the planet's climate caused directly or indirectly by emissions of greenhouse gases into the atmosphere. The next chapter discusses climate change and its relationship to sustainable development.

Scientists have observed that the global mean surface air temperature is rising, and have concluded that even a slight warming of the entire planet could cause severe climatic changes. Evidence points toward the *greenhouse effect* as the cause of global warming. This effect is caused by emissions of carbon dioxide (CO_2), methane (CH_4), nitrogen dioxide (NO_2), ozone (O_3), and the chlorofluorocarbons ($CFCl_3$ and $CF2Cl_2$, called CFCs).

Chlorofluorocarbons were first manufactured in large quantities in the 1940s, when wartime factories began producing household appliances. Production of CFCs increased because they are superb refrigerant gases (hence the common term *freons*). Only recently have we learned that as greenhouse gases, freons are much more potent than CO_2. One molecule of CFC has the same greenhouse effect as 10,000 molecules of CO_2. A more detailed mathematical analysis of these problems is found in the excellent text *Environmental Science: The Natural Environment and Human Impact.*[6]

Since the end of the 19th century, the amount of CO_2 in the atmosphere has increased significantly. Carbon dioxide is a vital factor in the earth's heat balance because it traps heat in the atmosphere. If there were substantially less CO_2 in the atmosphere, very little heat would be retained and the surface of the earth would be coated with ice. Conversely, if the amount of CO_2 were to double, the earth's average temperature might increase dramatically. How quickly the amount of atmospheric CO_2 could double depends on the rate at which we burn fossil fuel. Initial estimates ranged from 88 to 220 years,[7] but after the publication of the report by the IPCC (Intergovernmental Panel on Climate Change) in 2007, these estimates were reduced.

As the next chapter discusses in detail, researchers predict several climatic changes, including global temperature rises, sea level rises, decreases in the levels of inland lakes, and climate variations that may result in droughts, heavy precipitation, heat waves, melting of ice cover and glaciers, and more intense storms, hurricanes, and tropical cyclones.[8,9,10]

Chlorofluorocarbons play a sinister role in environmental degradation. They are greenhouse gases, but they are also the primary cause of ozone layer depletion. Ozone in the stratosphere helps screen out damaging ultraviolet rays (ozone at ground level is a pollutant and irritant). CFCs combine with ozone to create gaps in the stratospheric ozone layer, thus permitting ultraviolet rays to reach the earth's surface, where they harm plants and increase rates of skin cancer. Scientists from the British Antarctic Survey first observed this in 1985. Measurements since then show ozone depletion increasing in severity,[11] although some recovery has been observed in 2006 as the controls on CFCs take effect.

International agreements were successfully negotiated (known as the Montreal Protocol), and all major industrial countries passed laws to phase out the use of CFCs. These controls appear to be effective in protecting the stratospheric ozone and also lessen the greenhouse effect. Regrettably, a criminal black market in illicit CFCs has sprung up, mainly for servicing out-of-date refrigeration plants.[12,13]

As engineers and geoscientists, we must insist that any process that emits needless carbon dioxide or other greenhouse gases should be upgraded. In particular, refrigeration lines containing CFCs—which have been used in refrigeration systems for decades—should never be voided into the atmosphere. Criminal trade in CFCs should be reported to the police.

ENERGY CONSERVATION AND NUCLEAR POWER

During the 1970s, two oil shortages caused long lineups at gasoline stations. By the 1980s, there was an oil glut. Gasoline became cheaper, and we lost interest in energy conservation. However, another and more serious energy crisis looms in this century. Although coal reserves will last another three or four centuries,[14] the easily accessible oil and natural gas reserves are reaching their maximum productivity. The total global oil and gas production peak is predicted to occur in 2010.[15] Once the peak has passed, the supply of easily accessible oil will drop, but the global demand for oil will continue to increase—oil prices will rise, perhaps sharply, and may affect market stability. This "peak oil" problem is discussed in detail in Chapter 17.

Alternative energy sources (solar, wind, wave, geothermal) have added very little to the global energy supply. It seems that nuclear fission and nuclear fusion (should it ever be developed) will be essential if we are to maintain our present standard of living and extend it to the citizens of developing nations.[16] Between 1980 and 1989 the proportion of the world's energy needs satisfied by nuclear energy doubled from 2.5 to 5.0 percent.[17] The World Nuclear Association (WNA) reports that in 2008, more than 435 commercial nuclear reactors are operating in 30 countries, and 16 countries rely on nuclear energy for 25 percent or more of their electricity. For example, France obtains over 75 percent of its electricity from nuclear power; Japan obtains over 30 percent; the U.S. obtains about 20 percent; and Canada obtains 16 percent.[18] Although nuclear advocates sound aggressively self-serving in their predictions, it does appear that the world is going to need nuclear energy.[19]

Atomic Energy of Canada Ltd. (AECL), the designer of the CANDU nuclear reactor, contends that the CANDU is safer and more reliable than the Americans' light-water reactors, since it is fuelled by natural uranium and moderated by heavy water. In a recent letter to the *Ottawa Citizen*, Geoffrey Wasteneys, a retired consultant on electric power production, made the following comments about the CANDU system:

> The CANDU reactor . . . is an outstanding Canadian development that enabled the production of electricity by the use of natural uranium, thus avoiding the requirement for a vastly expensive process of enrichment.
>
> This enabled many countries that possessed uranium resources to produce electric power by a nuclear process without having to deal with the United States or the Soviet Union, which together possessed a virtual monopoly on the production of enriched uranium, a by-product of their production of nuclear bombs.
>
> There is no serious problem with the spent CANDU fuel: It is stored for possible later utilization. (It includes plutonium, and currently, salvaged plutonium from nuclear bombs is being used as fuel.)
>
> An aggressive U.S. marketing policy and propaganda by such firms as Westinghouse that derided the Canadian system as "a scientist's hobby" shut the CANDU out of all but marginal world markets such as India, Pakistan, Argentina, Korea, Romania and China. In consequence, heavy development costs were not fully supported by sales. The decline in the use of nuclear power stations in some countries relates to design faults in present systems and cheaper crude oil.
>
> There has never been a serious problem with a CANDU reactor. The breakdowns have related to the (non-nuclear) steam cycle. Most of the required renovations at Ontario Hydro stations are nearing completion. . . . The atrocious Darlington cost overruns were due to initial design faults, to some extent the outcome of political intervention.[20]

Nuclear energy has the great advantage of using fuel that is compact and plentiful. Even so, public concerns over operating safety and the disposal of radioactive waste are hindering plans to construct new nuclear generating plants. A meltdown at a nuclear plant would devastate surrounding cities and towns. The risk of it actually happening is extremely remote, but even a slight risk generates fear.

Another serious worry connected with nuclear power concerns the long-term disposal of highly radioactive waste. It will be necessary to keep the waste out of circulation for thousands of years because of the extremely long half-lives of some of the elements, such as plutonium. At present, the plan is to store such waste in stable geological underground layers from which water has been absent for millions of years. As with most matters related to nuclear energy, waste disposal is the subject of bitter debate. One issue has to do with the level of certainty regarding the future. Supporters of nuclear energy admit that there can be no absolute guarantee that humans will not be exposed to the waste for thousands of years into the future. However, they add that the risk of future exposure is very small. Opponents of nuclear energy insist on a guaranteed method for keeping nuclear waste permanently isolated. Since a guarantee is impossible, they declare that this is an adequate reason to phase nuclear energy out of existence.

Clearly, nuclear plants must be designed and managed carefully, professionally, and ethically—and this must be transparent to the general public. Nuclear plant designers and operators must recognize the public's anxiety and must demonstrate the safety of the CANDU reactor (and improve it even further).

Coal-fired power does not seem to be a reasonable alternative to nuclear energy because of the associated problems of air pollution. In the aftermath of the Three Mile Island and Chernobyl disasters, coal power has been compared with nuclear power to ascertain which is more dangerous. One writer asserts that their dangers are similar, even if we accept the high estimate of 39,000 cancer deaths (or future deaths) from Chernobyl. He estimates that the death toll from the use of coal in the former U.S.S.R. is between 5,000 and 50,000 per year. Many of these deaths result from the mining and transporting of coal, which requires 100 times as much material handling as uranium (with an equivalent energy output). In the United States, 100 or more coal miners die every year, and nearly 600 of the 1,900 deaths in railway accidents each year are the result of transporting coal. However, the big killer is air pollution, although it is impossible to say with certainty whether coal burning is the cause of a specific, given death. Even so, it has been estimated that 50,000 people in the United States die every year as a result of air pollution, mostly resulting from the burning of coal. Extrapolating these data, assuming similar populations and similar pollution conditions, we can achieve a rough estimate of from 5,000 to 50,000 deaths from burning coal in the former U.S.S.R.[21]

The debate over the safety of our energy sources will not be resolved in this textbook; however, the issue illustrates the importance of professional ethics, conserving energy, increasing efficiency, and avoiding waste. In all of these, ethical actions are essential.

WASTE DISPOSAL

The most common degradation of the environment involves the indiscriminate disposal of wastes—be they solid, liquid, or gaseous—as by-products of manufacturing, processing, or construction. Controlling waste disposal is usually within the authority of an engineer or geoscientist (or should be), and a professional must minimize harm to the environment.

In the early part of the 20th century, little was known about the insidious effects of heavy metals, dioxin, asbestos, pesticides, and other toxic substances.

People believed that the environment was a vast sink that could accept any amount of waste without becoming contaminated. Waste was tipped into dumps as cheaply and as quickly as possible, with little regard for the environment.

We now realize that in many parts of Canada, ill-considered methods of industrial and domestic waste disposal are creating a crisis. Large cities are transporting waste hundreds of kilometres for disposal, and small towns are beginning to realize that the local dump is a source of disease, a fire hazard, and a danger to groundwater. Dumps are gradually being replaced by closely monitored landfills, which accept only low-hazard solid waste and cover it every day with a layer of soil to reduce odour and pest problems. Some landfills are lined with plastic to prevent waste fluids from entering the groundwater, or at least reduce this problem. These "sanitary landfills" are a great improvement over town dumps, but they are not the ideal solution to waste disposal.

Incineration has one advantage: the volume of waste is reduced, so only the ash needs to be buried. However, the gases and particulate emissions released by the incinerators are still objectionable. Moreover, if chlorine-based organic compounds are burned, trace amounts of toxic chemicals may be dispersed.

The best way to solve the waste disposal problem is to reduce the volume of waste through more efficient use of resources, and to reuse or recycle waste materials whenever possible. For example, waste automobile tires are now being used as fuel to heat cement kilns; in this process, even the ash is consumed since it becomes part of the product.[22] Composting, pyrolysis, and density-based separation of organic and non-organic materials have been successful in various applications. Much more research is needed in the recycling of waste materials.

Every province now has environmental protection legislation (as discussed in Chapter 15) and has begun recycling programs to reduce solid waste. However, some liquid waste—especially toxic, flammable, explosive, radioactive, or hazardously reactive chemicals—is still being dumped illegally because of the shortage of proper incineration facilities and high-hazard disposal areas. Hazardous liquid waste can pose a serious threat to health if it leaks into the underground water table. This has happened many times in the past. The place names "Love Canal" and "Minamata" are now synonymous with the unethical disposal of industrial waste and with the human tragedy that followed. (Case histories of both are provided later in this chapter.) In the United States alone, 32,000 hazardous waste disposal sites had been identified by 1991; very few of them have been cleaned up.[23] Any solution to this problem will require both technical ability and political awareness.

AIR POLLUTION

Air pollution has many components, but the best known are sulphur oxides and nitrogen oxides. Sulphur oxides (such as SO_2) result from burning fossil fuels such as coal and petroleum, and by other industrial activities. SO_2 is a foul-smelling gas that reacts with oxygen in the atmosphere to form SO_3, which then combines immediately with water to yield sulphuric acid in the form of droplets. The highest SO_2 values have been reported in the northeastern U.S. and in Europe, where high-sulphur fossil fuels are burned in large quantities. In most large cities, SO_2 emissions have been reduced recently as a result of the shift from high-sulphur coal to low-sulphur natural gas.

Sulphur oxides are detrimental to plant life. They also corrode metals, discolour fabrics, and degrade building materials. Severe damage to plant life can be observed many miles downwind from certain smelting operations. It seems that a combination of sulphur oxides and air particles is especially damaging to human health, partly because of the action of small particles in conveying sulphuric acid into the lungs. SO_2 is a serious lung irritant, and dramatic episodes such as the "London smogs" have been attributed to the combination of SO_2 and particulates. The London smog of 1952 established the link between atmospheric pollution in smog and increased mortality.[24] The worst smog in Canada occurred in southern Ontario in 1962, and lasted five days. This episode is believed to have been London smog; however, it was named the "Grey Cup smog" because it caused the 1962 Grey Cup football game to be postponed as a result of poor visibility.[25]

Even in the absence of sulphur, the burning of fossil fuels causes serious air pollution in urban areas. Exhaust gases typically contain unburned hydrocarbons (HC), carbon monoxide (CO), nitrogen oxides (NOx), and "normal" combustion products such as SO_2 and water (H_2O). In the atmosphere, many of these products react chemically to produce new contaminants. Because these processes are stimulated by sunlight, the resulting products are referred to as photochemical oxidants. Two of the principal photochemical oxidants are ozone (a lung irritant) and peroxyacetyl nitrate (PAN, a lung and eye irritant). Ozone is constantly being created in the atmosphere by natural processes, but not to a degree great enough to constitute a pollution hazard.

Nitrogen oxides are also a problem in air pollution. There are several known oxides of nitrogen, but the important ones from the standpoint of air pollution are nitric oxide (NO) and nitrogen dioxide (NO_2). The term NOx is commonly used to refer to nitrogen oxides collectively. NOx is a product of almost any combustion process that uses air, since nitrogen is the chief component of air. To a great degree, the formation of NOx is the result of high combustion temperatures, principally from motor vehicles, which in industrialized urban areas account for 50 to 60 percent of atmospheric NOx. The Los Angeles type of photochemical smog is caused mainly by NO emissions from cars. It occurs on warm, sunny days when traffic is heavy and reaches a peak in the early afternoon. It therefore differs from the London smog, which forms on cold winter nights as a result of SO_2 produced by coal combustion.[26]

The nitrogen oxides participate actively in photochemical reactions with hydrocarbons, thus helping produce photochemical smog. NO_2 plays a double role in air pollution: it is a component in the formation of photochemical smog, and it is toxic in its own right. NO is much less toxic than NO_2, but NO is readily converted into NO_2 in the atmosphere: by reacting with oxygen in the presence of water, it becomes nitric acid, which is extremely toxic to any growing organism. The adverse effects of air pollution on humans and animals include serious lung disorders, reduced oxygen in the blood, eye and skin irritation, and damage to internal organs.[27] Damage to painted surfaces, cars, and buildings is mainly the result of acid rain, which is the topic of the following section.

Most provinces and the federal government have Clean Air Acts that specify emission standards and ambient air-quality standards. Air pollution control is mainly a provincial responsibility, although the federal government regulates trains, ships, and gasoline.[28] Professional engineers and geoscientists must follow government regulations—reducing harmful emissions, wherever possible, should always be a primary objective.

ACID RAIN

The problem of acid rain captured Canadians' attention in the 1980s. Both sulphur oxides and nitrogen oxides are implicated, because they form sulphuric and nitric acids in the atmosphere and cause rainfall to become more acidic than it otherwise would be. Neutral water ideally has a pH of 7.0, but "normal" rainfall in remote areas that are unpolluted has a pH of about 5.6 because of the presence of small amounts of acid of natural origin.[29] Rain is typically called "acid rain" when the pH falls below 5.0. In many areas of Canada and the northeastern United States, the rain has a pH value as low as 4.0.

When acid rain falls, it harms mainly fish, trees, farms, buildings, and cars. Aquatic life begins to be affected when the pH falls below 5.0, and most fish are killed when a pH of 4.5 is reached. The result is hundreds of lakes in northeastern North America (and in Scandinavia) that are devoid of fish, and thousands of other lakes that are threatened.[30] Lesions on plants caused by simulated acid rain have been observed when the pH drops lower than 3.4, although subtle effects may be occurring at higher pH levels. Humans may also be harmed, because the acidity leaches magnesium, aluminum, and heavy metals out of the soil and concentrates them in drinking water. Fish are especially susceptible to dissolved aluminum, and this may be a risk to humans as well. Acid rain does severe damage to limestone buildings and monuments, since it dissolves the key chemicals in limestone.[31]

Acid rain is an international problem. The heavier flow of pollutants is believed to be from the United States into Canada, because of the greater industrial activity south of the border. However, the smelters around Sudbury, Ontario, are major sources of the sulphur dioxides that cause acid rain. The federal governments in Canada and the United States have agreed to control acid rain, and the situation is improving. The most effective way to reduce acid rain is to reduce acid emissions, although adding lime to neutralize the acid can reverse low pH levels in lakes. Reducing emissions is costly, but an early study shows that in addition to a cleaner environment, the economic benefits are about equal to the cost of the controls.[32]

WATER POLLUTION

Some rivers are less polluted now than they were in the 19th century, when there was a serious water pollution crisis. In the mid-19th century, 20,000 people in London, England, died of cholera. As Donald Carr says, "in the Western world this was the greatest pollution disaster of history."[33] Typhoid and cholera epidemics stemming from water contaminated by sewage were widespread. Water pollution has at least six possible sources:

- disease-causing bacteria;
- organic waste decaying in the water, reducing the dissolved oxygen content;
- fertilizers that stimulate plant growth and also depress oxygen levels;
- toxic materials, such as heavy metals and chlorinated hydrocarbons (DDT, PCB);
- acidification, as mentioned earlier; *and*
- waste heat, which can also reduce dissolved oxygen levels.[34]

Cholera and typhoid are practically unknown in Canada today as a result of sewage treatment and the use of chlorine to kill bacteria in drinking water. Nevertheless, we must be vigilant in monitoring water quality. The pollution caused by lawn fertilizer in summer and highway salt in winter is said to come from "non-point" sources (that is, from everywhere) instead of from "point" sources such as sewage treatment plants, power plants, and factories, which can be more easily monitored.

Agriculture also creates dangerous pollution when the runoff from dairy or pig operations is too close to the intake of municipal water systems. Such a problem occurred in Walkerton, Ontario, in May 2000 (as discussed in Chapter 2). Before the cause was traced, 7 people died and more than 2,300 became ill.

Every summer, many beaches in Canada are closed to swimmers because of high bacteria and fecal content in the water. This is a scandal for a country that boasts the largest endowment of fresh water in the world. Laws to control pollution have been passed at all levels of government. In the 21st century, a key principle is that "the polluter pays" to remedy the problems. As a professional, you have a duty to society to ensure that environmental laws are followed. In most cases it is not enough to treat pollution at the "end of the pipe." Waste must be eliminated earlier in the process by increasing efficiency and by reducing, reusing, or recycling raw materials.

EXPONENTIAL POPULATION GROWTH

Population growth is generally viewed as an achievement, not a problem. Moreover, it is unlikely that any individual engineer or geoscientist is able to affect population growth. However, it must be mentioned, because the problems of consumption of non-renewable resources and environmental degradation are proportional to the population, and the world population is at record levels and growing quickly.

Around 1800, there were roughly 1 billion people on this planet. After the industrial revolution, new machines, medicines, and improvements in nutrition caused life expectancy to rise and infant mortality to fall. As a result, the world's population passed 2 billion in 1930, 3 billion in 1960, 4 billion in 1975, 5 billion in 1987, 5.6 billion in 1994,[35] and 6.5 billion in 2005.[36] A simple graph of these numbers indicates that the world's population is growing exponentially.

Engineers and geoscientists know that exponential growth cannot be sustained by finite resources—even when resources are as vast as this planet's. Moreover, the expectations of the citizens of developing countries are also increasing sharply. Satellite television is now showing everyone the conspicuous consumption of the developed world, and the world's poor want a share of this bounty. The pressure to consume resources will reach disastrous proportions within the lifetime of most readers. Obviously, we must use resources more efficiently if we hope to avoid catastrophe, or at least delay it.

Wildlife and plant populations do not increase in size indefinitely. Sooner or later they encounter environmental resistance that limits the population.[37] Every species in a given habitat has an equilibrium point for its population. For example, food shortages (such as dwindling prey) limit the population of wild animals. Less food leads to increased mortality—especially infant mortality—until the birth rate, the death rate, and the food supply are in equilibrium. Human population is a little different, since we can devise ways to improve our habitat and in this way raise the equilibrium point. But even for us there are limits, since resources are finite.

Many population forecasts have been made, using different assumptions. It is expected that the world's population will peak between 2050 and 2100, at which time an equilibrium point, estimated between 10.6 billion[38] and 12 billion,[39] will have been reached. Most population growth will occur in developing countries that do not have financial resources, stable governments, or infrastructure. The result will be anarchy, disease, and war. Many people will try to escape these conditions, and Canada will be faced with the ethical decision of sharing our prosperity by opening the doors to millions of refugees, or refusing to admit them in fear that the surging tide may lower our standard of life.

Most demographers call for a reduction in population growth to cope with the problem, although a few "cornucopians" insist that innovation, free enterprise, and international trade are the solutions and will let us develop methods of adapting to population growth. However, this optimistic view has been challenged for decades.

> . . . [W]hile the general optimism of the cornucopians is comforting, it conflicts with the rough consensus emerging among most demographers, scientists and policy analysts involved in population and resource research. Their view is that a high percentage of the planet's peoples are doomed to live with poverty and violence unless population growth is dramatically reduced. . . . [T]he Earth's biosphere can only produce enough renewable resources—food, fresh water and fish—to sustain [at most] two billion people at a standard of living equal to that in Europe. . . .[40]

Clearly, some action must be taken to avert these crises. In 1968, Garret Hardin, an American philosopher, published a paper on a syndrome observed about 140 years earlier called "the tragedy of the commons." Hardin applied this analogy to the problem of overpopulation and observed that, although our society has an ethical precept that everyone born has certain inalienable rights, it is impossible to divide finite resources among an indefinitely large number of people. Hardin concluded that the only way we can preserve the most precious freedoms is by the draconian measure of regulating the freedom to procreate, as China has already done. Several American groups, most notably Population Connection, are presently advocating similar solutions to the population problem.[41]

Engineers and geoscientists are not in a position to alleviate the problems of overpopulation, or even to affect it significantly. What we can do is ensure that we are not contributing to the inefficient use of resources or creating threats to the environment. In the long term, ethical actions are always in our own self-interest.

OPPORTUNISM AND THE TRAGEDY OF THE COMMONS

The tragedy of the commons is often mentioned in environmental discussions. William Forster Lloyd, a political economist at Oxford University, first described the degradation of common pastures in England in 1832. Lloyd observed that the cattle that grazed on these "commons" were smaller and stunted when compared to cattle on privately owned pastures. The privately owned pastures were obviously better kept and healthier environments for the cattle. His conclusions were republished in Hardin's paper in 1968.[42]

The concept can be explained as follows. Imagine a pasture open to all local cattle owners. Each owner, quite naturally, wants to increase his or her herd and thus tries to graze as many cattle as possible on the commons. Each animal added to the herd yields a significant benefit to the owner, but slightly reduces the food available to other animals. This process of increasing herd size may be sustainable for many years, especially if the commons is very large. Eventually, however, adding more animals will reduce the quality of the commons below an acceptable level. Yet even at this point, there is no way to stop the process. As Hardin says:

> [t]herein is the tragedy. Each man is locked into a system that compels him to increase his herd without limit—in a world that is limited. Ruin is the destination toward which all men rush, each pursuing his own best interest in a society that believes in the freedom of the commons. Freedom in a commons brings ruin to all.[43]

Hardin applied this concept as a motivating force for population limitation. However, the tragedy of the commons also applies to other environmental effects. For example, the inconsiderate citizen who dumps sewage along the roadway, or the negligent corporation that spills toxic or radioactive waste, or who emits heat, noise, or noxious or toxic fumes into the air, is like the cattle owner who adds one more animal to the commons. These selfish acts degrade the environment.

Whether we classify these acts as vandalism or simply as human nature, they are the philosophy of utilitarianism in reverse. Utilitarianism supports a small sacrifice by the general population in order to aid the public good. In the tragedy of the commons, a selfish individual imposes a general degradation upon society in order to reap a small personal benefit. Moreover, without intervention, human nature makes the outcome inevitable. As Hardin puts it:

> [t]he rational man finds that his share of the cost of the wastes he discharges into the commons is less than the cost of purifying his wastes before releasing them. Since this is true for everyone, we are locked into a system of "fouling our own nest," so long as we behave only as independent, rational, free enterprisers.[44]

Through thoughtful laws, regulations, and taxes, we can avoid the tragedy of the commons. The true owners—the public—must police the environment, and degradation must be linked to the cause. In other words, the polluter must pay. This monitoring creates the feedback loop that is required to protect the environment. This will not be an easy task, but as Case History 16.1 (below) shows, the cost of ignoring waste dumping is far more expensive than the cost of regulating it.

CASE HISTORY 16.1

TOXIC POLLUTION: LOVE CANAL, MINAMATA, BHOPAL, SUDBURY

The improper disposal of toxic waste is professional misconduct, whether it occurs deliberately, or through simple incompetence. The following four case histories of toxic pollution are so well known that their names—Love Canal, Minamata, Bhopal, and Sudbury—are now synonymous with environmental degradation or disaster.

Love Canal, New York—Dioxin

Love Canal is in Niagara Falls, in New York State, and is named after one of the early residents of the area. The canal was originally excavated for boats and barges, but it was never completed for navigational purposes, so it was more of a ditch than a canal. Between 1942 and 1953, the Olin Corporation and the Hooker Chemical Corporation saw it as a convenient hole for burying waste chemicals. Over 18,000 tons of chemical waste, including dioxins, were buried, and eventually the "canal" was again a flat plot of land.

In 1953 the Hooker Chemical Company sold the land to the Niagara Falls Board of Education for the nominal sum of one dollar. The conditions of the sale are not completely clear, but it appears that the Board needed the land to meet a growing school population boom, and allegedly insisted on the location, threatening expropriation. Hooker Chemical apparently attempted to warn the Board about the contamination by drilling test boreholes to show the Board the chemical contamination and including a clause in the deed that mentions the buried chemicals. In subsequent years, when the Board moved to sell parts of the land for real estate development, Hooker Chemical representatives publicly opposed such use because of the buried chemicals.[45]

In any case, the land eventually became a residential area, and homes, playgrounds, and a school were built on it. In 1976, after several seasons of heavy rains, people began to notice a terrible stench of chemicals. Homes reeked, children complained of chemical burns, and pets died or became sick. Yet these problems were minor when compared with the miscarriages, birth defects, and cases of cancer that occurred in the Love Canal area, more and more frequently, as the years passed. Residents soon demanded some action on the basis that these problems were far too frequent.

In 1978, a government study of the area exposed some remarkable—and frightening—statistics. More than 80 different chemicals had been detected, some of which were carcinogenic. The chemical pollution in the air was 250 to 5,000 times established safe levels. There was an unusually high (almost 30 percent) rate of miscarriage. Of 17 pregnant women in the Love Canal area in 1978, only 2 gave birth to normal children.

New York state authorities recognized the serious health threat posed by the buried chemicals and moved a few hundred families out of the area. The school was closed and surrounded by barbed wire. The area over the buried chemicals became a ghost town, with derelict houses, empty streets, and "No Trespassing" signs. Neighbouring residents who lived only a small distance from the chemicals were concerned about their health, but faced a dilemma: they wanted to move, but since "Love Canal" was now a synonym for hazardous waste, their homes were worthless. In 1980, residents' demands forced the government to carry out further testing. The tests showed high levels of genetic damage among the neighbouring residents. The area was declared a disaster area by the U.S. president, and 710 families were relocated. Many of the abandoned homes were demolished, and the chemical wastes were excavated for treatment and proper disposal. The total cost of the cleanup was estimated at US$250 million.[46]

The Love Canal tragedy revealed questionable ethics, although it is disputed whether the true fault lies with the Hooker Chemical Corporation, which buried much of the chemical waste, or with a careless or naive Board of Education. The agreement to transfer the land contained a clause that protected Hooker against future claims for liability, and should have alerted the Board of pollution. However, it is alleged that, after the extent of the disaster became public, chemical industry spokespeople ridiculed the residents as hypochondriacs.[47] This toxic secret brought tragedy to many families who risked their life savings in worthless homes, whose children were born deformed, and who even now may live in fear of contracting cancer.

Love Canal heightened awareness of the need for ethical conduct and environmental regulations. In the years following Love Canal, the U.S.

Environmental Protection Agency (EPA) discovered between 32,000 and 50,000 other toxic waste dumps scattered across the United States. Possibly 2,000 of them may still be a risk to the public.[48]

The resulting furor over Love Canal led to stricter laws and more severe penalties for improper disposal of waste. Unfortunately, the public awakening to the danger of improper and unethical dumping was too slow in coming. Lois Gibbs, the resident who took the leadership role in drawing public attention to the environmental disaster at Love Canal, recently stated: "As a society, we begin with this toxic thing and say: 'How much can we put in the environment before somebody is harmed?' This risk-assessment approach means there is a subset of our society that will always be sacrificed."[49] Dioxin was being released into the environment at Love Canal for years, and when housewives discovered that dioxin caused birth defects in their children, it still took 13 years for scientists to verify the facts.

Minamata Bay, Japan—Mercury Poisoning

Since 1953, thousands of residents of the Minamata Bay area of Japan have fallen ill as a result of organic mercury poisoning. Mercury (also called quicksilver) is the liquid metal often used in thermometers and barometers. Whether as a pure liquid or in compound form, mercury can cause serious renal and neurological dysfunction. Mild cases often mimic amyotrophic lateral sclerosis (ALS, also called "Lou Gehrig's disease"). The symptoms of severe poisoning include clumsiness, stumbling, severe mental or behavioural problems, and loss of speech, taste, and hearing.

The Chisso Company, a nitrogen fertilizer company in Minamata, the main city on Minamata Bay, first began producing acetaldehyde in 1932. Mercury was required as a catalyst in this process and for other chemicals the company would produce later, such as vinyl chloride. The mercury was used in liquid form, and during the production process, a portion of it was lost— washed into Minamata Bay with the wastewater. In the bay, microbes acted on the mercury and converted it into an organic (methyl or carbon-based) mercury compound. Shellfish absorbed the organic mercury, and since mammals do not excrete mercury, it becomes more concentrated as it moves up the food chain. Over time, the concentrations are sufficiently large that the toxic effect becomes apparent. The first humans affected were fishermen and their families, who had a diet rich in fish, including shellfish. The seriousness of the problem was recognized by physicians and health officials in the Minamata area around 1956, although cases were later traced back to 1953, about 20 years after mercury first began washing into the bay.[50]

The medical director of the hospital associated with the Chisso Company became sufficiently concerned about the problem in 1956 that he began a series of tests on cats. Since family cats ate fish, they were the first to exhibit this curious behaviour. He identified the manufacturing plant effluent as the cause of the problem, but when he reported his results to his superiors at

Chisso, he was ordered to stop his tests and forbidden to report his findings to the local health authorities.[51]

It was not until 1959 that medical authorities requested help from the Kumamoto medical school and an investigation was begun. By 1962, they were certain that the problem was caused by organic mercury and that the source of the mercury was the Chisso effluent. Initially it was estimated that about 2,900 people had contracted Minamata disease, as the debilitating neurological syndrome is now known. The government indicted Chisso, but it took years for the various cases to work their way through the legal system. The first decision was made in 1970, and the government also awarded some compensation. However, in 1995, the Japanese government reached a political settlement with about 11,000 unrecognized sufferers that called for lump-sum payments, and in 2007, an additional 5,000 people applied for official recognition as sufferers of Minamata disease. The lawsuits are still not settled.[52]

The ethical issues in this case are clear. The Chisso Company's actions—stopping the medical tests, suppressing knowledge of the problem, and continuing to permit mercury to be dumped in Minamata Bay—were unethical and inexcusable. Even though the damaging effects of mercury were not well known in 1932, the company knowingly inflicted personal tragedy on thousands of unwitting people in later years.

An outbreak of Minamata disease occurred in Grassy Narrows, Ontario, in 1970. Members of two Ojibwa bands living near the Wabigoon River began to show the debilitating symptoms of mercury poisoning, and the source of the pollution was traced to the Reed Paper Company in Dryden, just upstream from the Ojibwa reserves. The provincial government ordered Reed to reduce mercury usage, and the pollution was gradually eliminated. Although compensation was eventually paid to the two bands, the economic and social effects were devastating.[53]

Bhopal, India—Methyl Isocyanate

In the early morning hours of December 3, 1984, a poisonous cloud of methyl isocyanate gas escaped from the Union Carbide plant in Bhopal, India. It killed thousands of people up to 6 km (4 miles) away, many while they were asleep in their beds. This was probably the worst industrial accident in history, and its social and economic impact on Bhopal was devastating. It is estimated that between 3,000 and 12,000 people died in this catastrophe. Around 30,000 more suffered permanent injuries, 20,000, temporary injuries, and 150,000, minor injuries. And even these horrific numbers are disputed by victims' rights organizations, which say the real numbers were even higher.[54]

The Union Carbide plant was established in 1969 as a mixing factory for pesticides. Methyl isocyanate, which is used in large quantities in the production process, is highly volatile as well as highly toxic. Methyl isocyanate reacts vigorously with many common substances and must be maintained at very low temperatures to prevent uncontrolled reactions. The precise cause of the disaster is not known, but most explanations state that an employee closed a valve on a piping system so that a filter connected to the pipe could be washed. A metal disc should have been inserted to make certain that the valve could not leak, but this was not done. During the washing process, water leaked past the closed valve, entering piping that was connected to the methyl isocyanate holding tank. The water reacted with the methyl

Photo 16.1 — Union Carbide Factory, Bhopal. *A methyl isocyanate gas leak from a Union Carbide pesticide plant in Bhopal, India, killed thousands of people in neighbouring communities. The Union Carbide factory (in the background) looms over relatives and friends carrying a victim's body to cremation.*

Source: CP Photo/Associated Press.

isocyanate, generating intense heat. The pressure in the tank increased dramatically and pushed past pressure relief valves into the atmosphere. Safety measures were either inadequate or did not work. Over the next 90 minutes, about 40 tonnes of methyl isocyanate and other reaction products were released into the atmosphere. Since the vapour is heavier than air, it filled low-lying areas, crept into houses through windows and doors, and asphyxiated thousands of people while they slept.

Union Carbide India Limited (UCIL) operated the plant, but the plant was mainly owned (50.9 percent) by Union Carbide Corporation, a U.S. company. Union Carbide disputes the above explanation for the disaster. Based on a forensic study conducted by a consulting engineer, Union Carbide concluded that the gas emission was sabotage. They believe that an unknown plant employee deliberately added water to a storage tank, apparently intending to ruin the methyl isocyanate.[55]

An inquiry was held after the disaster. The construction, operation, and maintenance of the Bhopal plant were examined, as well as management decisions that permitted such a potentially dangerous plant to operate in such an unsafe manner, in an urban area, with no suitable emergency plan. The Indian government charged the company management with negligence, brought murder charges against its chief executive, and demanded US$3.3 billion to settle victims' claims. Then, in 1989, the Indian Supreme Court announced a settlement of all claims for US$470 million, conditional on the dropping of criminal charges.[56] Shortly after the settlement was announced, a new Indian government disallowed the claim and sought to reinstate the criminal charges.[57] While the litigation continues, the Indian government is supporting the survivors of the Bhopal disaster.

Sudbury, Ontario—Sulphur Dioxide

Canada is the world's second-largest producer of nickel, and the mines around Sudbury, Ontario, are the main source of this metal. Nickel was first discovered in the area in 1856, but it was not until the Canadian Pacific Railway reached Sudbury in 1883 that anyone realized the full extent of the ore body.[58] The nickel is in the form of sulphide ore, which cannot be converted directly into metallic form. It must first be smelted—that is, burned to remove the sulphur and convert the ore to nickel oxide, which can then be reduced to pure nickel. In the early 1900s the first conversion step was typically done in huge, open "roasts," where layers of timber were interspersed with layers of ore. The roasts burned around the clock and emitted a toxic cloud of sulphur dioxide.[59] The ecological impact of this process was ignored a century ago, when the Sudbury area was sparsely populated. Few people realized that the sulphur dioxide, when dissolved in rainwater, created acid. In 1928 the federal government became aware of the problem and banned the use of open roasts.[60] However, enclosed smelting is still done.

The problem was partly solved by the erection of taller smokestacks, which spread the pollutants over a wider area, thus reducing their concentration. The largest of these super stacks, at Sudbury's Copper Cliff mine, was built in 1972 and is about 380 m (1,247 ft.) high.

The environmental effects of acid rain on the Sudbury region were severe. In the area around the smelters, trees are stunted and sparse, lakes are devoid of fish, and only the hardiest species of birds can survive (where there are any birds left at all). In 1978 the Regional Municipality of Sudbury began an ambitious program to restore 10,000 hectares of barren land. By experiment, it was discovered that, in most locations, a combination of fertilizer, agricultural lime (to neutralize the acidity), and a seed mixture of grass and legumes would generate a healthy grass cover. Two years after grass began growing, crews would return to plant trees and shrubs. Fifteen different tree species were planted, and by 1995, many of these trees were more than 3 m (10 ft.) tall. Although the soil acidity is still very high, the levels of heavy metals have been reduced. Populations of insects, birds, and small mammals have increased, and successful tree growth has averaged 70 percent across all species. About 3,000 hectares have been reclaimed so far.[61] As a result of the extensive environmental damage and the efforts being made to remedy that damage, Sudbury has become one of the most closely studied ecological areas in the world.

Discussion of Toxic Pollution Cases

These environmental disasters were not of equal magnitude, nor were they equally unethical. Thousands of deaths and injuries were directly linked to Bhopal—the world's worst environmental disaster—and to Minamata. Hundreds of illnesses, miscarriages, and cancer cases were related to Love Canal, but no such human tragedies have been directly attributed to the Sudbury sulphur dioxide emissions, although the pollution is evident in the lakes and vegetation downwind of the smelters.

CASE HISTORY 16.2

NUCLEAR SAFETY: THREE MILE ISLAND AND CHERNOBYL

Nuclear-generated electrical power is a major source of environmental concern, although earlier American and British studies showed that power generation by coal is 250 times more hazardous than nuclear power and that generation by oil is 180 times more hazardous. Only natural gas poses fewer hazards than nuclear power to workers and the general population.[62] A few serious radiation releases occurred in the early experimental days of Canadian and American nuclear development, and a few deaths in North America have been attributed to power reactor accidents. However, the American reactor accident at Three Mile Island in 1979 and a later accident at Chernobyl in Ukraine (then part of the Soviet Union) in 1986 involved partial or complete meltdown of the nuclear reactor cores, and were therefore much more serious than any other nuclear accident (before or since). The public was horrified to learn that meltdowns—the most unthinkable nuclear accident—could occur.

Three Mile Island

Three Mile Island is located on the Susquehanna River in southern Pennsylvania. Construction on its two-unit nuclear power plant began in the late 1960s, and the second unit (unit TMI-2) was completed, tested, and brought online in December 1978. The two units were designed to produce a maximum of 880 megawatts of electrical power. The TMI-2 unit experienced many minor problems during its commissioning and early operation. It had been operating for only three months when it became the source of North America's worst nuclear accident.[63]

The Accident and Its Causes

The problem began shortly after 4 a.m. on March 28, 1979, and recovery efforts lasted for a month. The initial cause of the accident was a blocked feedwater line, which caused pumps to stop. Operator errors and other minor malfunctions then magnified the problem. James Carter, the American president at the time (and also a nuclear engineer), commissioned an inquiry into the accident. The inquiry report describes the first few minutes of the accident as follows:

> In the parlance of the electric power industry, a "trip" means a piece of machinery stops operating. A series of feedwater system pumps supplying water to TMI-2's steam generators tripped on the morning of March 28, 1979. The nuclear plant was operating at 97 percent power at the time. The first pump trip occurred at 36 seconds after 4:00 a.m. When the pumps stopped, the flow of water to the steam generators stopped. With no feedwater being added, there soon would be no steam, so the plant's safety system automatically shut down the steam turbine and the electric generator it powered. The incident at Three Mile Island was 2 seconds old. . . .
>
> When the feedwater flow stopped, the temperature of the reactor coolant increased. The rapidly heating water expanded. The pressurizer level (the level of the water inside the pressurizer tank) rose and the steam in the top of the tank compressed. Pressure inside the pressurizer built to 2,255 pounds per square inch, 100 psi more than normal. Then a valve atop the pressurizer, called a pilot-operated relief valve, or PORV, opened—as it was designed to do—and steam and water began flowing out of the reactor coolant system through a drain pipe to a tank on the floor of the containment building. Pressure continued to rise, however, and 8 seconds after the first pump tripped, TMI-2's reactor—as it was designed to do—scrammed: its control rods automatically dropped down into the reactor core to halt its nuclear fission.
>
> Less than a second later, the heat generated by fission was essentially zero. But, as in any nuclear reactor, the decaying radioactive materials left from the fission process continued to heat the reactor's coolant water. This heat was a small fraction—just 6 percent—of that released during fission, but it was still substantial and had to be removed to keep the core from overheating. When the pumps that normally supply the steam generator with water shut down, three emergency feedwater pumps automatically started. Fourteen seconds into the

accident, an operator in TMI-2's control room noted the emergency feed pumps were running. He did not notice two lights that told him a valve was closed on each of the two emergency feedwater lines and thus no water could reach the steam generators. One light was covered by a yellow maintenance tag. No one knows why the second light was missed.

With the reactor scrammed and the PORV [relief valve] open, pressure in the reactor coolant system fell. Up to this point, the reactor system was responding normally to a turbine trip. The PORV should have closed 13 seconds into the accident, when pressure dropped to 2,205 psi. It did not. A light on the control room panel indicated that the electric power that opened the PORV had gone off, leading the operators to assume the valve had shut. But the PORV was stuck open, and would remain open for 2 hours and 22 minutes, draining needed coolant water—a LOCA [loss of coolant accident] was in progress. In the first 100 minutes of the accident, some 32,000 gallons—over one-third of the entire capacity of the reactor coolant system—would escape through the PORV and out the reactor's let-down system. Had the valve closed as it was designed to do, or if the control room operators had realized that the valve was stuck open and closed a backup valve to stem the flow of coolant water, or if they had simply left on the plant's high pressure injection pumps, the accident at Three Mile Island would have remained little more than a minor inconvenience. . . .[64]

The combination of the initial blockage in the feedwater lines, plus the overlooked lights (warning that the valves were closed from the emergency feedwater pumps), plus the relief valve that was stuck open, created a confusing combination of effects that would lead to incorrect actions, exposure of the reactor core, and release of radioactive water. The series of events would end with the reactor destroyed and millions of people in fear for their health and safety.

In the days and weeks after the initial incident, the nuclear plant owners, the news media, and politicians from the local mayor to the president were involved in a debate over the need for evacuation and the correct information to be released to local residents. Control was not fully regained for another month. The plant went into a cold shutdown on April 27, 1979. The full report of the incident is available on the Internet and is a chilling story of confusion and fear.[65]

The Resulting Damage and Aftermath

During the accident, the loss of coolant caused the reactor core to be exposed. The extreme heat melted about one-third of the core, rendering the reactor useless—a billion-dollar loss. The site cleanup likely cost another billion dollars. A small amount of radioactive material was released from the reactor into the environment, raising the fear of radiation-induced health effects, principally cancer, in the neighbourhood of the reactor. Over the next 18 years the state government would trace the health of 30,000 people who had been living within five miles of Three Mile Island at the time of the accident. Fortunately, no unusual trends were found, and the registry was discontinued in 1997.[66]

Chernobyl

The Chernobyl nuclear power plant is located about 100 km (62 miles) north of Kiev, the capital of Ukraine. In 1986 the plant comprised four reactors constructed between 1977 and 1983. At full capacity, the plant generated 4,000 megawatts of electricity. The Chernobyl reactors were a Soviet design, designated by the acronym RBMK, which is a pressurized water reactor design that uses ordinary water as a coolant and graphite as a moderator. This type of Soviet reactor was intended for plutonium generation (for weapons) as well as electrical power production. No other power reactor design uses this combination of graphite and water.[67]

At 1:23 a.m. on April 26, 1986, the No. 4 nuclear reactor in Chernobyl exploded, releasing huge, sinister clouds of radioactive plutonium, cesium, and uranium dioxide into the atmosphere. It was the worst nuclear accident in history involving a nuclear generating plant. Ukraine was then part of the U.S.S.R., and regrettably, the Soviet authorities were slow to issue a warning or to release any details about the accident. In fact, two days later, Swedish experts, who had noticed the nuclear fallout over Scandinavian countries, released the first information about the accident.

The Accident and Its Causes

The reactor explosion had three basic causes: poor reactor design, inadequately trained reactor personnel, and unsafe operating procedures that permitted tests to be carried out at a low and unstable power. The reactor design flaws included an inadequate containment shell, poorly designed graphite control rods, and a feature called a "positive void coefficient," which refers to the tendency for voids (or steam pockets) to form in the cooling water. Although graphite is the main moderator, the cooling water also has a moderating effect. However, when steam pockets in the water get too large, then the moderating effect of the water begins to fluctuate, and the reactor may experience rapid and uncontrollable power surges.

The RBMK reactors were known to have a positive void coefficient, so other control features were in place to prevent the instability from occurring. However, the reactor's inexperienced operators overrode these features. By the time the problem was recognized, the heat had deformed the channels, and the control rods could not be reinserted. The World Nuclear Association (WNA) website describes the Chernobyl accident as follows:

> On 25 April [1986], prior to a routine shut-down, the reactor crew at Chernobyl-4 began preparing for a test to determine how long turbines would spin and supply power following a loss of main electrical power supply. Similar tests had already been carried out at Chernobyl and other plants, despite the fact that these reactors were known to be very unstable at low power settings.
>
> A series of operator actions, including the disabling of automatic shutdown mechanisms, preceded the attempted test early on 26 April. As the flow of coolant water diminished, power output increased. When the operator moved

to shut down the reactor from its unstable condition arising from previous errors, a peculiarity of the design [the positive void coefficient] caused a dramatic power surge.

The fuel elements ruptured and the resultant explosive force of steam lifted off the cover plate of the reactor, releasing fission products to the atmosphere. A second explosion threw out fragments of burning fuel and graphite from the core and allowed air to rush in, causing the graphite moderator to burst into flames.

There is some dispute among experts about the character of this second explosion. The graphite—there was over 1200 tonnes of it—burned for nine days, causing the main release of radioactivity into the environment. A total of about 14 EBq (10^{18} Bq) of radioactivity was released, half of it being biologically-inert noble gases. . . . Some 5000 tonnes of boron, dolomite, sand, clay and lead were dropped on to the burning core by helicopter in an effort to extinguish the blaze and limit the release of radioactive particles.[68]

Photo 16.2 — Chernobyl Nuclear Power Plant. *Nuclear reactor number 4 at the Chernobyl power plant exploded on April 26, 1986. The accident was caused by an unstable low-power operation test, which led to the meltdown of the nuclear reactor and the dispersion of radioactive contamination across several countries. However, the root causes were eventually traced to poor training, ignorance, and secrecy among bureaucrats in charge of the plant. This tragedy exposed the folly of making engineering design decisions for political or military reasons, and the danger of secrecy and information management in engineering.*

Source: © Reuters/CORBIS.

The Resulting Damage, Injuries, and Deaths

The WNA states that "30 people were killed, and there have since been up to ten deaths from thyroid cancer due to the accident." However, these numbers are challenged, and do not reveal the truly massive disruption of life that took place. A full list of damage and disruption was included in a 1988 UNSCEAR (United Nations Scientific Committee on the Effects of Atomic Radiation) report. It describes

- hundreds of direct injuries and deaths,
- evacuation of several cities, involving hundreds of thousands of people,
- extensive radiation monitoring,
- repeated decontamination of buildings and destruction of some buildings,
- creation of a prohibited area within a 5-km radius around Chernobyl,
- creation of a restricted movement area within a 30-km radius around Chernobyl,
- destruction of poisoned food and movement of thousands of cattle, and
- millions of protective medical treatments, radiation tests, and so on.[69]

Aftermath

Unit 4 at Chernobyl was enclosed in concrete shortly after the accident, to permit the other three reactors to continue operating. However, this concrete structure is not durable, and more repairs may be necessary. Much money was spent on improving the safety of the remaining reactors, and in view of the desperate need for electricity, they were kept running. Unit 2 was shut down after a turbine hall fire in 1991, and Unit 1 was shut down at the end of 1997. Unit 3 continued operating until December 2000.[70]

Controversial ideas and new secrets about the Chernobyl disaster were revealed by Grigori Medvedev in *The Truth about Chernobyl*,[71] first published in the Soviet Union in 1989 and translated into English in 1991. An excerpt from the book review written by Joseph Schull of *Maclean's* magazine follows and is reproduced here:

> . . . The first truth that emerges in Medvedev's book is that the Soviet nuclear industry was run by incompetents from top to bottom: Officials in charge of the construction and management of nuclear power stations simply had no training in the field, while their underlings at Chernobyl were no better prepared. Meanwhile, secrecy surrounded the industry and fostered utter ignorance about its potential dangers. Information about previous nuclear mishaps, including the 1979 accident at Three Mile Island, was reserved for high-placed officials unable to draw the appropriate lessons. A state bureaucracy that acknowledged successes but not setbacks was equally damaging. Attention to safety implied the possibility of accidents, and that could only mean that errors might be committed—a possibility that nearly everyone, from minister to technician, wanted to deny. Failure was not in the Soviet vocabulary. . . .
>
> The coverup continues even now. Soviet authorities have admitted to only 31 deaths in the immediate aftermath and have kept secret the numbers who have

died since then. But Vladimir Chernousenko, the scientific director now in charge of the 32-km exclusion zone surrounding the Chernobyl power station, recently estimated that fatal casualties to date number between 7,000 and 10,000. . . .[72]

Discussion of These Nuclear Accidents

It should be noted that nuclear power engineering has matured significantly in the decades since the accidents described above, and that regulatory procedures are more rigorous. Although the CANDU reactors have had problems with zircaloy pressure tubes that were prone to corrosion and required replacement,[73] the Canadian CANDU reactor was designed to be safer than the Soviet RBMK design and even safer than the American light-water reactor (LWR) design. Canadian nuclear power plants use heavy water, not graphite, for moderation. The heavy water is needed for the fission to proceed. Loss of heavy water causes the fission process to stop automatically.[74]

Nevertheless, the Three Mile Island and Chernobyl accidents warn us of the immense potential energy in nuclear plants. The possibility of disaster may be infinitesimally small, but it is not zero. These tragic accidents also expose the danger of making design decisions for political or military reasons, and the danger of secrecy and information management in engineering. Competence, clear communication, and high professional and ethical standards are essential in the design and operation of such facilities.

NOTES

[1]   Environment Canada, *National Pollutant Release Inventory*, available at <www.ec.gc.ca/inrp-npri> (June 15, 2009).

[2]   Environment Canada, *Canada's National Pollutant Release Inventory*, Fact sheet, available at <www.ec.gc.ca/inrp-npri> (June 15, 2009).

[3]   Environment Canada, *NPRI-Google Earth Map Layers (2006 Data)*, available at <www.ec.gc.ca/inrp-npri> (June 15, 2009).

[4]   Commission for Environmental Cooperation (CEC), *Taking Stock: 2004 North American Pollutant Releases and Transfers*, September 2007 (2004 data), available at <www.cec.org/home> (June 4, 2008).

[5]   Commission for Environmental Cooperation (CEC), *News Release: Toxic releases down from North American industry leaders, increasing from other facilities*, October 18, 2007, available at <www.cec.org/news> (June 4, 2008).

[6]   R.W. Jackson and J.M. Jackson, *Environmental Science: The Natural Environment and Human Impact*, Longman, Harlow, Essex, 1996.

[7]   P.H. Abelson, "Carbon Dioxide Emissions," Editorial, *Science*, November 25, 1983.

[8]   Jackson and Jackson, *Environmental Science*, p. 318.

[9]   "The Politics of Climate," *EPRI Journal*, June 1988, pp. 4–15.

[10]   M.F. Meier, "Contributions of Small Glaciers to Global Sea Level," *Science*, December 21, 1984, pp. 1418–1421; and *Carbon Dioxide and Climate: A Second Assessment*, National Academy Press, Washington, DC, 1982.

[11]   Jackson and Jackson, *Environmental Science*, p. 320.

[12]   M. Smith and M. Vincent, "Tanking a Killer Coolant," *Canadian Geographic*, September–October 1997, pp. 40–44.

[13] S. Tripp and B. Whiting, "The Illegal Trade in Chemicals That Destroy Ozone," in *TRIO—The Newsletter of the North American Commission for Environmental Cooperation* (CEC), Spring 2003, available at <www.cec.org/trio/> (June 4, 2008).

[14] E. Titterton, "Nuclear Energy: An Overview," in *Energy Alternatives: Benefits and Risks* (ed. H.D. Sharma), University of Waterloo, Waterloo, ON, 1990, p. 146.

[15] C.J. Campbell, "Oil depletion—Update through 2001," Hubbert Peak of Oil Production website at <www.hubbertpeak.com/campbell/update2002.htm> (June 4, 2008).

[16] Titterton, "Nuclear Energy."

[17] Jackson and Jackson, *Environmental Science*, p. 257.

[18] World Nuclear Association, *Information Papers,* available at <www.world-nuclear.org> (June 4, 2008).

[19] Allan Kupcis, Chairman, Canadian Nuclear Association (CNA), "Get Real: Nuclear Is in Your Future," *The Globe and Mail*, July 25, 2003, A11.

[20] G. Wasteneys, "CANDU Reactor Is Outstanding Canadian Feat," *Ottawa Citizen*, Letters, October 17, 2003, available at <www.fpinfomart.ca> (June 4, 2008). Excerpt reprinted with permission.

[21] "Letters: Chernobyl Public Health Effects," *Science*, October 2, 1987, pp. 10–11.

[22] Jackson and Jackson, *Environmental Science*, p. 330.

[23] Ibid., p. 334.

[24] Ibid., p. 311.

[25] R.E. Munn, "Air Pollution," *The Canadian Encyclopedia*, available at <www.canadianencyclopedia.ca> (June 4, 2008).

[26] Jackson and Jackson, *Environmental Science*, p. 312.

[27] Munn, "Air Pollution."

[28] Ibid.

[29] Jackson and Jackson, *Environmental Science*, p. 315.

[30] H.L. Ferguson, "Acid Rain," *The Canadian Encyclopedia*, available at <www.canadianencyclopedia.ca> (June 4, 2008).

[31] Ibid.

[32] Ibid.

[33] D.E. Carr, *Death of Sweet Waters*, Berkley, New York, 1971, p. 41.

[34] A.H.J. Dorcey, "Water Pollution," *The Canadian Encyclopedia*, available at <www.canadianencyclopedia.ca> (June 4, 2008).

[35] Jackson and Jackson, *Environmental Science*, p. 140.

[36] United Nations, *World Population to 2300,* Department of Economic and Social Affairs, New York, 2004, available at <www.un.org/esa/population> (June 4, 2008).

[37] Jackson and Jackson, *Environmental Science*, p. 140.

[38] "End of the Population Explosion?" *Discover*, July 1997, p. 14.

[39] Jackson and Jackson, *Environmental Science*, p. 163.

[40] P. Kaihla, C. Erasmus, J. Edlin, and B. Bethune, "Apocalypse When?: A United Nations Plan to Limit Global Population Growth Triggers an Acrid War of Words", *Maclean's*, September 5, 1994, p. 22. Excerpt reprinted with permission.

[41] *Population Connection* website at <www.populationconnection.org> (June 4, 2008).

[42] G. Hardin, "The Tragedy of the Commons," *Science* 162 (1968), pp. 1243–1248.

[43] Hardin, "The Tragedy of the Commons."

[44] Ibid.

[45] E. Zuesse, "Love Canal: The Truth Seeps Out," *Reason Magazine*, February 1981, available at <www.reason.com/news/show/29319.html> (June 3, 2008).

[46] L.G. Regenstein, "Love Canal Toxic Waste Contamination: Niagara Falls, New York," in N. Schlager, *When Technology Fails: Significant Technological Disasters, Accidents, and Failures of the Twentieth Century,* Gale Group, Detroit, 1994, pp. 354–360.

[47] Regenstein, "Love Canal Toxic Waste Contamination," p. 358.
[48] Ibid.
[49] L. Baker, "Be Safe! Lois Gibbs' New Campaign Urges Caution on Toxic Chemicals," *E-Magazine*, vol. XIV, no. 4, July–August 2003, available at <www.emagazine.com/view/?376> (June 4, 2008).
[50] J. Larson, "Mercury Poisoning: Minamata Bay, Japan," in Schlager, *When Technology Fails*, pp. 367–371.
[51] Larson, "Mercury Poisoning," p. 370.
[52] "Chisso to nix lump-sum redress package," *The Japan Times*, Friday, November 16, 2007.
[53] M. Bray, "Grassy Narrows," *The Canadian Encyclopedia*, available at <www.canadianencyclopedia.ca> (June 4, 2008).
[54] L. Ingals, "Toxic Vapour Leak: Bhopal India," in Schlager, *When Technology Fails*, pp. 403–410.
[55] "The Incident, Response, and Settlement," *Bhopal Information Center*, Union Carbide, available at <www.bhopal.com/irs.htm> (4-June 4, 2008).
[56] "Damages for a Deadly Cloud," *Time*, February 27, 1989, p. 53.
[57] "Haunted by a Gas Cloud," *Time*, February 5, 1990, p. 53.
[58] B. Sutherland, "Nickel," *The Canadian Encyclopedia*, available at <www.canadianencyclopedia.ca> (June 4, 2008).
[59] L.C. Ritchie, "Ecological Disaster: Sudbury Ontario," in Schlager, *When Technology Fails*, pp. 340–344.
[60] Ritchie, "Ecological Disaster," p. 341.
[61] International Council for Local Environmental Initiatives, Region of Sudbury, *Canada: Land Reclamation*, ICLEI Project Summary Series, Project Summary #22.
[62] R.D. Bott, "Nuclear Safety," *The Canadian Encyclopedia*, 1st ed., Hurtig, Toronto, 1985, p. 1302.
[63] D.E. Newton, "Three Mile Island Accident," in Schlager, *When Technology Fails*, p. 510.
[64] *Report of the President's Commission on the Accident at Three Mile Island*, U.S. Government document, October 30, 1979, available at <www.pddoc.com/tmi2/kemeny> (June 4, 2008).
[65] Ibid.
[66] World Nuclear Association, *Three Mile Island: 1979*, Information Papers, March 2001, available at <www.world-nuclear.org/info/inf36.html> (June 4, 2008).
[67] World Nuclear Association, *RBMK Reactors*, Information Papers, May 2007, available at <www.world-nuclear.org/info/inf31.html> (June 4, 2008).
[68] World Nuclear Association, *Chernobyl Accident*, Information Papers, March 2001, available at <www.world-nuclear.org/info/chernobyl/inf07.html> (4-June 4, 2008). Excerpt reprinted with permission.
[69] United Nations Scientific Committee on the Effects of Atomic Radiation (UNSCEAR), *UNSCEAR 2000 Report*, Annex J, Exposures and effects of the Chernobyl Accident, Vienna, Austria, 2000, available at <www.unscear.org/unscear/en/chernobyl.html> (June 4, 2008).
[70] World Nuclear Association, *Chernobyl Accident*.
[71] G. Medvedev, *The Truth about Chernobyl* (trans. E. Rossiter), HarperCollins, New York, 1991.
[72] J. Schull, "A Fatal Coverup: The Deadly Lies of Chernobyl Come to Light," *Maclean's*, May 13, 1991, p. 65.
[73] M. Nichols, "CANDU Flawed," *Maclean's*, August 25, 1997.
[74] J.A.L. Robertson, "Nuclear Power Plants," *The Canadian Encyclopedia*, available at <www.canadianencyclopedia.ca> (June 4, 2008).

Chapter 17
Environmental Sustainability

Environmental sustainability means sensible use of our huge (but finite) natural resources, so that they are not dissipated negligently, thus degrading the quality of life for future generations. Unfortunately, our current resource consumption is not sustainable. To avoid irreversible environmental degradation, we need a long-term strategy and aggressive action now. As this chapter shows, the topic is controversial and the proposed remedies are unpalatable, so achieving environmental sustainability may be the key battle of the 21st century.

This chapter defines *sustainability* and gives a brief overview of the history of sustainable thinking. The chapter then discusses the two greatest threats to sustainability at present: climate change and the depletion of fossil fuels, and suggests what we must do as engineers and geoscientists (and as citizens and society in general) to make our lifestyle sustainable. The chapter concludes with a discussion of the Ladyfern natural gas field in British Columbia, which was a disturbing example of uncontrolled exploitation, and illustrates why we must apply ethical principles to the development and use of our resources if we are to reach sustainability.

A DEFINITION OF SUSTAINABILITY[1]

Sustainability is a very simple concept, although it has serious implications for society. In 1987, the Brundtland Commission of the United Nations defined sustainable development in its report *Our Common Future:* "Sustainable development is development that meets the needs of the present, without compromising the ability of future generations to meet their own needs."[2]

Paul Hawken, in *The Ecology of Commerce,* defined sustainability more personally: "Sustainability . . . can also be expressed in the simple terms of an economic golden rule for the restorative economy: leave the world better than you found it, take no more than you need, try not to harm life or the environment, make amends if you do."[3]

Sustainability is a vital issue because the welfare of society is threatened, but most people do not understand (or do not believe) this crucial fact. The two main symptoms of unsustainability are climate change and peak oil. These are separate problems, but they are linked.

- **Climate change.** The careless disposal of waste, specifically carbon dioxide and similar fossil fuel emissions, is causing global warming and climate change, with predictions of dire global consequences within the next few decades.
- **Peak oil.** Our excessive consumption of oil has forced us to maximize the rate of production of this finite natural resource. Oil production is at or near its production peak and is beginning (or will soon begin) to decline. Within the next decade or two, easily accessible oil will become scarce and oil prices will rise, leading to energy shortages. In fact, sharply rising oil prices are already observable in 2008. Our lifestyle depends on the intense use of oil and its by-products for transportation, heating, electricity, and manufacturing, so a crisis lies ahead unless we move to a more sustainable exploitation of energy.

A BRIEF HISTORY OF SUSTAINABLE THINKING

The Industrial Revolution in the 1800s began the mechanization of industry, and increased society's living standards. The newly invented machinery was driven by steam power, fuelled by coal, and magnified the productivity of unskilled workers. By the turn of the century, machinery was gradually shifting from coal to oil as a source of energy. In the early 1900s, drilling for oil was common in the southern U.S. states. The need to control oil supplies was one of the causes of the two World Wars (1914–1918 and 1939–1945).

When the Second World War ended in 1945, Canada looked forward to an era of peace and prosperity. Industries, geared for war, converted their factories to appliances and vehicles. Petrochemical industries provided a selection of magical new plastics, and agro-chemical industries promised profitable farms and an end to world hunger, using new pesticides, herbicides, and fertilizers—all made from petroleum feedstock. The average person could fuel a car and heat a home effortlessly, using oil or natural gas drawn from massive reservoirs in the Canadian west.

In the 1950s, nuclear reactors were developed for power generation, and added to the electrical supply from hydroelectric and fossil fuel plants. Cheap electric power was available to almost everyone. Jet engines, which were in an early experimental stage before the Second World War, provided easy international travel, and low-cost vacations abroad became the norm. Air conditioning units and television sets, which were virtually nonexistent in pre-war days, became common. During this period, world population grew exponentially (from less than 1 billion in 1800 to about 6.5 billion in 2007, as discussed in Chapter 16). Everyone aspired to a lifestyle that consumed more resources in a month than our pre-war ancestors consumed in a year. Few observers saw any problem with this consumption-oriented lifestyle, for almost two decades.

SILENT SPRING   In 1962, Rachel Carson described the dangers associated with pesticides in her book *Silent Spring*.[4] Carson, a trained biologist, was trying to explain why songbirds did not return in the spring. She discovered that bird populations were dying because pesticides were being applied indiscriminately. Her book led to recognition that indiscriminate use of agricultural chemicals could be hazardous to bird, fish, animal, and human life. In our current terminology, such use was unsustainable.

THE LIMITS TO GROWTH   The Club of Rome, an organization concerned with the problems of humankind, published a report in 1972 titled *The Limits to Growth*, which warned that uncontrolled human activity had the potential to make our planet uninhabitable.[5] The book describes one of the earliest computer simulations of human system behaviour. A simple "world model" was created, which simulated the creation or consumption of five basic quantities as a function of time: population, capital, food, non-renewable resources, and pollution. Feedback loops and complex relationships between the quantities were included, yielding a mathematical model of the planet. The simulation was run many times, with varying inputs and controls; the goal was to find the levels that yielded sustainable equilibrium. Their "standard" run, which simulated the world of 1972 into the future, shows industrial output per capita peaking about the year 2000, with non-renewable resources depleting sharply thereafter. In the following decades, population and pollution peak, but food drops off rapidly, devastating the population and industrial output. In brief, the model showed that the 1972 lifestyle was unsustainable. The computer analysis is naive by today's standards, but it stimulated further research into global sustainability.

GAIA   In 1979, James Lovelock's book *Gaia: A New Look at Life on Earth* made a major contribution to sustainability. Lovelock's book views the earth as a self-regulating living being, with ability to adapt, and ability to heal, like other organisms.[6] Many challenged the Gaia concept, and the debate is still unsettled, but the message is important: if we damage the environment, we will regret our negligence.

THE BRUNDTLAND REPORT   A United Nations committee, commonly called the Brundtland Commission, was assigned to investigate growing concerns about the environment. In 1987, the Brundtland Commission issued their report titled *Our Common Future*. The Brundtland report defined the concept of sustainable development, and proposed that industrial development must not impair the ability of future generations to enjoy equal prosperity.

IPCC REPORTS   In 1988, the World Meteorological Organization (WMO) and the United Nations Environment Programme (UNEP) established the Intergovernmental Panel on Climate Change (IPCC). The IPCC is composed of scientists, and includes experts from all over the world. IPCC reports involve many researchers, from many countries, with the goal of being comprehensive, scientific, and balanced. IPCC does not conduct the research itself, but monitors world research on climate change, its causes, its consequences, and how to reduce these effects or adapt to them. IPCC reports are guiding documents for discussing global warming and climate change. The IPCC Fourth Assessment Report was released in 2007, and is discussed later in this chapter. The IPCC is a co-recipient of the 2007 Nobel Peace Prize.

MONTREAL PROTOCOL   The Montreal Protocol on Substances that Deplete the Ozone Layer was signed in 1987 to take effect in 1989.[7] The purpose of the Montreal Protocol is the eventual total elimination of chlorofluorocarbons. These compounds, commonly called *freons*, are effective refrigeration gases, but they interact with ozone. Although ozone is a pollutant at ground level, a layer of ozone in the stratosphere filters out harmful ultraviolet rays (as explained in Chapter 16). The ozone layer is essential to life on earth, and freons destroy it. Fortunately, the Montreal Protocol has been remarkably effective in restoring some of the ozone layer, and scientists predict that it will return to normal in a few decades. Developing countries have, on average, a 10- to 15-year grace period to match Canada's commitments under the Protocol. The Montreal Protocol is probably the most successful international environmental treaty, in terms of acceptance and compliance by all signatories. The struggle to implement the Kyoto Protocol (discussed below) is a stark contrast to the success of the Montreal Protocol.

THE EARTH SUMMIT IN RIO   In 1992, an "Earth Summit" was held in Rio de Janeiro, where 165 nations, including Canada and the United States, voluntarily agreed to reduce greenhouse gas (GHG) emissions, the main cause of global warming and climate change. This agreement, called the UN Framework Convention on Climate Change (UNFCCC), set a goal of reducing GHG emissions to 1990 levels by 2000. This goal was not achieved.

THE KYOTO PROTOCOL   In 1997, more than 160 countries met in Kyoto, Japan, to negotiate new GHG emission targets. More than 80 countries agreed to reduce their GHG emissions to an average level of 5.2 percent *below* 1990 levels by the year 2010. Each country was allotted a different target. Disagreements resulted over issues such as credits for carbon dioxide "sinks" such as forests (which absorb carbon dioxide), whether countries could pay credits instead of reducing emissions, and whether these rules were fair for developing nations. In March 2001, the United States announced that it would no longer participate in the Kyoto Protocol. In December 2002, Canada's Parliament voted to endorse the Kyoto Protocol. Canada's target was to reduce its emissions of greenhouse gases to 570 Mt (Megatonnes) of carbon dioxide by 2010. This amount is 6 percent lower than Canada's total greenhouse gas emissions in 1990. However, emissions have increased since 1990. In fact, Canada's predicted emissions for 2010 are about 810 Mt, which is about 42 percent higher than our 2010 goal. In 2007, Prime Minister Harper announced that Canada could not meet the Kyoto target, and would be negotiating a new international agreement with more realistic "aspirational" goals.[8]

BALI ROADMAP In 2007, the United Nations Framework Convention on Climate Change (UNFCCC) adopted the Bali "roadmap," which established a process for negotiating a post-Kyoto, international agreement on climate change to take effect in 2012.[9] Canada's position on future emissions targets in Bali was consistent with Canada's Clean Air Act, tabled in 2006, which sets *intensity-based* emissions targets. These targets are in sharp contrast with the total emissions targets in the Kyoto Protocol.[10] Intensity-based emissions targets require that the emissions per joule of energy consumed must decrease, but allow total emissions to increase. Canada also insisted that developing countries should not be exempt from emissions targets, as they were in the Kyoto Protocol. Negotiations are in progress to establish the terms that will apply in 2012.

CLIMATE CHANGE

Although evidence of global warming has been observed over the past three decades, full agreement on the cause of the warming has been elusive. Well-known scientists have postulated various causes for the observed evidence, and debate has been strong and often harsh on all sides. The Intergovernmental Panel on Climate Change (IPCC) tried to settle that debate when it issued its comprehensive Fourth Assessment Report (AR4) in 2007. The AR4 took six years to prepare, and its conclusions are based on research by over 800 authors, in 130 countries, reviewed by 2,500 scientific reviewers. The four-part report is long and technical, but summaries are freely available on the Internet.[11,12,13,14] The IPCC Report concludes with great certainty, based on extensive evidence, that climate change is directly linked to global warming, caused by greenhouse gases, resulting mainly from human use of energy, particularly fossil fuels. This four-step sequence is explained briefly below.

From Carbon Emission to Climate Change

- **Gas emission.** The combustion of coal, oil, natural gas, and other hydrocarbon fuels creates waste gas, mainly carbon dioxide, which is emitted into the atmosphere. Decaying foliage has always emitted carbon dioxide and methane, but prior to the Industrial Revolution, the process of photosynthesis (which absorbs carbon dioxide) was adequate for equilibrium. This equilibrium was disturbed by human activities, such as the burning of coal, oil, natural gas, or wood to heat homes, drive automobiles and engines, and generate electricity.

- **Greenhouse effect.** The atmosphere is transparent to most of the sun's radiation, which passes through to warm the earth's surface. The energy absorbed by the surface is eventually re-emitted, but this lower frequency thermal radiation is now partially blocked by the atmosphere. This creates a "greenhouse effect" in which the atmosphere acts as a thermal blanket. Certain "greenhouse gases" (GHGs) increase the heat absorbed. The most common is carbon dioxide (CO_2), which persists in the upper atmosphere for many decades. Other greenhouse gases are methane (CH_4), nitrogen dioxide (NO_2), ozone (O_3), and the chlorofluorocarbons (CFCs).[15] Some of these gases are more effective in absorbing radiation or persist longer than carbon dioxide.

- **Global warming.** The greenhouse effect is essential to human life, since it cushions the earth from the stark temperature extremes that exist on planets without an atmosphere. However, over the past two centuries, the concentration of greenhouse gases in the atmosphere has increased sharply and the earth's global mean temperature has increased. Research shows that the atmospheric carbon dioxide level was approximately constant at 280 ppm (parts per million) for two thousand years, until about the year 1800, when the level began to rise to the current level of approximately 380 ppm. Some of the effects of global warming, such as the melting of glaciers and the decrease of Arctic sea ice, are already observable, as shown in the 2005 documentary film *An Inconvenient Truth*. In 2007, a film titled *The Great Global Warming Swindle* was produced to refute the conclusions in the 2005 film. The debate over the existence and cause of global warming and the action needed to combat it are therefore still subject to vigorous debate.[16]

- **Climate change.** Human-caused global warming causes only a small increase in the earth's solar energy balance, but it has large nonlinear effects that can lead to potentially dramatic climate changes, such as severe storms, and changes in precipitation patterns, triggering droughts and severe flooding. Climate change will have serious negative effects on oceans, fisheries, agriculture, plants, and animals, as discussed below.

The Implications of Climate Change

A few conclusions from the 2007 IPCC Fourth Assessment Report (AR4) follow.

- **Human cause.** Global atmospheric concentrations of carbon dioxide, methane, and nitrous oxide have increased markedly as a result of human activities since 1750, and now far exceed pre-industrial values (determined from ice cores).

- **Temperature rise.** The global temperature over the next century is estimated to increase, in the range of 1.8°C to 4.0°C, depending on future fossil fuel use, technological change, economic development and population growth, but not including climate control initiatives, such as the Kyoto Protocol. The graph in Figure 17.1 shows the predicted range of temperatures for various scenarios of greenhouse gas emission. The graph shows that even if GHG concentrations could be kept constant (at year 2000 levels), a warming of about 1°C will occur in the next century as a result of GHG already emitted. The temperature rise may not appear to be great, but the effects are surprising.

- **Sea level rise.** If present trends (observed in the past decade) continue, sea level increases in the range of 18 cm to 59 cm (7 to 23 in.) are predicted. Sea level predictions are less certain, because of incomplete data on the melting of the Greenland ice sheet. (If the entire ice sheet were to melt, the sea level would rise 7 m (23 ft.), but that is not expected, even in the worst scenario, for a thousand years.)

- **Current observations.** Numerous long-term changes in climate have been observed. These include changes in Arctic temperatures and ice, widespread changes in precipitation amounts, ocean salinity, wind patterns, and aspects of extreme weather, including droughts, heavy precipitation, heat waves, and the intensity of tropical cyclones. Eleven of the last 12 years (1995–2006) rank among the 12 warmest years in the instrumental record of global surface temperature (since 1850).

- **Other consequences.** The IPCC lists many possible consequences of climate change, including, but not limited to, greater intensity and severity of inland floods, drought, glacier and sea ice melts, and possibly even major changes in ocean currents, such as the Gulf Stream, which moderates the temperature of Europe. The increased temperature has other unexpected effects, such as insect plagues, extinction of species, and bleaching of coral reefs. For example, the freezing winter temperatures in British Columbia and Alberta used to kill pine borer, but in recent years, the temperature rise permits them to destroy tens of millions of hectares of valuable trees.[17] Similarly, a small temperature rise in the south Pacific Ocean in 2004 bleached and killed part of the Great Barrier Reef, and by 2050 even the most protected coral reefs will suffer massive damage.[18] Other examples abound.

- **Feedback loops.** An important factor in the science of global warming is the role played by feedback loops. This is a familiar concept to engineers and geoscientists, but it is not well known to the average citizen. Positive feedback loops can cause rapid and nonlinear magnification of small changes. For example, forest fires release carbon dioxide and other GHGs into the atmosphere, leading to global warming and even higher temperatures and drier conditions, in which more forest fires occur. The melting of Arctic sea ice has an even more dramatic feedback effect: ice is very effective in reflecting sunlight back into space, but as the ice melts, the darker water absorbs more energy from the sun, thereby causing more ice to melt. Many such feedback loops exist. A book by George Monbiot warns that if the warming reaches 2°C compared to pre-industrial levels,

FIGURE 17.1 — IPCC Predictions for Temperature Change in the Next Century

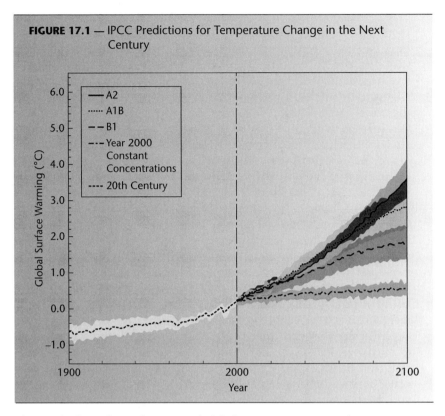

The graph above shows the expected global average temperature increase (relative to year 2000), due to global warming, until the year 2100. The line prior to 2000 is historical data, and the lines from 2000 to 2100 represent three scenarios (A2, A1B, and B1), shown as continuations of the 20th-century data. Scenario A2 represents slower economic growth, but continuously rising population, using fossil fuel energy. Scenario A1B represents very rapid economic growth and a population peak at mid-century, but rapid economic restructuring to use more diverse energy sources. Scenario B1 is similar to A1B, but includes the introduction of clean and resource-efficient technologies. The scenarios do not include climate initiatives to reduce emissions, such as the Kyoto Protocol. The lowest line is the scenario where concentrations are held constant at year 2000 values. Shading denotes the ±1 standard deviation range of annual averages.

Source: Intergovernmental Panel on Climate Change (IPCC), "Summary for Policymakers," Figure SPM-5, *Climate Change 2007: The Physical Science Basis,* February 2007, p. 14, available at <www.ipcc.ch> (June 6, 2008). Figure reproduced with permission.

positive feedback loops will be activated. For example, vast peat bogs in the subarctic, presently under permafrost, will begin to decay, leading to irreversible warming.[19]

The emission of greenhouse gases is changing our climate and damaging our environment, and unless serious efforts are made to reduce these emissions, climate change will harm future generations. Action is needed immediately.

Fighting Climate Change

To combat climate change, the IPCC recommends that we reduce our use of fossil fuels, but what goal we should aim for is debatable. Future goals will depend on international agreements that will be negotiated to take effect after 2012. However, we are presently about 42 percent above our agreed Kyoto Protocol goal for 2010, so reducing our GHG emissions by 42 percent would seem to be a good place to start. In the foreword to the Canadian edition of his book *Heat: How to Stop the Planet from Burning*, George Monbiot criticizes Canada's failure to reach its Kyoto goal, and our lack of "binding and immediate targets." Conversely, in a recent (February 2008) survey of members' opinions about climate change, an APEGGA report showed that, while almost all (99.4 percent of 1,077) respondents agreed that the climate is changing, the majority (68 percent) disagreed with the statement that "the debate on the scientific causes of recent climate change is settled."[20]

However, in the absence of agreement on the causes and action needed, engineers and geoscientists should follow a "precautionary principle." That is, we should take reasonable precautionary measures when harmful results may threaten human health or our environment, even if the precise "cause and effect" relationship is not yet clearly established. Therefore, in spite of the doubts and objections of a large part of society, we should make reasonable efforts to follow the IPCC recommendations to reduce GHG emissions.

But even if we apply the best *mitigation* (or reduction) methods, some *adaptation* is also essential, according to the IPCC Report. Adaptation means that we must learn to live with warmer temperatures, modify our building codes, and design our infrastructure (roads, drains, bridges, highways, and so forth) more robustly to withstand the more intense storms, hurricanes, floods, and droughts to come in future.

The IPCC AR4 report suggests emission reduction and energy conservation initiatives in seven key areas: energy supply, transport, residential and commercial buildings, industry, agriculture, forestry, and waste management. The following summaries give a brief insight into this comprehensive 94-page report, but you may wish to refer to the report itself to read the projections and recommendations for your area of practice.[21]

- **Energy supply.** Without effective policy changes, global CO_2 emissions from fossil fuel combustion are predicted to rise at least 40 percent from the year 2000 to 2030. A wide range of cost-effective energy saving is possible, including switching to low-carbon fuels, improving power-plant efficiency, as well as building nuclear power and renewable energy systems. Carbon capture and sequestration (CCS) is cost-effective at higher carbon prices. CCS uses existing technology to separate the carbon dioxide from exhaust gas and inject it into underground storage wells. Already, about 40 power generation plants are in operation using carbon capture and sequestration (about 10 in Australia). Other options still under development include advanced nuclear power, advanced renewable energy sources (hydro power, solar, wind, geothermal, and bio-energy) and, in the longer term, the possible use of hydrogen as an energy carrier.

- **Transport.** Global transportation activity is expected to grow robustly (unless there is a major shift away from current patterns) by about 2 percent per year, with energy use and carbon emissions about 80 percent above year 2002 levels by 2030. Significant improvements in efficiency are possible for light-duty vehicles and airplanes. Substituting bio-fuels for conventional fossil fuels can reduce emissions, but raises the ethical question of diverting food crops to transport. Clearly, we need more fuel-efficient vehicles; hybrid vehicles; cleaner diesel vehicles; shifts from road transport to railway and public transport systems; increased non-motorized transport (cycling, walking); better land-use and transport planning.

Photo 17.1 — Alberta Wind Farm. The strong winds in the Pincher Creek area of southwestern Alberta are harnessed by a growing number of wind farms. The wind turbines are connected to the electrical grid, significantly reducing the need to generate electricity using fossil fuels.

Source: Getty Images/Henry Georgi.

- **Buildings.** Energy consumption and GHG emission in buildings vary significantly, depending on the country and region. The rapid economic growth scenario expects all of the increase in CO_2 emissions (in buildings) to be in the developing world (Asia, Middle East and North Africa, Latin America, and sub-Saharan Africa). Between 2004 and 2030, the CO_2 emission growth is estimated at 1.5 percent to 2.4 percent annually, and this range is taken as the baseline. There is a global potential to reduce this projected baseline emission growth by about 30 percent, cost-effectively. In new buildings, it is possible to achieve 75 percent or more in energy savings compared with recent current practice, generally at little or no extra cost, with cooperation by architects, engineers, contractors, and clients, using passive methods to reduce energy needs. Almost all studies show that improved insulation in the colder climates, and greater efficiency in space cooling and ventilation in the warmer climates, come first in reducing emissions. Other measures with high savings potential are solar water heating, efficient appliances, and energy-management systems. Efficient cooking stoves rank second after efficient lighting (shift to fluorescent bulbs) in developing countries. In developed countries, appliance-related measures are most cost-effective, with upgrades of cooling-related equipment ranking high in warmer climates. Air-conditioning savings can be more expensive than other efficiency measures, but can still be cost-effective if they displace more expensive peak power. In general, there is a need for more efficient lighting and better use of daylight; more efficient electrical appliances and heating and cooling devices; improved cook stoves, improved insulation; better passive and active solar design for heating and cooling; alternative refrigeration fluids, recovery and recycling of fluorinated gases.
- **Industry.** Industrial CO_2 emissions (including electricity use) increased by 65 percent from 1971 to 2004. The projections for industrial CO_2 emissions for 2030 are a further increase of 40 percent over 2004, assuming no further action is taken to control these emissions. Energy-intensive industries (such as iron and steel, non-ferrous metals, chemicals and fertilizer, petroleum refining, cement, and pulp and paper) account for about 85 percent of this energy consumption. Much of this energy-intensive industry is now located in developing countries. Many older, inefficient facilities need investment to improve energy efficiency and reduce emissions. The IPCC report recommends reducing energy consumption and CO_2 (and other GHG) emissions through energy efficiency, fuel switching, power recovery, renewables, feedstock change, product change, and material efficiency.
- **Agriculture.** In 2005, agriculture accounted for an estimated 10 percent to 12 percent of total global anthropogenic emissions of GHGs, mainly methane (CH_4) and nitrous oxide (N_2O). Despite large annual exchanges of CO_2 between the atmosphere and agricultural lands, the net CO_2 is approximately balanced. However, assuming continued practices, annual agricultural N_2O emissions are projected to increase by about 35 to 60 percent and

CH_4 emissions by 60 percent, by 2030. Improved soil management can reduce net GHG emissions. About 90 percent of the total mitigation arises by sink enhancement or carbon sequestration (that is, increasing the content of organic material in the soil) and about 10 percent from emission reduction.

- **Forestry and forests.** Globally, forests cover 3,952 million hectares (9,765 million acres; about 30 percent of the world's land area). However, the forests are decreasing at an approximate rate of 13 million hectares per year, as forests are converted into agricultural land, cities, or highways; harvested by logging; or burned by forest fires. Forests store carbon but the process involves long periods of slow growth (carbon intake), interrupted by short periods of large, rapid releases of carbon during clearing, harvest, or fires. The largest short-term gains are always achieved by avoiding emissions (that is, by reducing deforestation or degradation, and increasing fire protection). In the long term, a sustainable forest-management strategy aimed at maintaining or increasing forest carbon stocks, while producing an annual yield of timber, fibre, or energy from the forest, will generate the largest sustained mitigation benefit.

- **Waste management.** Consumer waste contributes less than 5 percent of global GHG emissions, with landfill methane (CH_4) accounting for more than half of these emissions. Wastewater and sewage are secondary emission sources of methane and nitrous oxide (N_2O); in addition, incineration of waste containing fossil carbon causes minor emissions of CO_2. These emissions are small but significant, because they can easily be reduced or utilized using existing technology. In fact, waste can be a commercially valuable source of energy, through incineration, industrial co-combustion, landfill gas utilization, or anaerobic digester biogas. Waste has an economic advantage in comparison to many biomass resources, because it is regularly collected at public expense. Actions required include methane recovery from landfills; waste incineration with energy recovery; composting of organic waste; controlled wastewater treatment; recycling and waste minimization.

Climate change is a serious challenge, but several authors have suggested many other ways to fight it, adapt to it, or even profit from it.[22,23]

OIL AND GAS DEPLETION

The depletion of oil and natural gas is also a threat to sustainability, and is perhaps even more serious than climate change. Oil resources are approaching a "tipping point." Beyond the tipping point, they will become scarce and expensive, creating supply crises and economic turmoil. Oil and natural gas are essential for our lifestyle: they are needed to transport, refine, manufacture, and distribute the commodities that we need, and to generate the electrical power that heats and lights our homes, cooks our food, and even entertains us. Other resources are also affected, since they require oil to develop them.

Definition of "Peak Oil"

The term "peak oil" refers to the date when the global production rate of oil reaches its maximum; after that date, the production rate will remain constant for a while, but will then begin to decrease. This conclusion is based on the observation that oil and gas resources are large but finite, and some sources are more easily exploited than others. Obviously, cheaper oil is consumed first. In fact, in the past century, much of the easily accessible oil has already been consumed, and we must now exploit the less accessible oil. The global rate of production is now at or near its peak (and in some areas, the peak has passed).

Figure 17.2 shows the total global production rate since 1930, with predictions until 2050, based on the theory developed by M.K. Hubbert. Hubbert applied his theory in 1956 to predict, correctly, that United States oil production would peak between 1965 and 1970. As the figure shows, the United States (lower 48 states), the Middle East, Russia, and Europe have all reached their peak oil production rates. The total global oil and gas peak is predicted to occur

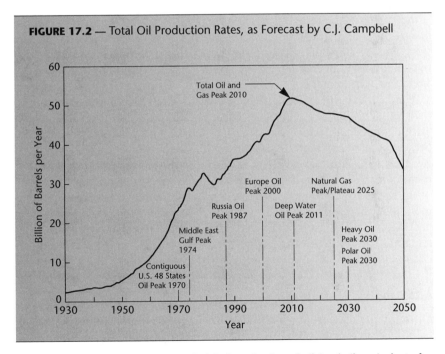

FIGURE 17.2 — Total Oil Production Rates, as Forecast by C.J. Campbell

This simplified figure shows the total global production of oil (and oil equivalent of other hydrocarbons) from 1930 to the present, with forecasts to 2050. The peak is predicted to occur in 2010, as well as the observed or predicted dates for productivity peaks in specific oil deposits.

Source: Figure adapted and re-drawn by Prof. Roydon Fraser, University of Waterloo, ON, from data provided by C.J. Campbell, Newsletter '89 for May 2008, Association for the Study of Peak Oil and Gas (www.aspo-ireland.org) (see Note 24). Figure reproduced with permission of Prof. Fraser and C.J. Campbell.

in 2010. Of course, this date is not certain, because developments in technology will make some wells more productive or more accessible, and will delay the date of the oil peak. However, once the peak has passed, the availability of easily accessible oil will drop and oil prices will rise, perhaps sharply.[24]

The Implications of Peak Oil

As Figure 17.2 illustrates, the global oil production rate is predicted to peak, then decrease in the near future. This drop will occur because the less accessible oil (such as Alberta's oil sands) cannot be produced as easily or as quickly as oil from the Persian Gulf. The less accessible oil fields will not be profitable unless oil prices rise. However, world demand for oil will not decrease. In fact, as China and India adopt higher living standards, the demand for oil will skyrocket. The gap between oil demand and oil production will widen, causing scarcity and even higher prices. We saw this problem before, in the oil crisis of 1973, when the Organization of Petroleum Exporting Countries (OPEC) reduced oil deliveries, and oil prices rose by a factor of 4 in about 6 months. Simultaneously, gasoline prices soared, cars lined up at gas stations, people hoarded gasoline, gas stations ran out of gasoline, and tempers flared. The stock markets lost billions of dollars, and in oil-importing Western nations, the combination of stagnant growth and inflation persisted for years, damaging their economies.[25]

The oil peak will affect us all. Our industries, electrical supply, transportation, and our lifestyle itself will be adversely affected by oil shortages or high prices. Natural gas, which has a separate but similar peak, is also included, because as oil becomes scarce, industries will switch, where possible, to natural gas, thus accelerating its peak. Oil and natural gas are essential for heating our homes and as the feedstock for most plastics, fertilizers, solvents, and adhesives.

In summary, the world may not run out of oil for another century or more, but the easily produced oil is almost gone, the production rate is peaking, and we will soon pay a very high price for the less accessible oil that remains. The price of oil doubled in 2007–08, reaching a record high of US$147 per barrel in July 2008, with even higher prices predicted.

Adapting to Oil Depletion

Although the date of global oil peak production is occasionally challenged, mainly because new techniques make wells more productive, most analysts believe that it will occur in the next few years, by 2010 or 2015.[26] This date is critical, because even a crash program to reduce our dependence on conventional (easily accessible) oil and move to synthetic oil, nuclear, bioenergy, or renewable energy, will take at least 20 years, according to a recent U.S. government report, commonly called the "Hirsch Report."[27] That is, if society, industry, and government cooperate fully to reduce energy consumption and to develop alternatives, we still need 20 years to avoid energy shortages and financial crises. Obviously, action is needed immediately!

To combat the peak oil crisis, we must reduce our fossil fuel consumption. Coincidentally, this is also the action recommended by the IPCC to mitigate the effects of climate change. Therefore, the IPCC recommendations for improving efficiency and reducing emissions (summarized earlier) also apply to the peak oil crisis. To achieve sustainability, we must be committed to energy efficiency.

A draft protocol (or international agreement) has been proposed to reduce consumption of fossil fuels, on a global scale, in a fair and orderly way.[28] However, to date, no major nation has seriously considered adopting it. Another bold proposal was made by M. Jaccard in his book *Sustainable Fossil Fuels: The Unusual Suspect in the Quest for Clean and Enduring Energy.* Jaccard proposes a strategy for sustainable energy in which fossil fuels continue to play a key role for many years, to avoid economic collapse.[29] By using carbon capture and sequestering (CCS) technology, we can continue to use fossil fuels without emitting GHGs. Fossil fuels are the cheapest source of clean energy for at least the next century. This book is a Canadian opinion on reaching a sustainable future, and won the prestigious 2006 Donner Prize for the top Canadian book in public policy. Jaccard's timely proposals deserve to be considered seriously.

THE ETHICS OF CLIMATE CHANGE AND PEAK OIL

The Brundtland Commission concluded that environmental degradation leads to poverty and economic disparities. The 2007 IPCC report shows that climate change makes the problem worse, because the economic loss caused by climate change will fall disproportionately on the poorest people. That is, the developing countries will carry most of the burden of climate change through droughts and floods. Although the peak oil crisis will affect rich countries because they are more dependent on oil and gas for their lifestyle, poorer members of those societies will suffer most, as factories close, prices rise, and the unemployed, the very young, and the old are unable to meet the rising costs.

This prospect raises some very basic questions of fairness for all citizens, and engineers and geoscientists in particular. Is it ethical or fair for richer countries to use fossil fuels indiscriminately, thus creating GHG emissions, and indirectly

- creating millions of "climate change" refugees when droughts reduce crops in Africa, or when sea levels rise and flood low-lying countries, such as Holland, Bangladesh, and the Pacific islands?
- creating an energy crisis for future generations, who will have to scrabble for hard-to-find, high-priced oil and gas after the easily accessible oil is depleted?

These ethical problems are very similar to the "tragedy of the commons," discussed in the previous chapter. However, when he first observed the syndrome in 1832, William Forster Lloyd could hardly have imagined the immensity of the problems we now face. Let us examine climate change and peak oil, using the four basic ethical theories discussed in Chapter 11:

- **Mill's utilitarianism.** The excessive use of fossil fuels, which lies at the root of both of these problems, is the philosophy of utilitarianism in reverse. Utilitarianism seeks the solution that produces the maximum benefit for the greatest number of people, with the benefit most widely distributed among the people affected. Therefore, utilitarianism would require a sacrifice by the general population, in the form of reduced use of fossil fuels, to benefit the quality of life for society in general. Reducing fossil fuel usage is not a small sacrifice, but it pales in comparison with the harmful effects, deprivation, and economic upheaval predicted by many sources in the next few decades.

- **Kant's formalism or duty ethics.** Kant observed that we should act only in ways that would be acceptable as universal principles for everyone to follow. In the case of climate change and peak oil, the wasteful or excessive use of fossil fuels is not a principle that everyone can follow. In fact, wasteful actions are the problem. The universal principle must be to reduce the consumption of fossil fuels and to encourage the development of alternative energy sources. Some people may challenge this ethical argument on the basis that we have no duty to future generations. However, civilized humans regard the lives of their children and grandchildren as highly as their own. Kant stated that respect for humanity was at the base of his formalism: life should always be treated as an end or goal, and never as a means to achieve some other purpose.

- **Locke's rights ethics.** This theory states that every individual has rights, simply because he or she exists. The right to life and the right to the maximum possible individual liberty and human dignity are fundamental; other people have a duty not to infringe on those rights. However, we recognize that rights are limited and can be taken away if justice is served by doing so. We already have laws against water and air pollution, as explained in Chapter 15, because people have no right to waste or destroy a resource that is in the public domain. If we reduce our dependence on oil through reasonable changes in behaviour, we can avoid serious infringement of personal rights.

- **Aristotle's virtue ethics.** Aristotle believed that humans would achieve true happiness by developing virtues (or qualities of character) using thought, reason, deduction, and logic. He visualized each virtue as a compromise between two extremes. In the case of climate change and peak oil, the extremes are the wasteful use of fossil fuels that our society has permitted (perhaps even encouraged) for the past century, and the future environmental degradation and scarcity of fossil fuels, which will lead to shortages, hardship, civil unrest, and possibly wars. In this case, the "golden mean" between these extremes is the recognition that wasteful practices must end, and that we must develop new energy sources and distribution systems.

In summary, sustainability is simple fairness. It is unfair and unethical to harm others, through inefficiency, negligence, greed, or abuse. In this case, the "others" are future generations, particularly those living in the poorer countries.

THE CRITICAL NEED FOR SOCIETAL CHANGE

Climate change and peak oil are not just technical problems; they are societal problems. Wasteful practices, accepted by society for centuries, need to change. Creating a sustainable lifestyle will require effort and sacrifice, and we must ensure that the sacrifice is spread evenly. Engineers and geoscientists are part of the solution, but the responsibility should not fall on them alone. Achieving sustainability is a duty for all of society.

REDUCING WASTEFUL CONSUMPTION   Examples of inefficient use abound, so the obvious first step is to reduce consumption and waste by conserving, re-using and recycling materials and consumer goods. Wasteful consumption must be replaced by the basic guideline: *Reduce; Re-use; Recycle*. Blue Box recycling systems are in place in most parts of Canada, and must be expanded and used more diligently. However, we must go beyond the Blue Box, and add another mantra: *Rethink and Reorganize*. We must examine our lifestyle, eliminate non-essential energy-consuming activities, and reorganize to be more independent of fossil fuels.

THE LIMITS TO EFFICIENCY   Efforts to reduce fossil fuel consumption are often stymied by society. In a 2007 study, a CIBC economist described an "efficiency paradox": For many decades, the United States has passed laws for more efficient automobiles, furnaces, and appliances, and great improvements were achieved. However, energy usage has still increased, because consumers are simply using energy more aggressively. The contradiction is most noticeable in fuel use for private vehicles. Although auto manufacturers have increased engine efficiency by about 30 percent in the past 25 years, American drivers are simply buying larger, more wasteful vehicles. For example, the sale of light trucks increased by 45 percent in the last decade (greatly in excess of passenger car sales), even though light trucks are much less fuel-efficient than cars. In addition, the distance driven by each vehicle has increased by about 25 percent since 1970. A similar trend was reported in home heating and cooling. Energy efficiency of air conditioners increased by 17 percent since 1990, but air conditioner use increased by 36 percent. Furnaces are now far more fuel-efficient, but house size has doubled in the past 60 years, requiring more heat.[30] This trend indicates that, to reduce energy consumption, we must have limits to growth, not intensity levels based on efficiency.

GOVERNMENT ACTION   Unfortunately, very few people will eliminate wasteful or unhealthy habits voluntarily. People need encouragement, as tobacco smoking bans have proved. Society must pressure governments to limit or reduce GHG emissions, and governments (at municipal, provincial, and federal levels) must pass laws, regulations, and standards, or impose taxes or charges to limit fossil fuel use and GHG emissions. Governments must stimulate research and development to find new efficiencies, and perhaps even new energy sources. A discussion of these methods and a comparison of their effectiveness are in the IPCC report.[31]

In summary, if we are to maintain our standard of living without depriving future generations of vital resources, we must begin to reduce fossil fuel consumption now. Ethics and justice require the burdens of climate change and peak oil to be shared equally. Engineers and geoscientists must inform the public of the consequences of inaction. Although the remedies may be difficult to apply, ethical reasoning insists that customs and habits must change; legislation must be passed; regulations, permits, subsidies, and incentives may be essential.

SUSTAINABILITY FOR PROFESSIONAL ENGINEERS AND GEOSCIENTISTS

Several Associations publish environmental guidelines, as discussed in Chapter 15. However, following guidelines is not enough. A British engineer, long involved in the oil and gas industry, expressed the following comment on sustainability:

> It is important to distinguish sustainability from environmental compliance. Compliance with environmental regulations is an essential, daily, operational requirement, but sustainability is a long-range strategy to combat a slow-moving, planet-wide terminal condition. Sustainability requires innovative thinking, persistence and possibly some sacrifice. Decades may pass before the social acceptance, legislation and treaties are devised to deal effectively with sustainability. Engineers must, of course, ensure that regulations are followed, but in planning projects and activities, engineers must aim their conceptual thinking at the higher hurdle of sustainability.[32]

To reach sustainability, engineers and geoscientists must, of course, strive to increase efficiencies in the areas identified in the IPCC AR4 report mentioned earlier. However, increased efficiency is not enough. We must *Rethink and Reorganize* our use of energy to

- explore new and innovative ways to reduce our energy consumption, drastically,
- explore new methods of clean consumption of fossil fuels, such as CCS,
- develop renewable energy sources such as solar, wind, and geothermal energy,
- exploit nuclear energy (which is also non-renewable, and will eventually peak), and
- develop entirely new energy sources and/or distribution systems, such as the hydrogen economy, or clathrate technology (mining the methane embedded in ice, found in abundance at great depths around the world, according to T. Flannery).[33] Innovation and new ideas are essential!

Moreover, we, as professionals, must be able to say "no" when environmental degradation is proposed. As responsible engineers and geoscientists, we have a duty to assess projects and activities for their sustainability, and decline to participate in projects or activities that can be shown to be clearly unsustainable. However, engineers and geoscientists must not be required to be martyrs in the battle against climate change and peak oil; we need objective methods for evaluating sustainability and legislation to protect whistle-blowers. For example, life-cycle analysis (LCA), which is still in development, attempts to measure the environmental impact of a project at the conceptual stage. Life-cycle analysis requires an inventory of all the energy and material inputs and outputs from a project, and determines the net result: the energy and resources consumed, and the products and wastes produced. LCA is usually carried out for specific emissions, such as CO_2, SOx, NOx, phosphates, or other chemicals, and the amount of water and energy consumption. Specific toxic emissions, such as arsenic and lead, are also sometimes considered, and efforts have been made to include mass and economic considerations. Recent efforts to create an international LCA standard by the International Standards Organization (ISO) have been hampered by serious restrictions on data and theoretical limitations about defining the input and/or output boundaries for an LCA.[34] Although its value is not yet fully realized, engineers and geoscientists are urged to adopt LCA or to develop better objective tools for evaluating sustainability of major projects.

In summary, although climate change and peak oil are different problems, they are linked; they will have a seriously negative effect on our lives; and their main cause is the same: the excessive consumption of fossil fuels. The solution is therefore the same: to maintain our standard of living, we must reduce our consumption of fossil fuels drastically, and move to other energy sources. The wasteful use of energy is not just unethical; it is a hazard to our way of life. Society must implement regulations and fund research so that we can avoid the worst effects of climate change, and so that we can develop alternative energy sources that are sustainable.

Finally, we professionals must speak out in the political and ethical debate over climate change to help change society's attitudes, laws, and way of life.

CASE HISTORY 17.1

THE LADYFERN NATURAL GAS DEPLETION

In 2000, a huge natural gas discovery in the Ladyfern field in northeastern British Columbia was announced. The find was touted as British Columbia's solution to the continent's gas shortage, and promised to provide about 25 percent of the province's gas production and to make many people rich. However, Ladyfern is almost depleted. Many of the wells that promised such riches are now pumping saltwater. Ladyfern is a key example of wasteful development—exactly the opposite of sustainable development. In his comprehensive summary of the Ladyfern debacle, Andrew Nikiforuk states:

> What many analysts initially talked up as an "elephant play" of one trillion cubic feet has been downsized to a 400-billion-cubic-foot warthog—or less. And the find that was supposed to ease the supply/demand crunch for gas, well, has not. In fact, the rapid production and depletion of Ladyfern played a central role in setting up this [2002/2003] winter's gas price shock . . . .
>
> That's just the beginning of Ladyfern's many disappointments. . . . In a clear-cut case of deadly competition and "value destruction," [several] firms spent more than half a billion dollars in less than a year to build three times the infrastructure needed to drain one very finite pool of gas. It was, as one local resident put it, "like watching 20 spoiled kids all going at the same milk shake."
>
> But Ladyfern also highlighted the BC government's greed for fast money—as well as the dubious competence of the province's energy regulator. "The people who truly got burned were the owners of the resource: the people of British Columbia," says Ian Doig, editor of *Doig's Digest*, a monthly Calgary-based oil and gas newsletter. "Ladyfern was supposed to be a monster play, but with regulatory neglect it became a national embarrassment."[35]

The Ladyfern story is a long, twisted tale in which greed and espionage turned a natural gas treasure into lawsuits, waste, and loss. Although the companies involved showed a profit, it was a fraction of what they could have made, and the real losers were the public and our next generation. A few highlights from Nikiforuk's narrative are summarized below.

Initial Drilling and Discovery

The initial seismic data was obtained by Shell Canada, and later "farmed out" to three firms who became drilling partners: Murphy, Beau Canada Exploration, and Apache. Two exploratory wells were drilled; the first well was dry, but the second Ladyfern well found natural gas at a depth of 3,000 m (9,840 ft.) in early 2000. Tests showed that the well was "one of the top-10 onshore natural gas discoveries in Canadian history." The timing of the find was also significant, as explained by Nikiforuk:

> Although industry had been drilling at record rates throughout the Western Canada Sedimentary Basin, it hadn't been finding much gas. In fact, no one had found a really big pool since 1986. That geological fact, combined with rising US appetite for natural gas, was about to push prices from a decade-long low of $1.50 per thousand cubic feet to record highs of $12 per thousand cubic feet.

Although gas discoveries are required to be kept confidential, two of the partners announced the find precipitously in February 2000. Nikiforuk takes up the tale at this point:

> Predator Corp., a three-man Calgary firm, read the reports with keen interest. The aptly named company specializes in buying and drilling on land next to proved oil and gas fields in Alberta and BC. It smelled a big find at Ladyfern, and quickly

learned that the BC government was about to sell off two drill sites within 600 metres of the discovery well. Utilizing something called "Predator methodology" to assess Ladyfern's wealth potential, the company prepared to outbid Murphy on an 842-hectare parcel of land adjacent to the discovery well . . . .

. . . Predator successfully outbid Murphy by laying out a hefty $8.65 million in a sealed bid process. That amounted to $10,268.20 per hectare—more than any oil and gas company has ever paid for land in Western Canada. At the time, an average hectare went for $323 in BC. (Just to keep things in perspective, Murphy purchased its original discovery lease for $808 a hectare.)

Land Rush, Frantic Drilling, and Industrial Espionage

News of the auction prices immediately started a wild land rush. To prevent Predator and other competitors—such as Canadian Natural Resources (CNR) and the Alberta Energy Co. (AEC)—from closing in on its gas find, Murphy then paid more than $15.7 million for another six and a half sections of land, or $7,515 per hectare. Meanwhile, Predator made secret presentations about Ladyfern's potential to 16 different Calgary companies in an attempt to find a drilling partner flush with cash. Ricks Nova Scotia, a Calgary firm owned by Ricks Exploration in Oklahoma City, bought into the gamble. Along with Predator, it also purchased another two sections of land for $8.56 million, or $15,240 a hectare . . . .

. . . Ladyfern's multiplying suitors had a 100-day winter window to find what Murphy had already struck. Muskeg generally isn't drilled in the summer because bogs and swamps have a habit of swallowing equipment, if not cash. As such, AEC, the biggest gas player in BC, prepared to invade the forest south of Murphy's big find in early winter, while CNR readied its drilling crews to assault land just north of Murphy in early spring. Predator's partner, Ricks Nova Scotia, was so eager to get at Ladyfern that it didn't want to wait for winter freeze-up. Instead of trucking across frozen muskeg, it proposed dropping in its drilling rigs by helicopter.

As natural gas prices rose to $6 per thousand cubic feet in the fall of 2000, the industry unleashed its scouts. Scouts act as intelligence spies, and most come equipped with high-tech communications gear, powerful scopes and listening devices. Their job is to coolly collect information on competitors—and nearly every company in the patch employs at least one or two. For the next year, scouts eavesdropped on cell phone conversations, talked up seismic crews and spied on drilling rigs. "It was like the Gulf War out there," recalls one helicopter pilot. "They were running around in camouflage fatigues through the cut lines doing cloak-and-dagger stuff." . . .

Lawsuits, Waste, and Duplication

Murphy was now in a tight position. It had lots of gas but no pipeline. In order to construct a line from BC to a nearby Alberta gas plant, it needed the blessings of the National Energy Board (NEB), the federal government regulator. But

Ricks Nova Scotia had already secured a hearing for its own pipeline plans without having drilled a single well. Over a two-day hearing held in Calgary in December 2000, a Ricks lawyer told the NEB that his client needed a $3-million conduit because no one wanted to share existing lines. He also held up copies of the Daily Oil Bulletin and Apache's press releases as proof that there was gas in the area. Ricks executives noted that Murphy and its partners were already sucking the great Ladyfern milk shake dry—and that time was money.

Murphy and Apache, of course, objected to Ricks's application and questioned the firm's motives. Murphy's lawyer, Kemm Yates, even charged that Ricks's application made a mockery of the national regulatory process for oil and gas. "You might as well turn out the lights, lock the doors and go home," if the Ricks pipeline was approved, Yates told the NEB. "You would be acceding to complete deregulation and to a position that says anybody who is prepared to spend a lot of money can do what they want."

Yates then explained that regulation arose in the oil patch specifically to prevent gold rushes—a pattern of big discoveries followed by frantic development, flush production, great waste and then sharp decline. In reply, a Ricks lawyer noted that Murphy and Apache were rough-and-ready competitors and had just sued Ricks and Predator for allegedly using stolen data. "Here we have the Apache wolf donning obviously ill-fitting sheep's clothing," he argued. It wasn't your normal sleepy NEB hearing.

Shortly afterward, the NEB approved Ricks's pipeline, but the company never built it. Just weeks after the decision, Ricks settled with Murphy out of court (the amount is unknown) and surrendered its 75% interest in the Ladyfern play to Murphy and Apache. Much to Predator's dismay, its 25% interest then entered a trust until the lawsuit is settled. [See Note 36. This lawsuit was settled in favour of Murphy, in 2006.]

Murphy's next hurdle came at its own pipeline hearing before the NEB. Although Predator was now pretty much out of the picture, AEC's $65-million drilling campaign had already hit pay dirt: a monster well that yielded 60 million cubic feet just five miles south of Murphy's discovery. Without quick approval, Murphy would miss the winter construction deadline. And that meant that AEC, which had smartly bought Ricks's pipeline approval, would be in a position to gulp most of the Ladyfern milk shake.

Although Murphy expected a routine and quick hearing before the NEB, it got some surprises. A local trapper whose business had been ruined by gas wells and roads hired Mike Sawyer, a combative Calgary-based environmentalist, to raise some very legitimate concerns about how the industry had busted up the surrounding forest. Sawyer's questions drew out the Calgary hearing for days. (BC's Ministry of Energy and Mines supported Murphy's bid, arguing that a year's delay might cost the government between $15 million and $30 million in royalties.) [Murphy president Harvey Doerr] still calls Sawyer's performance a nerve-wracking "filibuster." Just two weeks before a caribou migration threatened to shut down all construction along the Alberta border, the NEB gave Murphy's

17-kilometre pipeline the green light. "We built it in 11 days and then drilled all the wells we needed," says Doerr.

By now, Ladyfern looked like a boreal war zone. Helicopters bearing construction equipment and work crews crowded the sky. Some of the camps held as many as 400 men and women. Drilling rig crews working for Murphy could eyeball AEC's contractors and vice versa. "It was in-your-face competition," recalls one service contractor. And all of it was unfolding in a patch of the forest many dubbed "the middle of nowhere."

In the race to drain Ladyfern, industry spared no expense. Murphy and AEC rented helicopters from as far afield as Vancouver Island and Grand Prairie. Chinooks and Sikorskys ferried tractors, trucks, men and pipes at a cost of $12,000 an hour. Some helicopters hauled wood for roads, while others hauled water produced from gas wells—something most oil patchers had never seen before, or since. Trucks normally do that job.

Because CNR came a bit late to the Ladyfern party, it continued drilling throughout the summer of 2001—an almost prohibitively expensive development in sink-or-swim northern muskeg. In the process, it spent in excess of $120 million on helicopters, roads, pipelines and other infrastructure. But the wild spending spree didn't really strengthen the foundation of the service industry, says Brian Churchill, a former longtime city councillor in Fort St. John. "It was the antithesis of stable development beneficial to a community," he argues. "It was a gold-rush mentality."

To support all the wells being drilled, industry needed an all-season road. Doerr built one at an estimated cost of $25 million, thinking there would be only one. But BC's Oil and Gas Commission (OGC), which most oil patchers describe as a "rubber-stamp agency," then turned around and gave CNR and AEC approvals to build another $16 million in roads. "I doubt if anyone at the OGC knew that all three roads went to the same place," notes one local observer caustically. Confirms Bob Fedderly, president of the Northern Society of Oilfield Contractors & Service Firms: "We now have three times the environmental footprint we needed out there."

The same deadly competition infected all aspects of Ladyfern's development. "We didn't need a well on every section of land to drain this pool," notes Doerr. "But if you didn't drill them, you didn't get credit proportional to your share of the field. So the field was vastly over-drilled."

Instead of then determining the life and potential of Ladyfern reservoir to allow for rational exploitation, the OGC let the companies work out their own production schedules. Having spent oodles of money, they simply wanted to earn their investments back quickly. To do so, they jointly set a withdrawal rate of 785 million cubic feet per day. That figure astounded many oil patch veterans because it meant Ladyfern would be exhausted quickly.

As a result, Ladyfern accounted for about half the increase in Alberta's gas production in the summer of 2001. In other words, about 20 of Ladyfern's wells poured as much gas onto the market as more than 10,000 normal wells. This flood of gas, combined with a slowdown in the US economy, brought gas prices from a high of $12 per thousand cubic feet in the winter of 2001 to $3.50 last summer [2001]. The price collapse dropped AEC's stock value by 31.4% between April and July 2001.

By March 2002, so many wells were sucking on Ladyfern that production declined from 700 million cubic feet per day to 400 million cubic feet. Production today stands at about 300 million cubic feet—and is expected to soon become a trickle. As a result, there wasn't much gas left to cushion this winter's price shock. . . .

Closing Comments

With the exception of the BC government, just about everybody involved in Ladyfern now offers some sobering hindsight. "The tragedy was the amount of money industry spent to get the gas out of there," says Murphy's Doerr. "We could have developed resources elsewhere and been much more profitable. . . . Ladyfern was over-drilled—it's an atrocity," says [Predator] company president Robert Shields. The play was produced so fast it brought continental gas prices down, he adds, and "now it's depleted, and gas prices are up. It wasn't handled properly." AEC, now EnCana, has consistently described Ladyfern as a profitable experience, even though several of its wells are now producing water. However, CNR, which has substantially downgraded the size of its reserves, acknowledges that Ladyfern wasn't as big as it thought—and was definitely overcapitalized. Three of its wells are now sucking water.

The OGC is understandably reluctant to acknowledge that Ladyfern was an out-and-out case of natural resource abuse. Rather, the entire play proved that "competition is alive and well," says Commissioner Derek Doyle. He notes that northeast BC desperately suffers from a lack of infrastructure and needed more roads. He admits, however, that the OGC faced a major challenge "in having competent personnel" assess the wells' geology and life span. Concludes Doyle: "I think Ladyfern was an urgent cry for us to stay with our business as a regulator and not be captured by the enthusiasm of industry or those that promote the industry."

To service workers and consultants who competed at Ladyfern, the whole debacle remains a familiar study in human nature. "The BC government needed the money," notes one Grand Prairie oil and gas consultant. "They should have controlled production and limited the competition. But it was greed on all parts. Companies that should have made 200% profit only made 20%."

Rob Woronuk, a Calgary-based analyst with the Canadian Gas Potential Committee, an independent natural gas resources watchdog, pointedly describes Ladyfern as a model of how things shouldn't be done. For a find as precious as that, a competent regulator should have paid attention to reservoir geology, conservation and the public interest, he argues. A proper plan for depleting the reservoir over time in an economic manner would have maximized returns to the taxpayers of BC, Woronuk adds. "You are supposed to do that," he says. "That's the public's gas. We own that." Woronuk also points out that most of Ladyfern's gas was sold when gas prices abruptly dived in 2001. "Wouldn't it have been nice if that gas sold at $12 a thousand cubic feet?"

To Woronuk, Ladyfern just illustrates the short-sightedness that has sadly defined Canada's gas supply strategy for the past decade. "Conservation is an economic parameter that has been forgotten," he maintains. The country is now operating under a just-in-time approach to gas production, he adds, which guarantees high prices for years to come. "It's not economic," says Woronuk. "It's not efficient. And it doesn't make any sense. . . . "

The Ladyfern development is an example of clearly unsustainable development. The British Columbia (APEGBC) *Guidelines for Sustainability* were apparently ignored in this case. Similar stories are occurring in resource development around the world. In particular, the Alberta oil sands development is cited frequently as an example of excessively rapid and unsustainable development.[37] For example, David Olive, Business reporter for *The Toronto Star*, described the development of Alberta's oil sands in the Athabasca region northeast of Edmonton as a "moonscape of strip mines." He goes on to say:

The Athabasca River is suffering rapid depletion given that huge amounts of water are required to process heavy oil. And the oil sands operators have created some of Alberta's largest lakes, consisting of post-production toxic water.[38]

The tragedy of the commons is being reinvented in the modern world by resource developers. It is time that we insisted on conformance to a philosophy of sustainable development.

DISCUSSION TOPICS AND ASSIGNMENTS

1. Sustainability is based on the axiom that we have a duty to future generations. However, some philosophers challenge that axiom, arguing that future generations will likely have unimagined technological resources, and will be "better off" than we are. To illustrate: past generations could never have imagined nuclear energy, computers, cell phones or the Internet and the benefits associated with them. If future generations will be better off than we are, why should we sacrifice for them? Discuss this concept (sometimes called *prioritarianism*), and state whether (and why) you agree or disagree.

2. Population growth is a critical factor in sustainability. China's population of 1.322 billion people (as of 2007) exists on a land area that is slightly smaller than Canada, which has just over 33 million people. China has implemented a severe one-child program to limit its birth rate. Couples who agree to have only one child receive free health care for the birth and for the child, whereas couples that do not agree must pay their own expenses. Under this policy, the birth rate declined to about 1.2 percent in 1981, but by 1991, it had risen to 1.4 percent. As well, there is recent evidence that people refuse to comply with the one-child policy. For

example, more unwanted children are ending up in orphanages. Discuss the ethics of government policies that limit birth rates, using the ethical theories described in Chapter 11. Which should take precedence: personal freedom to bear children, or the duty-based concept that, in cases like this, everyone must share the responsibility and limit growth? How would you evaluate the greatest good for the greatest number? Is there a virtue, a golden mean, or a compromise that can be identified in this case? Can you support the Chinese policy on an ethical basis? Review the paper by G. Hardin, *The Tragedy of the Commons* (cited in Chapter 16), in preparing your answer.

3. Satellite television, recently available in many less developed countries, reveals the disparity in consumption between rich and poor countries. People throughout the world have begun to expect improved standards of living. Even if China is able to control population growth, it is likely to experience an increase in consumption patterns such as those seen in South Korea and Singapore. Estimate the effect on global warming if China consumed resources at the same rate as South Korea or Canada. (Hint: China's development is currently in the news, so statistics are easily found by Internet search.) How can the world meet the resource needs of China's present population if they adopt North American consumption traits?[39]

NOTES

[1]   The first four sections of this chapter were originally contributed to "Environmental Sustainability," Chapter 21 in G.C. Andrews, J.D. Aplevich, R.A. Fraser, and C. MacGregor, *Introduction to Professional Engineering in Canada,* 3rd ed., Pearson Education Canada, Inc., Toronto, Ontario, 2008. These sections are adapted and reproduced here with permission from Pearson Education.

[2]   *Our Common Future,* Oxford University Press, 1987, also called the "Brundtland Report." Many Internet sources; for example, <www.worldinbalance.net/agreements/1987-brundtland.html> (June 6, 2008).

[3]   P. Hawken, *The Ecology of Commerce: A Declaration of Sustainability,* HarperBusiness, New York, 1994.

[4]   R. Carson, *Silent Spring,* Fawcett World Library, New York, 1962.

[5]   D.H. Meadows, D.E. Meadows, J. Randers, and W.W. Behrens, *Limits to Growth,* Universe Books, New York, 1972. Research sponsored by The Club of Rome's Project on the Predicament of Mankind.

[6]   J. Lovelock, *Gaia: A New Look at Life on Earth,* Oxford, New York, 1987.

[7]   United Nations Environment Programme (UNEP), *Handbook for the Montreal Protocol on Substances that Deplete the Ozone Layer,* 7th ed., Nairobi, Kenya, 2006.

[8]   A. Freeman, "Canada gets its way on climate change," *The Globe and Mail,* November 24, 2007, and Canadian Press, available at <www.theglobeandmail.com> (June 6, 2008).

[9]   UN Framework Convention on Climate Change (UNFCCC), Bonn, Germany, available at <http://unfccc.int> (June 6, 2008).

[10] Government of Canada, *Clean Air Act,* Environment Canada, Ottawa, ON, 2006, available at <www.ec.gc.ca/cleanair-airpur/Clean_Air_Act-WS1CA709C8-1_En.htm> (June 6, 2008).

[11] Intergovernmental Panel on Climate Change (IPCC), "Summary for Policymakers," in *Climate Change 2007: The Physical Science Basis,* February 2007, Contribution of Working Group I to the Fourth Assessment Report of the Intergovernmental Panel on Climate Change (ed. Solomon, Qin, Manning, Chen, Marquis, Averyt, Tignor, and Miller), Cambridge University Press, Cambridge, U.K., and New York, U.S.A., available at <www.ipcc.ch> (June 6, 2008).

[12] Intergovernmental Panel on Climate Change (IPCC), "Summary for Policymakers," in *Climate Change 2007: Impacts, Adaptation and Vulnerability,* April 2007, Contribution of Working Group II to the Fourth Assessment Report of the Intergovernmental Panel on Climate Change (ed. Parry, Canziani, Palutikof, van der Linden, and Hanson), Cambridge University Press, Cambridge, U.K., available at <www.ipcc.ch> (June 6, 2008).

[13] Intergovernmental Panel on Climate Change (IPCC), "Summary for Policymakers," in *Climate Change 2007: Mitigation of Climate Change,* May 2007, Contribution of Working Group III to the Fourth Assessment Report of the Intergovernmental Panel on Climate Change (ed. Metz, Davidson, Bosch, Dave, Meyer), Cambridge University Press, Cambridge, U.K., and New York, NY, U.S.A., available at <www.ipcc.ch> (June 6, 2008).

[14] Intergovernmental Panel on Climate Change (IPCC), "Summary for Policymakers," in *Climate Change 2007: The AR4 Synthesis Report,* November 2007, available at <www.ipcc.ch> (June 6, 2008).

[15] R.W. Jackson and J.M. Jackson, *Environmental Science: The Natural Environment and Human Impact,* Longman, Harlow, Essex, 1996, p. 317.

[16] A. Gore, *An Inconvenient Truth: A Global Warning,* Motion picture, Paramount Pictures, Hollywood, CA, 2006; and M. Durkin, *The Great Global Warming Swindle,* Motion picture, WAGtv, London, UK, 2007. The Gore film received a dozen film awards, but was condemned by the U.S. Senate Committee on Environment and Public Works, then chaired by Senator Jim Inhofe (R-OK), in a press release stating that "global warming is the greatest hoax ever perpetrated on the American people," as reported by Wikipedia at <http://en.wikipedia.org/wiki/An_Inconvenient_Truth> (August 5, 2008). The Durkin film, intended to refute the Gore film, was "criticised heavily by many scientific organisations and individual scientists," who argued that it had "misused and fabricated data, relied on out-of-date research, employed misleading arguments, and misrepresented the position of the Intergovernmental Panel on Climate Change" as reported by Wikipedia at <http://en.wikipedia.org/wiki/The_Great_Global_Warming_Swindle> (August 5, 2008). Clearly the debate over the existence of climate change, and the action needed to combat it, is far from over.

[17] CBC News, "Cold snap won't wipe out pine beetle in B.C." January 28, 2008, available at <http://cbc.ca> (June 6, 2008).

[18] T. Flannery, *The Weather Makers,* Canadian Edition, HarperCollins Publishers Ltd, New York, 2006, p. 110.

[19] G. Monbiot, *Heat: How to Stop the Planet from Burning,* Anchor Canada, Random House of Canada, Toronto, 2006, p. 11.

[20] Association of Professional Engineers, Geologists and Geophysicists of Alberta, *Member Consultation on Climate Change: Report to Council,* a survey by APEGGA's Environment Committee, February 2008, available at <www.apegga.org/Environment/reports/ClimateChangesurveyreport.pdf> (August 4, 2008).

[21] Intergovernmental Panel on Climate Change (IPCC), "Summary for Policymakers", in *Climate Change 2007: Mitigation,* pp. 43–76.

[22] T. Homer-Dixon, *The Upside of Down,* Knopf, New York, 2006.

[23] J. Simpson, M. Jaccard, and N. Rivers, *Hot Air: Meeting Canada's Climate Change Challenge,* McClelland & Stewart, Toronto, 2007.

[24] C.J. Campbell, Personal communication reported in keynote paper by J. D. Hughes, "The Energy Sustainability Dilemma: Powering the Future in a Finite World," in *Proceedings of Plug-in Hybrid Electric Vehicle (PHEV) Conference: Where the Grid Meets the Road,* Winnipeg, Manitoba, 2007, Keynote address, available at <www.pluginhighway.ca/proceedings.php> (June 3, 2008). Data also found in C.J. Campbell, "Oil Depletion—Update through 2001," available at <www.hubbertpeak.com/campbell/update2002.htm> (June 3, 2008).

[25] R. Heinberg, *The Oil Depletion Protocol: A Plan to Avert Oil Wars, Terrorism and Economic Collapse,* Clairview Books, U.K., and New Society Publishers, BC, 2006, p. 38.

[26] Heinberg, *The Oil Depletion Protocol,* p. 23.

[27] R.L. Hirsch, *Peaking of World Oil Production: Recent Forecasts,* U.S. Department of Energy, February 2007, available at <www.netl.doe.gov> (June 6, 2008).

[28] Heinberg, *The Oil Depletion Protocol,* p. 153.

[29] M. Jaccard, *Sustainable Fossil Fuels: The Unusual Suspect in the Quest for Clean and Enduring Energy,* Cambridge University Press, New York, 2006.

[30] S. McCarthy, "Dim prospects that 'energy efficient' will pay off: CIBC," *The Globe and Mail,* November 27, 2007, available at <www.theglobeandmail.com> (June 6, 2008).

[31] Intergovernmental Panel on Climate Change (IPCC), "Summary for Policymakers," in *Climate Change 2007: Mitigation,* p. 87.

[32] Dennis Burningham, MSc., chemical engineer and U.K. industry consultant, personal correspondence. Mr. Burningham provided his personal insight into sustainability, and many of his comments are woven into this chapter. The author would like to acknowledge this personal assistance and express his thanks.

[33] Flannery, *The Weather Makers,* p. 199.

[34] M. Raynolds, M.D. Checkel, and R.A. Fraser, "The Relative Mass-Energy Economic (RMEE) Method for System Boundary Selection—A Means to Systematically and Quantitatively Select LCA Boundaries," *Int'l Journal of Life Cycle Assessment,* vol. 5, no. 1 (2000), pp. 37–46.

[35] This case study has been adapted, with permission, from A. Nikiforuk, "Northern greed," *Canadian Business Magazine,* May 2003, available at <www.canadianbusiness.com/article.jsp?content=20030512_53695_53695&page=1> (June 6, 2008).

[36] L. Schmidt, "Judge awards $27M, property to Murphy: Gas field legal battle nears end," *Calgary Herald,* Final Edition, Saturday, September 16, 2006, C4.

[37] P. Cizek, "Scouring Scum and Tar from the Bottom of the Pit," *Energy Security and Climate Change: A Canadian Primer,* Copyright C. Gonick (editor), Fernwood Publishing, Black Point, NS, 2007.

[38] D. Oliver, "Alberta's Inconvenient Truths," *The Toronto Star,* October 14, 2007, A12.

[39] Assignment problems 17.2 and 17.3 were inspired by Dr. Jerry M. Whiting (Professor Emeritus, University of Alberta). The author thanks Dr. Whiting for permission to adapt these questions.

Chapter 18
Writing the Professional Practice Exam

To obtain a licence to practise engineering or geoscience in Canada, every new applicant must write a Professional Practice Examination (PPE). The exam is set by the provincial or territorial Association of professional engineers and/or geoscientists. Ten of the 16 licensing Associations presently use the National Professional Practice Exam (NPPE) set by Alberta (APEGGA), although some Associations add questions and others set and administer their own exams. The exam ensures that the applicant is familiar with the principles of practice, including a basic knowledge of the profession, ethics, liability, and Canadian law as it applies to engineers and geoscientists in the applicant's province or territory of practice.

This chapter describes the examination and shows typical solved questions from previous exams. It should be especially useful to readers who are preparing to write it.

EXAMINATION SYLLABUS AND FORMAT

The first step in preparing for the PPE is to determine the syllabus and format of the exam in your province or territory.

- **Syllabus:** The Professional Practice Examination in each jurisdiction follows a general syllabus set by Engineers Canada, although most jurisdictions include minor changes in content to suit the local legislation. The topics include professionalism, ethics, liability, and Canadian law. This textbook covers most of these topics, although other textbooks are recommended for the legal concepts in the syllabus.
- **Format:** The type and duration of the exam are important. The Alberta (National) Professional Exam is typically a multiple-choice exam, two hours in length, although the Ontario PPE is a short-answer and essay-type exam, three hours in length. Other jurisdictions have variations on these formats.

It is important that you contact your Association as soon as possible to determine the requirements for the PPE, and the format for the exam. You need to know: What topics are covered by the exam? What is the duration and format of the exam? Are any aids or references permitted or provided?

Syllabus—Engineers Canada

Engineers Canada (formerly called the Canadian Council of Professional Engineers—CCPE) publishes a general syllabus for the exam, which is accepted in principle by most Associations. The Engineers Canada syllabus includes both professionalism and engineering law, defined as follows:

Professionalism

Topics to be covered by the examination in the general areas of engineering practice and ethics should include, but need not be limited to: the definition of professional engineering; the role of the association and the responsibilities associated with self-governance; professional accountability, conduct and ethics, the professional engineer's responsibility to the public and duty to report illegal or unethical engineering practice; the ethical use of the engineer's seal; continuing competence; and the social and environmental impacts of engineering on society.

Engineering Law

Topics to be covered by the examination in the area of the law as it relates to engineers and to the practice of engineering should include, but need not be limited to: the basic structure of the Canadian legal system, common law, Quebec civil law, statute law and the provincial court system; tort law, liability and liability issues; business organizations; contract law, specifications and tendering, discharge and breach of contract, bonding, estoppel and construction lien legislation; intellectual property, patents, technology transfer, copyrights, trademarks, industrial designs and trade secret; fiduciary responsibility; professional advertising, unfair competition and merchandising rights; dispute resolution, negotiation and arbitration; litigation and the engineer as expert witness; the Canadian Human Rights Act; environmental legislation; worker's compensation and occupational health and safety legislation.[1]

NATIONAL PROFESSIONAL PRACTICE EXAMINATION (NPPE)

Since 1998 the Association of Professional Engineers, Geologists, and Geophysicists of Alberta (APEGGA) has provided a National Professional Practice Examination (NPPE). It has been adopted by 10 jurisdictions, from the Northwest Territories to Manitoba to Newfoundland. The NPPE consists of a set of one hundred multiple-choice questions, administered in a two-hour closed-book format. All questions are common to engineering, geology, geophysics, and geoscience. Some Associations extend the examination in their jurisdictions by adding multiple-choice and/or essay questions to the NPPE, and in those cases, candidates are allowed additional time to complete the examination. Contact your Association for specific details.

The pass mark is 65 percent, with no penalty for wrong answers. The grade is final and cannot be appealed. The major subject areas on the exam (as of June 5, 2008) are as follows:

A. Professionalism (30%)

1) Definition and interpretation of professional status
2) The role and responsibilities of a professional in society
3) The role and responsibilities of a professional to management
4) Professional conduct, ethical standards and codes
5) Safety and loss management—the professional's duties
6) Environmental responsibilities

B. Professional Practice (20%)

1) Professional accountability for work, workplace issues, job responsibilities and standards of practice
2) Continuing competence
3) Quality management and standards of skill in practice
4) Business practices as a professional
5) Insurance and risk management
6) Professional and technical societies
7) Non-statutory standards and codes of practice

C. Regulatory Authority Requirements (9%)

1) Safety and loss management—regulatory aspects
2) Environmental regulations
3) Occupational health and safety
4) Workers compensation
5) Other statutory standards of practice

D. Law and Legal Concepts (25%)

1) Canadian legal system and international considerations, basics of business organizations
2) Contract Law—elements, principles, types, discharge, breach, interpretation etc.
3) Tort Law—Elements, application of principles, interpretation, liabilities of various kinds
4) Intellectual Property—patents, trademarks, software issues, copyright
5) Arbitration and Alternative Dispute Resolution (ADR)
6) Expert Witness

E. The Act (16%)

1) Definitions of the professions and scopes of practice
2) Structure and functions of a Provincial Association
3) Regulations and By-Laws
4) Registration
5) Discipline and enforcement
6) Use of seals and stamps[2]

ONTARIO AND BRITISH COLUMBIA

In Ontario, the PPE is a three-hour essay-type written examination, typically containing from eight to ten questions, half of these devoted to professionalism (practice and ethics), and half to Canadian engineering law. The syllabus is similar to the Engineers Canada syllabus, but not identical, and may be obtained from the PEO website.[3] In Ontario, candidates are usually provided copies of the Code of Ethics and the definition of professional misconduct (Ontario Regulation 941, sections 72 and 77), but all candidates should read the exam instructions carefully to find out what aids are permitted during the exam.

British Columbia has a slightly different approach: the PPE administered by APEGBC is a three-hour examination consisting of the National Professional Practice Exam (a two-hour multiple-choice section) plus an additional one-hour essay question. The exam syllabus and a Law & Ethics Seminar are available from the APEGBC website.[4]

THE EGAD! STRATEGY FOR ETHICS QUESTIONS

The ethics questions on the PPE specifically test your knowledge of the Code of Ethics, so review this code before the examination. Questions on multiple-choice exams are usually brief and direct, although essay-type questions usually describe a hypothetical situation and ask you to suggest the proper course of action. Whenever the answer involves a clause in the Code of Ethics, you should state the clause (or at least refer to the code). Many questions can be answered easily if you have a good grasp of the code.

Some questions may be "open-ended," and may require a more general discussion of ethics. These questions do not occur often in the exam setting (although they are common in real life). In these cases, the strategy for solving ethical problems, described in Chapter 11, may be useful, and it is simplified below for easier use on the exam. This six-step strategy has been renamed the "EGAD!" method and is similar to a solution method taught to law students.[5] It is remembered easily by these three words:

READ—EGAD!—WRITE

The term "EGAD!" (an old English exclamation of surprise) is an acronym or mnemonic for the four key steps in the solution strategy: **E**thical issues, **G**eneration of alternatives, **A**nalysis, and **D**ecision. The six-step strategy is explained as follows:

STEP 1: READ

Read the problem thoroughly and gather information.

Exam questions contain less information than real problems, so read each question thoroughly! Highlight or underline key facts, but do not copy the question into the exam book, since this wastes valuable time. Ask yourself the typical reporter's questions:

- *Who is involved?* (i.e., Who has caused harm to whom?)
- *What harm or damage has occurred* (or may potentially occur)?
- *How has this harm occurred* (or may potentially occur)?

STEP 2: E—ETHICAL ISSUES

Identify the basic ethical issues.

The exam question may state the ethical problem directly. For example: "Has Mr. Smith broken the Code of Ethics?" However, some exam questions may say simply "Explain and discuss this case." You must then imagine which ethical issues should apply, and if possible compare the similarities (and differences) of the case with previous cases. If the ethical issue is not obvious, then ask yourself: "What, exactly, is wrong in this situation? Do any actions contravene the law or the Association's Code of Ethics? What is unfair?" Once you identify the ethical problem, the proper course of action may be obvious. If so, write down your answer. However, in some problems you may have to suggest or imagine ("generate") a proper course of action.

STEP 3: G—GENERATION OF ALTERNATIVES

Generate (or suggest) possible courses of action.

Some exam questions simply ask: "What should you do?" You must suggest the proper action. This step requires creative thought, so it may be difficult. Creative techniques such as brainstorming may be useful. You might suggest a compromise, or a totally new idea. You might also try to imagine yourself as one of the participants. What would you do in this situation? The goal is to find a new course of action that is ethically correct and that has a minimum of side effects.

Sometimes the exam question involves an ethical dilemma with only two alternatives, both of which are very nasty. In this case, you should try to suggest a third possibility that is better. In your answer, you might list the reasons why the two choices are undesirable; do not, however, assume that you must choose one of them.

STEP 4: A—ANALYSIS

Analyze the possible courses of action.

When two or more courses of action have been suggested (in the previous step), you must examine each one to find the best. You want the simplest course of action that solves the problem without nasty side effects. You should test each course of action as follows:

- Is this course of action legal?
- Is it consistent with human rights, employment standards, and design standards?
- Does it obey the Code of Ethics and maintain the ideals of the profession?
- Can the solution be published and withstand the scrutiny of your colleagues and the public?
- What benefits will result, and are they equally distributed? (Utilitarianism)
- Can this solution be applied to everyone uniformly? (Kant)

- Does this solution respect the rights of all participants? (Locke)
- Does the solution develop or support moral virtues and/or is it a golden mean between unacceptable extremes? (Aristotle)
- Is it fair? Does it have any unfair side effects on those concerned?

STEP 5: D—DECISION

Make a logical decision.

The previous step should yield at least one acceptable course of action. However, in some cases all of the alternatives may be unacceptable, or the alternatives may be so equally balanced that none is clearly superior. In this case, you may need to review the above steps. If you still face a dilemma, with two equal choices, you must decide which is better (or least negative). If the choices are equally balanced, select the course of action that does not yield a benefit to the person making the decision. This will help you to defend the decision.

STEP 6: WRITE

Write a professional summary of your answer.

Finally, you must explain your answer clearly, logically, and neatly. You can't afford to waste time, so you must practise writing good answers. Start by stating your decision, which answers the question asked by the examiner. Then explain why you came to that conclusion. It is very important to cite sub-section numbers (from the Code of Ethics or regulations) for the ethics questions, just as you would cite past cases (or precedents) for the law questions. Do not copy clauses from the Code of Ethics; identify them by number, if possible. It is also important to write neatly and legibly. The examiners greatly appreciate this courtesy.

An Important Hint

An examiner who sets one of the essay-type PPEs says that candidates using the EGAD! method spend too much time on the EGA steps (Ethical Issues, Generation of Alternatives, and Analysis) and not enough time on step D (explaining the Decision). These short or incomplete answers get lower grades. Remember that the EGAD! process is intended merely to help you think about the problem in an orderly way. Do not write out all of the steps; your exam grade is based only on your written decision (step D), so explain it thoroughly.[6]

PREVIOUS EXAMINATION QUESTIONS

This section contains about 30 examination questions selected from previous PPEs in several provinces.[7] Readers are encouraged to attempt all of the following questions, regardless of their province of residence. Ethics concepts are universal, problems are similar, and answers will differ only slightly from province to province. The questions have been chosen to show the various exam formats: essay-type, short-answer, multiple-choice. Solutions are suggested for most questions. An asterisk (*) indicates where the specific clause number(s) from your provincial Act (or Code of Ethics) should appear, if appropriate.

Essay-Type Examination Questions

In the essay-type examination, the applicant is typically asked to answer four or five ethics questions and would be permitted about twenty minutes per question. A copy of the Code of Ethics may be provided for reference during the exam.

1—COMPETING WITH AN EMPLOYER

Professional Engineer A takes a job with a manufacturing company and almost immediately thereafter is given responsibility for preparing the draft of a bid for replacement turbine runners for a power corporation. While working on preparing the bid for the manufacturing company, Engineer A, as president and shareholder of his own company, which he runs privately from his home, writes to the power corporation requesting permission to submit a tender on the same project. A few days later, and while continuing to work on the bid for the manufacturing company, he receives word from the power corporation that a bid from his company would be considered. The day after learning this, he resigns his position with the manufacturing company and proceeds to finalize and submit a bid on behalf of his own company.

QUESTION *Discuss Engineer A's actions from an ethical point of view.*

SUGGESTED ANSWER Engineer A is clearly unethical in his actions. By running a private company in competition with his employer, he is not being fair or loyal to his employer, as required by the Code of Ethics.(*) He has taken advantage of inside information, betrayed the trust of his employer, and yielded to a conflict of interest. If his private company was unknown to his employer, then he has failed to disclose his conflict of interest as required by the Code of Ethics.(*) By his actions, he has failed to show the necessary devotion to professional integrity required by the Code of Ethics.

In his defence, it could be said that since he resigned before actually signing the contract, he did not compete with his employer, but this is irrelevant; the serious conflicts of interest occurred during the bid preparation stage. Engineer A has exposed himself to the serious possibility of disciplinary action for conflict of interest under the provincial or territorial Act.

2—FORMING A PRINTING COMPANY

You are a Professional Engineer with XYZ Consulting Engineers. You have become aware that your firm subcontracts nearly all the work associated with the setup, printing, and publishing of reports, including artwork and editing. Your wife has some training along this line and, now that your children are at school, is considering going back into business. You decide to form a company to enter this line of business together with your neighbours, another couple. Your wife will be the president, using her maiden name, and you and your neighbours will be directors.

Since you see opportunities for subcontract work from your company, you reason that there must be similar opportunities with other consulting firms. You are aware of the existing competition and the rates they charge for services and see this as an attractive sideline business.

QUESTION   *Can you do this ethically, and if so, what steps must you take?*

SUGGESTED ANSWER   You can do this ethically, but there is a potentially serious conflict of interest unless you follow the Code of Ethics for your province or territory. You can undertake the sideline business, provided it does not interfere with your regular employment and provided your employer is fully informed, as required by the Code of Ethics.(*) Your wife, of course, is free to use any legal name in her business affairs; however, if the sole reason for using her maiden name is to conceal your participation in the company's ownership and operations, your cooperation could be considered unethical. If you tell your employer all the details about your wife's company, then you have disclosed your conflict of interest as required by the Code of Ethics.(*) However, your wife's other clients may worry about a possible loss of confidentiality. A publishing company often receives confidential reports and must not reveal the contents of those reports to others. Therefore, your wife must not allow you to see sensitive engineering information submitted to her by other clients for publishing. Obviously, confidentiality must be guaranteed, or conflicts of interest may arise in the future.

3—CHEMICAL POLLUTION

Brenda MacDonald, a Professional Engineer, is manager of a chemical plant in a northern Canadian town. Early this summer she noticed that the plant was creating slightly more water pollution in the lake into which its waste line drains than is legally permitted. If she contacts the province's environment ministry and reveals the problem, the result will be a considerable amount of unfavourable publicity for the plant. The publicity will also hurt the lakeside town's resort business and may scare the community. In addition, solving the problem will cost her company well over $100,000. If she tells no one, it is unlikely that outsiders will discover the problem, because the violation poses no danger whatever to people. At the most, it will endanger a small number of fish.

QUESTION   *Should MacDonald reveal the problem despite the cost to her company, or should she consider the problem little more than a technicality and disregard it? Discuss the ethical considerations affecting her decision.*

SUGGESTED ANSWER   MacDonald must, legally and ethically, take action to remedy this situation. She is obligated under the Code of Ethics to consider the public welfare as paramount.(*) The legal limit for pollution has been exceeded, and failure to take action could be considered professional misconduct under the Act.(*) If she has known about the excess for some time, she may already be considered negligent and therefore subject to discipline under the Act.(*)

MacDonald must abide by the ministry's regulations, which would probably require her to submit a complete, factual report to inform officials about the pollution. Before sending the report, she should discuss it fully with her employer. If the employer reacts adversely, MacDonald must, nevertheless, forward the report to the ministry as required by law and by the Code of Ethics.(*) If the employer attempts to dismiss her, MacDonald may find it useful to ask the provincial Association to mediate and to inform her employer of the requirements under the Act. Should MacDonald be dismissed while acting properly and in good faith, she would have grounds for a suit against the employer for wrongful dismissal to recoup lost wages and costs. It would be advisable for her to consult a lawyer in that event.

The engineer's concern over adverse publicity and the cost to the company must not obscure the requirement to act within the law. If the situation is permitted to continue unabated, the long-term consequences will be much more serious. The pollution could ruin the neighbouring resort industry, and MacDonald could find herself subject to disciplinary action for negligence or professional misconduct.

4—INVOICE ERRORS

You are a Professional Engineer employed by a consulting engineering firm. Your boss, who is also a Professional Engineer, is the project manager. You examine a recent invoice that your boss sent to the client for work done by you and by members of your staff. You are surprised to see how much of your time and your staff's time has been charged to the job. You check further by reviewing the time sheets, which show that time that should be charged to other clients has been deliberately transferred to this job. You try to raise the subject with your boss, but are rebuffed. You are quite sure something is wrong, but you are not sure where to turn.

QUESTION   *Which articles in the Code of Ethics are relevant to this situation? What action must you take, according to the Code of Ethics?*

SUGGESTED ANSWER   The first step in solving any problem (as discussed in Chapter 11) is to get all the information. If no suitable answer is given, you have an ethical dilemma. The Code of Ethics says that you must be loyal to the employer,(*) but the code also states that you must be fair and loyal to the client.(*) You would solve the dilemma in two steps: First, you ask your boss, again, for an explanation or justification for this transfer. (For example, it may be a simple accounting error.) As a professional engineer, your boss is subject to the Code of Ethics, and has a duty to you as an employee.(*) If your boss is completely unwilling to explain the reasons for this action, then the over-billing may be fraud or theft, which is illegal. If you ignore the discrepancy, you may be implicated in your boss's apparently unprofessional conduct. In this case, your responsibility to the client outweighs the duty to the employer. Second, you try to resolve the discrepancy internally, by contacting your boss's boss for clarification of the discrepancy and assurance that you are not implicated in criminal activity. If the problem can be resolved internally, it is likely not essential to inform the client who is being overcharged. If your boss retaliates by dismissing you, you would consult a lawyer about suing for wrongful dismissal.

5—CHEMICAL SPILL CLEANUP

A consulting engineering firm is preparing to submit a proposal to clean up an area contaminated by a chemical spill during a train derailment. From past experience, the engineers and geoscientists in the firm know the amount of work involved in doing the job properly. The experts will include people with training in ecology, water quality, ground water, soils, air pollution, and other areas. The methodology that they feel must be followed will result in an expenditure of about $5 million. Before their proposal is submitted, however, the federal government, which is the potential client, issues a news release saying that it has budgeted only $1 million for this work.

QUESTION   *What can the consulting firm do? To reduce the level of work to one-fifth of what it thinks is necessary would infringe on the firm's perceived ethical responsibilities to the environment.*

SUGGESTED ANSWER   The question implies that the consulting firm feels pressure to submit a bid to do a partial or inadequate job, within the $1 million limit, simply to get the work. This behaviour contravenes every Code of Ethics (either directly or indirectly), which states that the professional engineer must act competently in providing engineering services.(*) Moreover, most Codes of Ethics require the professional engineer to uphold the principle of adequate compensation for engineering work.(*) It is therefore unethical for the consulting firm to submit a bid to perform an inadequate job, or to perform a $5 million job for $1 million.

However, the question asks what the consulting firm should do. Every Code of Ethics requires a professional engineer to explain the consequences to be expected if engineering judgment is overruled by non-technical authority.(*) In this case, the financial officials who set the budget are likely unaware of the required engineering work. This fact must be communicated to the federal department that issued the call for tenders, which must be advised to correct its specifications. If the problem cannot be resolved by simple communication, the consulting firm must evaluate the seriousness of the matter. (This requires details not provided in the question.) For example, if an inadequate job will endanger the public, the consulting firm has an obligation to put the public interest first, by publicizing the issue.(*) Furthermore, if the consulting firm submits a proper ($5 million) bid, and a competing firm obtains the job for less and thereby creates a dangerous situation, the consulting firm has an obligation to expose any unprofessional or unethical conduct by the other practitioner.(*)

6—ROAD DEFICIENCIES

Engineer A enters into a consulting contract with a client to provide design and construction supervision of road surfaces in a partially completed land development project. He has taken over from another consultant, who was discharged partway through the job. Before Engineer A can finish the project, his contract also is terminated. Shortly thereafter, it becomes obvious that there are deficiencies in the work done under A's supervision. Investigation shows that hastily paved road surfaces, completed under adverse late-fall weather conditions, are not up to specifications. It seems that A is aware of this. He intended to require remedial work by the contractor in the spring, but his termination occurred before that time. Engineer A did not advise his client that he was expecting to re-inspect in the spring and to have deficiencies corrected, nor did he inform his client of the existing state of the roads after he was released from his contract.

QUESTION *Did Engineer A act in an ethical way in his dealings with his client, even though he may feel that he was unfairly terminated? Discuss the articles of the Code of Ethics that have a bearing on this case.*

SUGGESTED ANSWER Engineer A, as a professional engineer, is required to act as a faithful agent for the client, in spite of other problems that might interfere.(*) Therefore, even if Engineer A felt that he was unfairly terminated, it would be unethical for him to neglect his responsibilities, such as listing the deficiencies (known only to Engineer A) so that the work could continue. This is especially important if the deficiencies might lead to endangering other workers or the general public. The Code of Ethics requires the professional engineer to put public safety first.(*)

7—CONTRACT CONTINGENT ON BOND ISSUE

An engineer enters into a contract with a public body [city or town] whereby he agrees to conduct such field investigations and studies as may be necessary to determine the most economical and proper method of designing and constructing a water supply system. He also agrees to prepare an engineering report, including an estimate of the cost of the project, and to estimate the amount of bond issue required. The contract provides that if the bond issue passes, the engineer will be paid to prepare plans and specifications and supervise the construction, and he will be paid a fee for his preliminary services. If the bond issue should fail, the public body will not be obligated to pay for the preliminary work. The public body's bylaws prohibit it from committing funds for preliminary work until the bond issue is approved.

QUESTION *May an engineer ethically accept a contingent contract under these conditions?*

SUGGESTED ANSWER   At first glance, this may appear to be a simple entrepreneurial activity. However, the project is structured to create a massive conflict of interest for the engineer. Any public project must be carried out such that the public interest is protected (and is seen to be protected). The water supply system for a city or town is an especially sensitive matter, since the life of the community depends on a reliable water supply.

In this case, the contract proposes that the engineer will not be paid for the report if the bond issue is not approved (presumably by city and/or financial officials). Such a proposal is contrary to the Code of Ethics, which requires the engineer to uphold the principle of adequate compensation for engineering work.(*) Moreover, the contract proposes that the engineer reap a double reward if the bond issue passes. He will be paid both for the report and for future work—to prepare plans and specifications and supervise the construction. This creates pressure on the engineer to put the bond issue ahead of the engineering quality of the project. This is a clear conflict of interest, and in this case the conflict is insurmountable, since disclosing the conflict would discredit his report in the eyes of the public, and concealing the conflict is unacceptable. Therefore, since such a conflict of interest is contrary to the Code of Ethics(*), he must not accept the contract as proposed.

The proper way to organize a project such as this is to split it into two parts. The first contract would be for preparing the report on the most economical water supply system. The engineer should be paid for this work, regardless of the success of the bond issue. The second contract—to prepare plans and specifications and supervise the construction—would be arranged only if the bond issue passed. Such a contract should be offered for tender, and payment would follow the standard procedures (lump sum, percentage, per diem, and so on) recommended by the provincial Association.

8—DEFICIENCIES IN BUILDING DESIGN

Engineer X, a civil engineer and an employee of ABC Consultants Ltd., is designated under the company's Certificate of Authorization (C of A; also called a Permit to Practice) as the engineer taking responsibility for seeing that the Professional Engineers Act, its bylaws, and its regulations are complied with.

ABC Consultants Ltd. prepared the electrical and mechanical designs for a multistorey building. Although Engineer X had very little to do with this project, he permitted his seal to be applied to the design drawings. These designs were found to be deficient in a number of respects. Contrary to the Building Code, firewalls were omitted, fire dampers were not shown, and sprinklers were improperly connected, among other things. On investigation, it was found that other professional engineers working for ABC did both the electrical work and the mechanical work.

QUESTION   *What is Engineer X's ethical position in this matter?*

SUGGESTED ANSWER   As explained in the Act(*), corporations that practise engineering must identify the individuals who personally supervise the work performed by the corporation. These people are required to be experienced engineers, and their names are designated on the corporation's Certificate of Authorization (C of A or Permit to Practise). Since Engineer X was designated on the C of A, and permitted his seal to be applied to the design drawings, he is responsible for the work. Clearly, Engineer X has been negligent in permitting deficient or unsafe work to be carried out and has failed to carry out his responsibility to supervise and direct the work. Engineer X will likely be subjected to a disciplinary action for negligence (as explained in Chapter 4 of this text). Designation on a C of A is not a formality, nor is it a meaningless title.

The other ABC engineers, who actually designed the deficient electrical and mechanical work, might also be subject to disciplinary action for incompetence. Although they work under the supervision of Engineer X, they are responsible for producing competent work, and their designs were apparently inadequate.

9—BRAKE SYSTEM SAFETY
Auto-Tran, Inc., a manufacturing firm, has contracted to develop and produce a fully automated mass-transportation system to serve residents of a large city. A failure could have catastrophic consequences. One of Auto-Tran's mechanical engineers, who is a licensed professional engineer (P.Eng.), is concerned that, during the installation and testing phase, a major part of the control system did not appear to function satisfactorily. The P.Eng. reported the apparent malfunction to the project manager and recommended that the firm engage a licensed software engineering practitioner to look into the problem. However, the project manager advised the P.Eng. that there was no budget available and that it was important for Auto-Tran, Inc. to make delivery in order to meet its contractual commitments to the client.

QUESTION   *The P.Eng. strenuously expressed his concerns to the project manager and learned subsequently that shipment to the client had already been made. Does the P.Eng. have any obligation to take further actions under the circumstances?*

10—PROFESSOR'S RESEARCH
Professor Nu, a professional engineer, divides his time between teaching and research on a project in association with the University. Nu is also associated with a private research and development company, where Nu provides part-time consulting services. Another federal government agency has invited proposals for a new project from a number of organizations. This project may be viewed as an extension of work previously done with the University by Professor Nu.

QUESTIONS   *Can Professor Nu ethically participate in a proposal preparation for the University, Nu's development company, or both? If one of these proposals is successful, can Nu participate in executing the project? In each case, discuss the basis for your answer.*

11—PATENT QUESTION

Multicommon Tires Ltd. ("Multicommon") designs and manufacturers automobile tires. Recently, Multicommon developed a new polymer, which would greatly improve the expected life of its tires. Multicommon has not yet started using the new polymer in its tires. The development of the technology is still in its early stages and the company has not yet obtained a patent.

Multicommon has retained Chem Engineering Inc. ("Chem") to help Multicommon develop a process for putting the new polymer into large-scale production. Chem assigned one of its professional engineers, Ben Evolent, to Multicommon's project. Ben Evolent is also a member of several trade and professional associations, including the International Building Materials Institute (the "IBMI"), an association of designers, manufacturers, sellers, and users of building materials. IBMI's mission statement is to improve the building materials industry by the mutual cooperation of its members. Ben Evolent often volunteers to serve on IBMI's committees.

Ben Evolent is very intrigued about the new technology and soon realizes that the new polymer could improve the durability of building materials made of synthetic rubber. At a recent committee meeting of the IBMI, Ben Evolent suggested to a group of materials designers that they use the new polymer in their products. Ben Evolent was happy to share the information and remembered reading something in [the Association's] Code of Ethics about practitioners being required to "extend the effectiveness of the profession through the interchange of engineering information and experience." In addition, Ben Evolent believed "it would be good for the environment and the public interest in general to have more durable building materials" and "besides, Multicommon doesn't make building materials anyway."

QUESTIONS *Comment on Ben Evolent's conclusions. What do you think of Ben Evolent's conduct? Explain.*

12—FOUNDATION INSPECTION NEGLIGENCE

J. Doe, the owner of a house in the City of Alpha, Ontario, was notified by the City that the condition of the foundation walls of the house violated the standards set out in the City's property standards bylaw. The City, concerned that the foundation walls had deteriorated to the point of being structurally unsafe, ordered Doe to obtain a written report by a professional engineer as to the walls' condition. Doe retained Omega, a licensed member of the PEO, to prepare the report. Omega prepared a report stating that Omega had inspected the foundation and that the foundation walls appeared to be "structurally sound and capable of safely sustaining the house for many more years."

Doe submitted Omega's report to the City. In response, the City sent a letter to Omega pointing out the City's observations regarding the deterioration of the walls, including evidence of significant water permeation, together with photographs taken by the City's inspector. In the letter, the City requested that Omega reassess the condition of the foundation and respond to the City within two weeks. Omega does not respond to that request.

Several months later, the City asks Doe to obtain a response from Omega to the City's request for Omega's reassessment of the foundation. One week after that, Omega finally responds by letter to the City, advising that Omega never examined the interior of the walls and admitting that the photographs provided by the City indicate that the foundation is not structurally sound.

QUESTIONS

a. *Comment on the engineering services provided by Omega, in relation to [the Act in your province]. In your answer, also discuss Omega's conduct regarding Omega's dealings with the City.*

b. *Omega does not have a Certificate of Authorization [or Permit to Practise, needed to provide services to the public, in many provinces]. Does Omega need one under the facts described above? Explain why or why not. What are the possible consequences to a professional engineer of acting without a certificate of authorization when one is required?*

13—UNDER-THE-COUNTER JOB

SmallBox Corp. operates a small chain of three retail stores that specialize in selling lumber and other home improvement products. In order to improve its inventory and distribution efficiencies, SmallBox would like to build a central warehouse that would serve all of its stores. SmallBox contacted Engco, a large engineering firm, to inquire about hiring them to design the facility.

Eager is employed as a professional engineer by Engco. At the request of Honcho, the head of Eager's division, Eager accompanied Honcho to a meeting at Engco's offices with some representatives of SmallBox to discuss how Engco might be able to assist SmallBox with the potential project. At the meeting, SmallBox's representatives described to Honcho and Eager the attributes that SmallBox was looking for in the proposed new warehouse. They also asked about the fees that Engco proposed to charge for its services. Upon being advised of Engco's standard rates, SmallBox's representatives stated that, unfortunately, they could not afford to hire Engco for this project. Honcho was not prepared to discount Engco's quoted rates, which Honcho described as being "extremely competitive." Although everyone was disappointed, the meeting ended pleasantly, and both sides politely thanked each other for attending.

The next day, Eager received a telephone call from Frugal, one of the representatives of SmallBox. Frugal was wondering if Eager would be interested in preparing the design for the warehouse "on the side," after work in the evenings and on weekends. SmallBox was prepared to pay Eager at an hourly rate that was 50 percent of the hourly rate that Engco would have charged for Eager's time. In a hushed voice, Eager undertook to give the proposal some consideration and get back to Frugal.

Eager thought about Frugal's offer. It had been three long years since Eager had last received a salary increase from Engco. The extra money would be nice. Even at rates discounted by 50 percent from those charged by Engco, this would be a very profitable opportunity for Eager. The money Eager would earn from SmallBox would be more, on an hourly basis, than the rate on which Eager's current salary was based, and unlike Engco, Eager didn't have to worry about big overheads and other expenses. Eager then thought about how Engco might react to the arrangement, but decided that since Engco wouldn't be getting this work anyway, there shouldn't be a problem. Besides, Eager thought, there was no reason why they even needed to know about it. Eager called Frugal back the next day to accept the engagement and enthusiastically began working on the project that evening.

QUESTION   *Please comment on the appropriateness of Eager's conduct.*

14—CONTRACT DISPUTE

You are a professional engineer employed by FirstConcept Ltd., a consulting engineering company. One of the firm's clients, ShopCo Developments Ltd. ("ShopCo"), has hired your firm to provide the engineering design for a new elevated walkway for one of ShopCo's major shopping malls. You have been assigned the responsibility of preparing the design for ShopCo.

You develop the design and meet with R. Epp, a representative of ShopCo, to discuss your design. R. Epp is not a professional engineer, but disagrees with your design. The representative gives you some suggestions on how to simplify it. You listen politely, but realize that the design would be compromised if the suggestions were incorporated.

Because of the disagreement, R. Epp fires FirstConcept and demands that you turn the design drawings over to A.L. Ternate, a professional engineer employed by SecondConcept Ltd., another engineering firm. R. Epp had worked with A.L. Ternate on a previous project, and the two of them got along quite well. A.L. Ternate has agreed to complete the design of the new elevated walkway as R. Epp wishes.

You refuse to turn over the design drawings, even when R. Epp offers to pay for all of FirstConcept's services to date.

QUESTIONS   *Do you have any obligation to turn over the drawings? Do you have any other responsibilities? Comment on A.L. Ternate's agreement with R. Epp.*

Short-Answer Examination Questions

The Ontario PPE often includes a few short-answer questions in the Ethics part, similar to those that follow:

1.   Provide a definition of "ethics."

     ANSWER   Ethics is the study of right and wrong, good and evil, obligations and rights, justice, and social and political ideals. (See textbook, Chapter 11.)

2.   In a few sentences, describe what a "profession" is.

     ANSWER   A profession is an occupation that requires specialized knowledge and skills, obtained by intensive learning and practice, and that is organized or regulated to ensure that its practitioners apply high standards of performance and conduct, commit themselves to continuing competence, and place the public good ahead of narrow personal interests. (See Chapter 1.)

3. Is your province's Code of Ethics for engineers enforceable under your professional engineering Act? Explain.

   ANSWER   Every province and territory (except Ontario) states or implies that the Code of Ethics is enforced under the Act, and violations of the code may result in disciplinary action. In Ontario the code is not enforceable, and a separate clause defines professional misconduct. (See Chapter 11.)

4. Explain what "conflict of interest" means.

   ANSWER   A conflict of interest occurs whenever a practitioner receives a benefit from more than one source when performing professional duties. For example, if an engineer supervising a construction contract for a client also accepts a commission from a supplier, the engineer has a conflict of interest (whether or not the engineer purchases the supplier's materials for the contract). Concealing such a conflict of interest is unethical. (See Chapter 12.)

5. Does your province's professional engineering Act explicitly restrict an engineer to practise in his or her branch of registration only? How does the Code of Ethics deal with the problem of practising outside of one's branch of registration?

   ANSWER   Although a few specialties are regulated in some provinces (especially the SER—the Structural Engineer of Record), professional practice is not generally limited to the branch of registration. However, every Code of Ethics forbids practitioners from accepting or performing work for which they are not qualified. Therefore, practitioners must obtain appropriate preparation—such as academic studies, on-the-job experience, and/or assisted practice—to develop adequate skill and knowledge in the new area. The question of competence is left to the judgment of the practitioner. In the event of a complaint, the practitioner would be expected to demonstrate evidence of adequate preparation in the new area. Failure to show adequate preparation would be a basis for disciplinary action as either negligence or incompetence. (See your Code of Ethics in Appendix B, or on your Association's website.)

6. The Association of Professional Engineers and/or Geoscience is the self-regulating organization responsible for the practice of engineering and/or geoscience in your province. What is the principal objective of this organization?

   ANSWER   The principal objective of every Association is to regulate the profession, and to protect the public interest. This is usually stated or implied near the start of the Act.

7. To become licensed to practise professional engineering and/or geo-science in your province, you must meet certain requirements. Discuss briefly the five most significant of these.

   ANSWER Since age is almost never a problem, the five most important are:
   (1) Adequate education—an accredited university degree (or equivalent);
   (2) Examinations—typically the professional practice exam must be written;
   (3) Adequate experience—typically four years;
   (4) Good character as determined from references; and
   (5) Citizenship (or permanent resident status) in Ontario, although this is not required in all provinces. (See Chapter 2.)

8. Are there any restrictions on how professional engineering services may be advertised? Explain.

   ANSWER Yes, it must be factual, clear, and dignified. (See Chapter 14.)

9. Is a civil engineer allowed to perform services that are normally within the scope of mechanical engineering? Explain.

   ANSWER Yes, providing that the civil engineer can show experience, fur-ther education, or training in the area of mechanical engineering. (See discussion of competence in Chapter 3.)

Multiple-Choice Examination Questions

The National Professional Practice Examination, set by Alberta (APEGGA) and used by 10 Associations across Canada, uses a multiple-choice format, as illus-trated in the following questions. A typical exam is two hours long, consists of one hundred multiple-choice questions, and is closed-book (that is, no aids are permitted).[8]

1. According to the Code of Ethics, which of the following activities by a professional member would be considered UNETHICAL?
   A. Not charging a fee for presenting a speech.
   B. Signing plans prepared by an unknown person.
   C. Reviewing the work of another member with that member's consent.
   D. Providing professional services as a consultant.

   ANSWER B. is correct. It is unethical for professionals to sign plans not prepared by themselves or under their direct supervision.

2. Which of the following is an example of a fraudulent, contractual mis-representation?
   A. A party is coerced into signing a contract by means of intimidation.
   B. A party knowingly makes false statements to induce another party into a contract.
   C. A party induces his son-in-law to sign an unfair contract.
   D. A party unknowingly provides false information about a portion of a contract.

   ANSWER B. is correct. Knowingly providing false information to induce a contract is fraudulent misrepresentation.

3. Contractual disputes of a technical nature may be most expeditiously and effectively solved through:
   A. a lawsuit.
   B. court appeals.
   C. contract renegotiations.
   D. arbitration.

   ANSWER   D. is correct. Arbitration provides an effective, expeditious resolution to technical disputes.

4. Which type of original work below is automatically protected by copyright upon creation?
   A. Paintings.
   B. Inventions.
   C. Clothing designs.
   D. Signatures.

   ANSWER   A. is correct. Of the works listed, only a painting is protected by copyright law.

5. In order for compensation to be awarded to a plaintiff in a tort liability case, the defendant must have:
   A. Caused injury to the plaintiff.
   B. Been willfully negligent.
   C. Signed a contract of performance.
   D. Performed under supervision.

   ANSWER   A. is correct. Injury is one of three criteria that must be met for compensation to be awarded in a tort liability case.

6. Which of the following is the most common job activity of top-level managers?
   A. Writing and reading corporate financial reports.
   B. Developing and testing new products.
   C. Designing and implementing production systems.
   D. Directing and interacting with people.

   ANSWER   D. is correct. Most top managers spend most of their time interacting with other people.

7. The professional's standard of care and skill establishes the point at which a professional:
   A. may or may not charge a fee for services.
   B. has the duty to apply "reasonable care."
   C. may be judged negligent in the performance of services.
   D. has met the minimum requirements for registration.

   ANSWER   C. is correct. The standard of care is used to judge whether or not a professional has been negligent in the performance of services.

8. To effectively reduce liability exposure the professional engineer, geologist, or geophysicist should:
   A. pursue continuing educational opportunities.
   B. work under the supervision of a senior engineer, geologist, or geophysicist.

C. maintain professional standards in practice.

D. provide clients with frequent progress reports.

ANSWER   C. is correct. Maintaining professional standards of practice is the most effective way of reducing liability exposure.

Additional Questions

Additional exam questions are in Appendix E-5 and in Chapters 12 to 14. Additional solved case studies are in Chapters 12 to 14, and 25 cases are located in Appendix F.

NOTES

[1]   Engineers Canada, *Guideline on the Professional Practice Examination,* Ottawa, ON, available at <www.engineerscanada.ca/e/pu_guidelines.cfm> (June 15, 2009). Excerpt reproduced with permission.

[2]   Association of Professional Engineers, Geologists and Geophysicists of Alberta (APEGGA), *National Professional Practice Examination* (Revised May 2007), Edmonton, AB, available at <www.apegga.org/Applicants/faqs/scope.html> (June 5, 2008). Excerpt reproduced with permission. Please be advised that the NPPE syllabus changes over time to reflect changes occurring in the professions. Candidates should visit the APEGGA website, or the website of their association to ensure they are using the most current information.

[3]   Professional Engineers Ontario (PEO), *PPE Syllabus,* <www.peo.on.ca> (June 5, 2008).

[4]   Association of Professional Engineers and Geoscientists of British Columbia (APEGBC), *Professional Practice Examination,* available at <www.apeg.bc.ca/reg/ppe.html> (June 5, 2008).

[5]   J. Delaney, *How to Do Your Best on Law School Exams,* J. Delaney Publ'ns, Bogota, NJ, 1990.

[6]   G.C. Andrews, *Study Guide for the PEO Professional Practice Exam,* 6th ed., Distance Education Department, University of Waterloo, Waterloo, ON N2L 3G1, 2004.

[7]   The essay-type and short-answer examination questions are taken mainly from Professional Practice Exams administered by Professional Engineers Ontario (PEO) over the past decade. The author would like to express his appreciation to PEO for their assistance in obtaining these questions and permission to publish them.

[8]   The multiple-choice examination questions are sample questions for the National Professional Practice Exam administered by the Association of Professional Engineers, Geologists and Geophysicists of Alberta (APEGGA), Edmonton, AB. The author would like to express his appreciation to APEGGA for permission to publish them.

APPENDIX A

PROVINCIAL AND TERRITORIAL ENGINEERING/ GEOSCIENCE ASSOCIATIONS
[**NOTE:** All Web addresses are valid as of August 1, 2008.]

Engineers Canada
(Formerly the Canadian Council of Professional Engineers—CCPE)
180 Elgin St., Suite 1100
Ottawa, ON K2P 2K3

Tel: (613) 232-2474 / Fax: (613) 230-5759
E-mail: info@engineerscanada.ca
Website: www.engineerscanada.ca

Canadian Council of Professional Geoscientists (CCPG)
Suite 200, 4010 Regent Street
Burnaby, BC V5C 6N2

Tel: (604) 412-4888 / Fax: (604) 433-2494
E-mail: info@ccpg.ca
Website: www.ccpg.ca

Association of Professional Engineers, Geologists and Geophysicists of Alberta (APEGGA)
1500 Scotia One
10060 Jasper Avenue NW
Edmonton, AB T5J 4A2

Tel: (780) 426-3990 / 1-800-661-7020
Fax: (780) 426-1877
E-mail: email@apegga.org
Website: www.apegga.org

Association of Professional Engineers and Geoscientists of British Columbia (APEGBC)
200–4010 Regent Street
Burnaby, BC V5C 6N2

Tel: (604) 430-8035 / 1-888-430-8035
Fax: (604) 430-8085
E-mail: apeginfo@apeg.bc.ca
Website: www.apeg.bc.ca

Association of Professional Engineers and Geoscientists of the Province of Manitoba (APEGM)

870 Pembina Highway
Winnipeg, MB R3M 2M7

Tel: (204) 474-2736 / 1-866-227-9600
Fax: (204) 474-5960
E-mail: apegm@apegm.mb.ca
Website: www.apegm.mb.ca

Association of Professional Engineers and Geoscientists of New Brunswick (APEGNB)

183 Hanwell Road
Fredericton, NB E3B 2R2

Tel: (506) 458-8083 / Fax: (506) 451-9629
E-mail: info@apegnb.com
Website: www.apegnb.com

Professional Engineers and Geoscientists of Newfoundland and Labrador (PEG-NL)

Courier: Suite 203, Baine Johnston Centre
10 Fort William Pl., St. John's, NL
Mail: P.O. Box 21207
St. John's, NL A1A 5B2

Tel: (709) 753-7714 / Fax: (709) 753-6131
E-mail: main@pegnl.ca
Website: www.pegnl.ca

Association of Professional Engineers, Geologists and Geophysicists of the Northwest Territories & Nunavut (NAPEGG)

Bowling Green Building
201, 4817–49th Street
Yellowknife, NT X1A 3S7

Tel: (867) 920-4055 / Fax: (867) 873-4058
E-mail: napegg@tamarack.nt.ca
Website: www.napegg.nt.ca

Engineers Nova Scotia
Association of Professional Engineers of Nova Scotia (APENS)

Mail: P.O. Box 129, Halifax, NS B3J 2M4
Courier: 1355 Barrington Street
Halifax, NS B3J 1Y9

Tel: (902) 429-2250 / 1-888-802-7367
Fax: (902) 423-9769
E-mail: info@apens.ns.ca
Website: www.apens.ns.ca

**The Association of Professional Geoscientists
of Nova Scotia (APGNS)**
P.O. Box 8541
Halifax, NS B3K 5M3

Tel.: (902) 420-9928
Office e-mail: nkeeping@dal.ca
Website: www.geoscientistsns.ca

**Professional Engineers Ontario (PEO)
Association of Professional Engineers of Ontario**
25 Sheppard Avenue West
Suite 1000
Toronto, ON M2N 6S9

Tel: (416) 224-1100 / 1-800-339-3716
Fax: (416) 224-8168 / 1-800-268-0496
E-mail: (See directory on PEO website)
Website: www.peo.on.ca

Association of Professional Geoscientists of Ontario (APGO)
60 St. Clair Avenue East
Suite 913
Toronto, ON M4T 1N5

Tel: (416) 203-2746 / 1-877-557-2746
Fax: (416) 203-6181
E-mail: info@apgo.net
Website: www.apgo.net

**Engineers PEI
The Association of Professional Engineers
of Prince Edward Island**
549 North River Road
Charlottetown, PE C1E 1J6

Tel: (902) 566-1268 / Fax: (902) 566-5551
E-mail: info@EngineersPEI.com
Website: www.engineerspei.com

Ordre des ingénieurs du Québec (OIQ)
Gare Windsor, bureau 350,
1100, rue De La Gauchetière West,
Montreal, QC H3B 2S2

Tel: (514) 845-6141 / 1-800-461-6141
Fax: (514) 845-1833
Website: www.oiq.qc.ca

Ordre des Géologues du Québec (OGQ)
500 rue Sherbrooke Ouest
Bureau 900
Montréal, QC H3A 3C6

Tel: (514) 278-6220 / 1-888-377-7708
Fax: (514) 278-7591
E-mail: info@ogq.qc.ca
Website: www.ogq.qc.ca

**Association of Professional Engineers and Geoscientists
of Saskatchewan (APEGS)**
2255–13th Avenue
Suite 104
Regina, SK S4P 0V6

Tel: (306) 525-9547 / 1-800-500-9547
Fax: (306) 525-0851
E-mail: apegs@apegs.sk.ca
Website: www.apegs.sk.ca

Association of Professional Engineers of Yukon (APEY)
312B Hanson Street
Whitehorse, YT Y1A 1Y6

Tel: (867) 667-6727 / Fax: (867) 668-2142
E-mail: staff@apey.yk.ca
Website: www.apey.yk.ca

WEB APPENDIXES
The website accompanying this textbook (www.andrews4e.nelson.com) consists of over 300 pages of additional information in the following appendixes:

**APPENDIX A—Provincial and Territorial Engineering/Geoscience
          Associations**
**APPENDIX B—Excerpts from the Acts, Regulations, and Codes of
          Ethics**
APPENDIX C—Codes of Ethics for Various Technical Societies
**APPENDIX D—NSPE Guidelines to Employment for Professional
          Engineers**
APPENDIX E—Additional Assignments and Discussion Topics
APPENDIX F—Additional Case Studies
APPENDIX G—Additional Articles of Interest

INDEX